Beginning and Intermediate Algebra

Edited by
Tucker Watson

Larsen & Keller
www.larsen-keller.com

Beginning and Intermediate Algebra
Edited by Tucker Watson
ISBN: 978-1-63549-023-7 (Hardback)

☰ Larsen & Keller

Published by Larsen and Keller Education,
5 Penn Plaza,
19th Floor,
New York, NY 10001, USA

Cataloging-in-Publication Data

Beginning and intermediate algebra / edited by Tucker Watson.
    p. cm.
Includes bibliographical references and index.
ISBN 978-1-63549-023-7
1. Algebra. 2. Mathematics. I. Watson, Tucker.
QA155 .B44 2017
512--dc23

The publisher's policy is to use permanent paper from mills that operate a sustainable forestry policy. Furthermore, the publisher ensures that the text paper and cover boards used have met acceptable environmental accreditation standards.

Printed and bound in the United States of America.

For more information regarding Larsen and Keller Education and its products, please visit the publisher's website www.larsen-keller.com

Table of Contents

Preface

This book is a compilation of chapters that discuss the most vital concepts in the field of algebra. It provides in depth knowledge of this field of study. As a part of mathematics, algebra includes the study of groups and fields and their manipulation. It is generally referred to as the backbone of mathematics. Some of the diverse topics covered in this book address the varied branches that fall under this category. The book studies, analyses and upholds the pillars of algebra and its utmost significance in modern times. This textbook is an essential guide for students who are seeking detailed information in this area.

Given below is the chapter wise description of the book:

Chapter 1- Algebra is one of the major parts of mathematics. It is the study of symbols that represent numbers. The basic part of algebra is called elementary algebra. This chapter is on overview of the subject matter incorporating all the major aspects of algebra.

Chapter 2- Algebra has a number of branches; some of these branches are commutative algebra, elementary algebra, operator algebra, linear algebra, universal algebra and von Neumann algebra. Elementary algebra is the basic part of algebra and one of the main branches of mathematics. This text is a compilation of the various branches of algebra that forms an integral part of the broader subject matter.

Chapter 3- The major concepts of algebra are discussed in this text. Identity element, inverse element, commutative property, associative property and isomorphism are some of the key concepts of algebra. The section strategically encompasses and incorporates the major components and key concepts of algebra, providing a complete understanding.

Chapter 4- This text provides a plethora of the major theorems of algebra. The main theorems in algebra are fundamental theorem of algebra, Abel-Ruffini theorem, binomial theorem, Chevalley-Warning theorem and Boolean prime ideal theorem. The topics discussed in the chapter are of great importance to broaden the existing knowledge on the theorems of algebra.

Chapter 5- The chapter serves as a source to understand the basic functions related to algebra. Some of the basic functions are algebraic function, cubic function, quantic function and quartic function. Algebraic functions are functions that can be expressed as the root of a polynomial equation. The following text will provide an integrated understanding of algebra.

Chapter 6- Algebraic structures are sets of finitary operations defined on it that satisfies a list of axioms. Associative algebra is an algebraic structure with compatible functions of addition and multiplication. This chapter helps the reader in developing an in-depth understanding of algebraic structures.

Chapter 7- In mathematics, any equation which is represented in the form of P=Q is an algebraic equation. Linear equation is an algebraic equation in which either a term is constant of a product or a constant. This section is an overview of the subject matter incorporating all the major aspects of algebraic equation.

Chapter 8- Algebraic geometry is the branch of mathematics, mainly studying zeros whereas the algebraic construction is used in geometry to study areas and volumes is known as exterior algebra. Some other applications of algebra that have been discussed in the following chapter are symbolic computation, permutation, algebraic number theory, algebraic K-theory etc. The diverse applications of algebra in the current scenario have been thoroughly discussed in this text.

Chapter 9- Algebra is one of the basic branches of mathematics. It emerged in the 16th century and since then has continuously been advanced as a branch of mathematics. In order to completely understand algebra, it is vital to understand the evolution of algebra. The aspects elucidated in this chapter are of vital importance, and provide a better understanding of algebra.

At the end, I would like to thank all those who dedicated their time and efforts for the successful completion of this book. I also wish to convey my gratitude towards my friends and family who supported me at every step.

Editor

Introduction to Algebra

Algebra is one of the major parts of mathematics. It is the study of symbols that represent numbers. The basic part of algebra is called elementary algebra. This chapter is on overview of the subject matter incorporating all the major aspects of algebra.

Algebra

Algebra (from Arabic *"al-jabr"* meaning "reunion of broken parts") is one of the broad parts of mathematics, together with number theory, geometry and analysis. In its most general form, algebra is the study of mathematical symbols and the rules for manipulating these symbols; it is a unifying thread of almost all of mathematics. As such, it includes everything from elementary equation solving to the study of abstractions such as groups, rings, and fields. The more basic parts of algebra are called elementary algebra, the more abstract parts are called abstract algebra or modern algebra. Elementary algebra is generally considered to be essential for any study of mathematics, science, or engineering, as well as such applications as medicine and economics. Abstract algebra is a major area in advanced mathematics, studied primarily by professional mathematicians. Much early work in algebra, as the Arabic origin of its name suggests, was done in the Middle East, by Persian mathematicians such as al-Khwārizmī (780–850) and Omar Khayyam (1048–1131).

$$x = \frac{-b \pm \sqrt{b^2 - 4ac}}{2a}$$

The quadratic formula expresses the solution of the degree two equation $ax^2 + bx + c = 0$ in terms of its coefficients $a, b, c,$ where a is not zero.

Elementary algebra differs from arithmetic in the use of abstractions, such as using letters to stand for numbers that are either unknown or allowed to take on many values. For example, in $x + 2 = 5$ the letter x is unknown, but the law of inverses can be used to discover its value: $x = 3$. In $E = mc^2$, the letters E and m are variables, and the letter c is a constant, the speed of light in a vacuum. Algebra gives methods for solving equations and expressing formulas that are much easier (for those who know how to use them) than the older method of writing everything out in words.

The word *algebra* is also used in certain specialized ways. A special kind of mathematical object in abstract algebra is called an "algebra", and the word is used, for example, in the phrases linear algebra and algebraic topology.

A mathematician who does research in algebra is called an algebraist.

Etymology

The word entered the English language during the fifteenth century, from either Spanish, Italian, or Medieval Latin. It originally referred to the surgical procedure of setting broken or dislocated bones. The mathematical meaning was first recorded in the sixteenth century.

Different Meanings of "Algebra"

The word "algebra" has several related meanings in mathematics, as a single word or with qualifiers.

- As a single word without an article, "algebra" names a broad part of mathematics.

- As a single word with an article or in plural, "an algebra" or "algebras" denotes a specific mathematical structure, whose precise definition depends on the author. Usually the structure has an addition, multiplication, and a scalar multiplication. When some authors use the term "algebra", they make a subset of the following additional assumptions: associative, commutative, unital, and/or finite-dimensional. In universal algebra, the word "algebra" refers to a generalization of the above concept, which allows for n-ary operations.

- With a qualifier, there is the same distinction:

  - Without an article, it means a part of algebra, such as linear algebra, elementary algebra (the symbol-manipulation rules taught in elementary courses of mathematics as part of primary and secondary education), or abstract algebra (the study of the algebraic structures for themselves).

  - With an article, it means an instance of some abstract structure, like a Lie algebra, an associative algebra, or a vertex operator algebra.

  - Sometimes both meanings exist for the same qualifier, as in the sentence: *Commutative algebra is the study of commutative rings, which are commutative algebras over the integers.*

Algebra as a Branch of Mathematics

Algebra began with computations similar to those of arithmetic, with letters standing for numbers. This allowed proofs of properties that are true no matter which numbers are involved. For example, in the quadratic equation

$$ax^2 + bx + c = 0,$$

a, b, c can be any numbers whatsoever (except that a cannot be 0), and the quadratic formula can be used to quickly and easily find the values of the unknown quantity x which satisfy the equation. That is to say, to find all the solutions of the equation.

Historically, and in current teaching, the study of algebra starts with the solving of equations such as the quadratic equation above. Then more general questions, such as "does an equation have a solution?", "how many solutions does an equation have?", "what can be said about the nature of the solutions?" are considered. These questions lead to ideas of form, structure and symmetry. This development permitted algebra to be extended to consider non-numerical objects, such as vectors, matrices, and polynomials. The structural properties of these non-numerical objects were then abstracted to define algebraic structures such as groups, rings, and fields.

Before the 16th century, mathematics was divided into only two subfields, arithmetic and geometry. Even though some methods, which had been developed much earlier, may be considered nowadays as algebra, the emergence of algebra and, soon thereafter, of infinitesimal calculus as subfields of mathematics only dates from the 16th or 17th century. From the second half of 19th century on, many new fields of mathematics appeared, most of which made use of both arithmetic and geometry, and almost all of which used algebra.

Today, algebra has grown until it includes many branches of mathematics, as can be seen in the Mathematics Subject Classification where none of the first level areas (two digit entries) is called *algebra*. Today algebra includes section 08-General algebraic systems, 12-Field theory and polynomials, 13-Commutative algebra, 15-Linear and multilinear algebra; matrix theory, 16-Associative rings and algebras, 17-Nonassociative rings and algebras, 18-Category theory; homological algebra, 19-K-theory and 20-Group theory. Algebra is also used extensively in 11-Number theory and 14-Algebraic geometry.

History

Early History of Algebra

The roots of algebra can be traced to the ancient Babylonians, who developed an advanced arithmetical system with which they were able to do calculations in an algorithmic fashion. The Babylonians developed formulas to calculate solutions for problems typically solved today by using linear equations, quadratic equations, and indeterminate linear equations. By contrast, most Egyptians of this era, as well as Greek and Chinese mathematics in the 1st millennium BC, usually solved such equations by geometric methods, such as those described in the *Rhind Mathematical Papyrus*, Euclid's *Elements*, and *The Nine Chapters on the Mathematical Art*. The geometric work of the Greeks, typified in the *Elements*, provided the framework for generalizing formulae beyond the solution of particular problems into more general systems of stating and solving equations, although this would not be realized until mathematics developed in medieval Islam.

By the time of Plato, Greek mathematics had undergone a drastic change. The Greeks created a geometric algebra where terms were represented by sides of geometric objects, usually lines, that had letters associated with them. Diophantus (3rd century AD) was an Alexandrian Greek mathematician and the author of a series of books called *Arithmetica*. These texts deal with solving algebraic equations, and have led, in number theory to the modern notion of Diophantine equation.

Earlier traditions discussed above had a direct influence on the Persian Muhammad ibn Mūsā al-Khwārizmī (c. 780–850). He later wrote *The Compendious Book on Calculation by Completion and Balancing*, which established algebra as a mathematical discipline that is independent of geometry and arithmetic.

The Hellenistic mathematicians Hero of Alexandria and Diophantus as well as Indian mathematicians such as Brahmagupta continued the traditions of Egypt and Babylon, though Diophantus' *Arithmetica* and Brahmagupta's *Brāhmasphu⬚asiddhānta* are on a higher level. For example, the first complete arithmetic solution (including zero and negative solutions) to quadratic equations was described by Brahmagupta in his book *Brahmasphutasiddhanta*. Later, Persian and Arabic mathematicians developed algebraic methods to a much higher degree of sophistication. Although Diophantus and the Babylonians used mostly special *ad hoc* methods to solve equations, Al-Khwarizmi's contribution was fundamental. He solved linear and quadratic equations without algebraic symbolism, negative numbers or zero, thus he had to distinguish several types of equations.

In the context where algebra is identified with the theory of equations, the Greek mathematician Diophantus has traditionally been known as the "father of algebra" but in more recent times there is much debate over whether al-Khwarizmi, who founded the discipline of *al-jabr*, deserves that title instead. Those who support Diophantus point to the fact that the algebra found in *Al-Jabr* is slightly more elementary than the algebra found in *Arithmetica* and that *Arithmetica* is syncopated while *Al-Jabr* is fully rhetorical. Those who support Al-Khwarizmi point to the fact that he introduced the methods of "reduction" and "balancing" (the transposition of subtracted terms to the other side of an equation, that is, the cancellation of like terms on opposite sides of the equation) which the term *al-jabr* originally referred to, and that he gave an exhaustive explanation of solving quadratic equations, supported by geometric proofs, while treating algebra as an independent discipline in its own right. His algebra was also no longer concerned "with a series of problems to be resolved, but an exposition which starts with primitive terms in which the combinations must give all possible prototypes for equations, which henceforward explicitly constitute the true object of study". He also studied an equation for its own sake and "in a generic manner, insofar as it does not simply emerge in the course of solving a problem, but is specifically called on to define an infinite class of problems".

Another Persian mathematician Omar Khayyam is credited with identifying the foundations of algebraic geometry and found the general geometric solution of the cubic equation. Yet another Persian mathematician, Sharaf al-Dīn al-Tūsī, found algebraic and numerical solutions to various cases of cubic equations. He also developed the concept of a function. The Indian mathematicians Mahavira and Bhaskara II, the Persian mathematician Al-Karaji, and the Chinese mathematician Zhu Shijie, solved various cases of cubic, quartic, quintic and higher-order polynomial equations using numerical methods. In the 13th century, the solution of a cubic equation by Fibonacci is representative of the beginning of a revival in European algebra. As the Islamic world was declining, the European world was ascending. And it is here that algebra was further developed.

History of Algebra

François Viète's work on new algebra at the close of the 16th century was an important step towards modern algebra. In 1637, René Descartes published *La Géométrie*, inventing analytic geometry and introducing modern algebraic notation. Another key event in the further development of algebra was the general algebraic solution of the cubic and quartic equations, developed in the mid-16th century. The idea of a determinant was developed by Japanese mathematician

Seki Kōwa in the 17th century, followed independently by Gottfried Leibniz ten years later, for the purpose of solving systems of simultaneous linear equations using matrices. Gabriel Cramer also did some work on matrices and determinants in the 18th century. Permutations were studied by Joseph-Louis Lagrange in his 1770 paper *Réflexions sur la résolution algébrique des équations* devoted to solutions of algebraic equations, in which he introduced Lagrange resolvents. Paolo Ruffini was the first person to develop the theory of permutation groups, and like his predecessors, also in the context of solving algebraic equations.

Italian mathematician Girolamo Cardano published the solutions to the cubic and quartic equations in his 1545 book *Ars magna.*

Abstract algebra was developed in the 19th century, deriving from the interest in solving equations, initially focusing on what is now called Galois theory, and on constructibility issues. George Peacock was the founder of axiomatic thinking in arithmetic and algebra. Augustus De Morgan discovered relation algebra in his *Syllabus of a Proposed System of Logic.* Josiah Willard Gibbs developed an algebra of vectors in three-dimensional space, and Arthur Cayley developed an algebra of matrices (this is a noncommutative algebra).

Areas of Mathematics with the Word Algebra in their Name

Some areas of mathematics that fall under the classification abstract algebra have the word algebra in their name; linear algebra is one example. Others do not: group theory, ring theory, and field theory are examples. In this section, we list some areas of mathematics with the word "algebra" in the name.

- Elementary algebra, the part of algebra that is usually taught in elementary courses of mathematics.

- Abstract algebra, in which algebraic structures such as groups, rings and fields are axiomatically defined and investigated.

- Linear algebra, in which the specific properties of linear equations, vector spaces and matrices are studied.

- Commutative algebra, the study of commutative rings.

- Computer algebra, the implementation of algebraic methods as algorithms and computer programs.

- Homological algebra, the study of algebraic structures that are fundamental to study topological spaces.

- Universal algebra, in which properties common to all algebraic structures are studied.

- Algebraic number theory, in which the properties of numbers are studied from an algebraic point of view.

- Algebraic geometry, a branch of geometry, in its primitive form specifying curves and surfaces as solutions of polynomial equations.

- Algebraic combinatorics, in which algebraic methods are used to study combinatorial questions.

Many mathematical structures are called algebras:

- Algebra over a field or more generally algebra over a ring. Many classes of algebras over a field or over a ring have a specific name:

  - Associative algebra

  - Non-associative algebra

  - Lie algebra

  - Hopf algebra

  - C*-algebra

  - Symmetric algebra

  - Exterior algebra

  - Tensor algebra

- In measure theory,

  - Sigma-algebra

  - Algebra over a set

- In category theory

  - F-algebra and F-coalgebra

  - T-algebra

- In logic,

  - Relational algebra: a set of finitary relations that is closed under certain operators.

  - Boolean algebra, a structure abstracting the computation with the truth values *false* and *true*. The structures also have the same name.

- Heyting algebra

Elementary Algebra

$$3x^2 - 2xy + c$$

Algebraic expression notation:
1 – power (exponent)
2 – coefficient
3 – term
4 – operator
5 – constant term
$x\,y\,c$ – variables/constants

Elementary algebra is the most basic form of algebra. It is taught to students who are presumed to have no knowledge of mathematics beyond the basic principles of arithmetic. In arithmetic, only numbers and their arithmetical operations (such as +, −, ×, ÷) occur. In algebra, numbers are often represented by symbols called variables (such as a, n, x, y or z). This is useful because:

- It allows the general formulation of arithmetical laws (such as $a + b = b + a$ for all a and b), and thus is the first step to a systematic exploration of the properties of the real number system.

- It allows the reference to "unknown" numbers, the formulation of equations and the study of how to solve these. (For instance, "Find a number x such that $3x + 1 = 10$" or going a bit further "Find a number x such that $ax + b = c$". This step leads to the conclusion that it is not the nature of the specific numbers that allows us to solve it, but that of the operations involved.)

- It allows the formulation of functional relationships. (For instance, "If you sell x tickets, then your profit will be $3x - 10$ dollars, or $f(x) = 3x - 10$, where f is the function, and x is the number to which the function is applied".)

Polynomials

A polynomial is an expression that is the sum of a finite number of non-zero terms, each term consisting of the product of a constant and a finite number of variables raised to whole number powers. For example, $x^2 + 2x - 3$ is a polynomial in the single variable x. A polynomial expression is an expression that may be rewritten as a polynomial, by using commutativity, associativity and distributivity of addition and multiplication. For example, $(x - 1)(x + 3)$ is a polynomial expression, that, properly speaking, is not a polynomial. A polynomial function is a function that is defined by a polynomial, or, equivalently, by a polynomial expression. The two preceding examples define the same polynomial function.

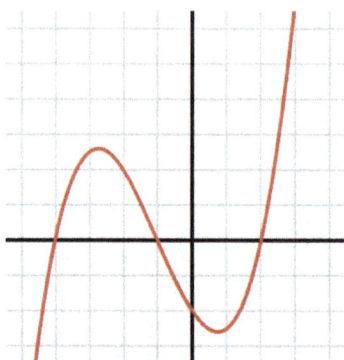

The graph of a polynomial function of degree 3.

Two important and related problems in algebra are the factorization of polynomials, that is, expressing a given polynomial as a product of other polynomials that can not be factored any further, and the computation of polynomial greatest common divisors. The example polynomial above can be factored as $(x - 1)(x + 3)$. A related class of problems is finding algebraic expressions for the roots of a polynomial in a single variable.

Education

It has been suggested that elementary algebra should be taught to students as young as eleven years old, though in recent years it is more common for public lessons to begin at the eighth grade level ($\approx$ 13 y.o. $\pm$) in the United States.

Since 1997, Virginia Tech and some other universities have begun using a personalized model of teaching algebra that combines instant feedback from specialized computer software with one-on-one and small group tutoring, which has reduced costs and increased student achievement.

Abstract Algebra

Abstract algebra extends the familiar concepts found in elementary algebra and arithmetic of numbers to more general concepts. Here are listed fundamental concepts in abstract algebra.

Sets: Rather than just considering the different types of numbers, abstract algebra deals with the more general concept of *sets*: a collection of all objects (called elements) selected by property specific for the set. All collections of the familiar types of numbers are sets. Other examples of sets include the set of all two-by-two matrices, the set of all second-degree polynomials ($ax^2 + bx + c$), the set of all two dimensional vectors in the plane, and the various finite groups such as the cyclic groups, which are the groups of integers modulo n. Set theory is a branch of logic and not technically a branch of algebra.

Binary operations: The notion of addition (+) is abstracted to give a *binary operation*, $*$ say. The notion of binary operation is meaningless without the set on which the operation is defined. For two elements a and b in a set S, $a * b$ is another element in the set; this condition is called closure. Addition (+), subtraction (−), multiplication (×), and division (÷) can be binary operations when defined on different sets, as are addition and multiplication of matrices, vectors, and polynomials.

Identity elements: The numbers zero and one are abstracted to give the notion of an *identity ele-*

ment for an operation. Zero is the identity element for addition and one is the identity element for multiplication. For a general binary operator ∗ the identity element e must satisfy $a * e = a$ and $e * a = a$, and is necessarily unique, if it exists. This holds for addition as $a + 0 = a$ and $0 + a = a$ and multiplication $a \times 1 = a$ and $1 \times a = a$. Not all sets and operator combinations have an identity element; for example, the set of positive natural numbers (1, 2, 3, ...) has no identity element for addition.

Inverse elements: The negative numbers give rise to the concept of *inverse elements*. For addition, the inverse of a is written $-a$, and for multiplication the inverse is written a^{-1}. A general two-sided inverse element a^{-1} satisfies the property that $a * a^{-1} = e$ and $a^{-1} * a = e$, where e is the identity element.

Associativity: Addition of integers has a property called associativity. That is, the grouping of the numbers to be added does not affect the sum. For example: $(2 + 3) + 4 = 2 + (3 + 4)$. In general, this becomes $(a * b) * c = a * (b * c)$. This property is shared by most binary operations, but not subtraction or division or octonion multiplication.

Commutativity: Addition and multiplication of real numbers are both commutative. That is, the order of the numbers does not affect the result. For example: $2 + 3 = 3 + 2$. In general, this becomes $a * b = b * a$. This property does not hold for all binary operations. For example, matrix multiplication and quaternion multiplication are both non-commutative.

Groups

Combining the above concepts gives one of the most important structures in mathematics: a group. A group is a combination of a set S and a single binary operation ∗, defined in any way you choose, but with the following properties:

- An identity element e exists, such that for every member a of S, $e * a$ and $a * e$ are both identical to a.

- Every element has an inverse: for every member a of S, there exists a member a^{-1} such that $a * a^{-1}$ and $a^{-1} * a$ are both identical to the identity element.

- The operation is associative: if a, b and c are members of S, then $(a * b) * c$ is identical to $a * (b * c)$.

If a group is also commutative—that is, for any two members a and b of S, $a * b$ is identical to $b * a$—then the group is said to be abelian.

For example, the set of integers under the operation of addition is a group. In this group, the identity element is 0 and the inverse of any element a is its negation, $-a$. The associativity requirement is met, because for any integers a, b and c, $(a + b) + c = a + (b + c)$

The nonzero rational numbers form a group under multiplication. Here, the identity element is 1, since $1 \times a = a \times 1 = a$ for any rational number a. The inverse of a is $1/a$, since $a \times 1/a = 1$.

The integers under the multiplication operation, however, do not form a group. This is because, in general, the multiplicative inverse of an integer is not an integer. For example, 4 is an integer, but its multiplicative inverse is ¼, which is not an integer.

The theory of groups is studied in group theory. A major result in this theory is the classification of finite simple groups, mostly published between about 1955 and 1983, which separates the finite simple groups into roughly 30 basic types.

Semigroups, quasigroups, and monoids are structures similar to groups, but more general. They comprise a set and a closed binary operation, but do not necessarily satisfy the other conditions. A semigroup has an *associative* binary operation, but might not have an identity element. A monoid is a semigroup which does have an identity but might not have an inverse for every element. A quasigroup satisfies a requirement that any element can be turned into any other by either a unique left-multiplication or right-multiplication; however the binary operation might not be associative.

All groups are monoids, and all monoids are semigroups.

Examples										
Set	Natural numbers N		Integers Z		Rational numbers Q (also real R and complex C numbers)				Integers modulo 3: Z_3 = {0, 1, 2}	
Operation	+	× (w/o zero)	+	× (w/o zero)	+	−	× (w/o zero)	÷ (w/o zero)	+	× (w/o zero)
Closed	Yes	Yes	Yes	Yes	Yes	Yes	Yes	Yes	Yes	Yes
Identity	0	1	0	1	0	N/A	1	N/A	0	1
Inverse	N/A	N/A	−a	N/A	−a	N/A	1/a	N/A	0, 2, 1, respectively	N/A, 1, 2, respectively
Associative	Yes	Yes	Yes	Yes	Yes	No	Yes	No	Yes	Yes
Commutative	Yes	Yes	Yes	Yes	Yes	No	Yes	No	Yes	Yes
Structure	monoid	monoid	abelian group	monoid	abelian group	quasigroup	abelian group	quasigroup	abelian group	abelian group (Z_2)

Rings and Fields

Groups just have one binary operation. To fully explain the behaviour of the different types of numbers, structures with two operators need to be studied. The most important of these are rings, and fields.

A ring has two binary operations (+) and (×), with × distributive over +. Under the first operator (+) it forms an *abelian group*. Under the second operator (×) it is associative, but it does not need to have identity, or inverse, so division is not required. The additive (+) identity element is written as 0 and the additive inverse of a is written as $-a$.

Distributivity generalises the *distributive law* for numbers. For the integers $(a + b) \times c = a \times c + b \times c$ and $c \times (a + b) = c \times a + c \times b$, and × is said to be *distributive* over +.

The integers are an example of a ring. The integers have additional properties which make it an integral domain.

A field is a *ring* with the additional property that all the elements excluding 0 form an *abelian group* under ×. The multiplicative (×) identity is written as 1 and the multiplicative inverse of a is written as a^{-1}.

The rational numbers, the real numbers and the complex numbers are all examples of fields.

Algebraic Number

An algebraic number is any complex number that is a root of a non-zero polynomial in one variable with rational coefficients (or equivalently – by clearing denominators – with integer coefficients). All integers and rational numbers are algebraic, as are all roots of integers. The same is not true for all real and complex numbers because of transcendental numbers such as π and e. Almost all real and complex numbers are transcendental.

Examples

- The rational numbers, expressed as the quotient of two integers a and b, b not equal to zero, satisfy the above definition because $x = a/b$ is the root of $bx - a$.

- The quadratic surds (irrational roots of a quadratic polynomial $ax^2 + bx + c$ with integer coefficients a, b, and c) are algebraic numbers. If the quadratic polynomial is monic ($a = 1$) then the roots are quadratic integers.

- The constructible numbers are those numbers that can be constructed from a given unit length using straightedge and compass. These include all quadratic surds, all rational numbers, and all numbers that can be formed from these using the basic arithmetic operations and the extraction of square roots. (Note that by designating cardinal directions for 1, −1, i, and −i, complex numbers such as $3 + \sqrt{2}i$ are considered constructible.)

- Any expression formed from algebraic numbers using any combination of the basic arithmetic operations and extraction of nth roots gives another algebraic number.

- Polynomial roots that *cannot* be expressed in terms of the basic arithmetic operations and extraction of nth roots (such as the roots of $x^5 − x + 1$). This happens with many, but not all, polynomials of degree 5 or higher.

- Gaussian integers: those complex numbers $a + bi$ where both a and b are integers are also quadratic integers.

- Trigonometric functions of rational multiples of π (except when undefined): that is, the trigonometric numbers. For example, each of cos π/7, cos 3π/7, cos 5π/7 satisfies $8x^3 − 4x^2 − 4x + 1 = 0$. This polynomial is irreducible over the rationals, and so these three cosines are *conjugate* algebraic numbers. Likewise, tan 3π/16, tan 7π/16, tan 11π/16, tan 15π/16 all satisfy the irreducible polynomial $x^4 − 4x^3 − 6x^2 + 4x + 1$, and so are conjugate algebraic integers.

- Some irrational numbers are algebraic and some are not:

- The numbers $\sqrt{2}$ and $\sqrt[3]{3}/2$ are algebraic since they are roots of polynomials $x^2 - 2$ and $8x^3 - 3$, respectively.

- The golden ratio φ is algebraic since it is a root of the polynomial $x^2 - x - 1$.

- The numbers π and e are not algebraic numbers; hence they are transcendental.

Properties

Algebraic numbers on the complex plane colored by degree (red=1, green=2, blue=3, yellow=4)

- The set of algebraic numbers is countable (enumerable).

- Hence, the set of algebraic numbers has Lebesgue measure zero (as a subset of the complex numbers), i.e. "almost all" complex numbers are not algebraic.

- Given an algebraic number, there is a unique monic polynomial (with rational coefficients) of least degree that has the number as a root. This polynomial is called its minimal polynomial. If its minimal polynomial has degree n, then the algebraic number is said to be of *degree n*. An algebraic number of degree 1 is a rational number. A real algebraic number of degree 2 is a quadratic irrational.

- All algebraic numbers are computable and therefore definable and arithmetical.

- The set of real algebraic numbers is linearly ordered, countable, densely ordered, and without first or last element, so is order-isomorphic to the set of rational numbers.

- For real numbers a and b, the complex number $a + bi$ is algebraic if and only if both a and b are algebraic.

The Field of Algebraic Numbers

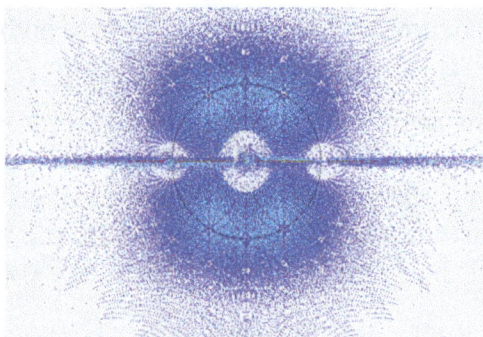

Algebraic numbers colored by degree (blue=4, cyan=3, red=2, green=1). The unit circle is black.

The sum, difference, product and quotient (if the denominator is nonzero) of two algebraic numbers is again algebraic (this fact can be demonstrated using the resultant), and the algebraic numbers therefore form a field Q (sometimes denoted by A, though this usually denotes the adele ring). Every root of a polynomial equation whose coefficients are *algebraic numbers* is again algebraic. This can be rephrased by saying that the field of algebraic numbers is algebraically closed. In fact, it is the smallest algebraically closed field containing the rationals, and is therefore called the algebraic closure of the rationals.

The set of *real* algebraic numbers itself forms a field.

Related Fields

Numbers Defined by Radicals

All numbers that can be obtained from the integers using a finite number of integer additions, subtractions, multiplications, divisions, and taking nth roots where n is a positive integer (i.e., radical expressions) are algebraic. The converse, however, is not true: there are algebraic numbers that cannot be obtained in this manner. All of these numbers are roots of polynomials of degree ≥ 5. This is a result of Galois theory. An example of such a number is the unique real root of the polynomial $x^5 - x - 1$ (which is approximately 1.167304).

Closed-form Number

Algebraic numbers are all numbers that can be defined explicitly or implicitly in terms of polynomials, starting from the rational numbers. One may generalize this to "closed-form numbers", which may be defined in various ways. Most broadly, all numbers that can be defined explicitly or implicitly in terms of polynomials, exponentials, and logarithms are called "elementary numbers", and these include the algebraic numbers, plus some transcendental numbers. Most narrowly, one may consider numbers *explicitly* defined in terms of polynomials, exponentials, and logarithms – this does not include all algebraic numbers, but does include some simple transcendental numbers such as e or $\log(2)$.

Algebraic Integers

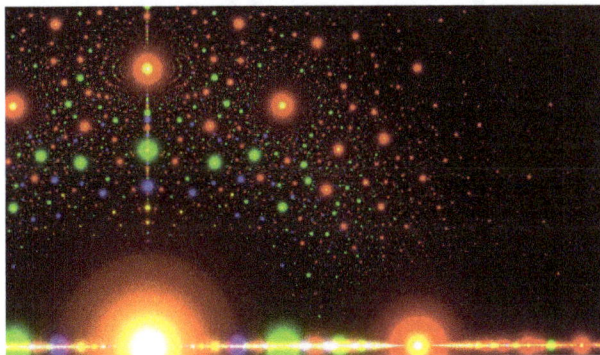

Algebraic numbers colored by leading coefficient (red signifies 1 for an algebraic integer)

An algebraic integer is an algebraic number that is a root of a polynomial with integer coefficients with leading coefficient 1 (a monic polynomial). Examples of algebraic integers are $5 + 13\sqrt{2}$, $2 -$

$6i$, and $1/2(1 + i\sqrt{3})$. Note, therefore, that the algebraic integers constitute a proper superset of the integers, as the latter are the roots of monic polynomials $x - k$ for all $k \in \mathbb{Z}$. In this sense, algebraic integers are to algebraic numbers what integers are to rational numbers.

The sum, difference and product of algebraic integers are again algebraic integers, which means that the algebraic integers form a ring. The name *algebraic integer* comes from the fact that the only rational numbers that are algebraic integers are the integers, and because the algebraic integers in any number field are in many ways analogous to the integers. If K is a number field, its ring of integers is the subring of algebraic integers in K, and is frequently denoted as O_K. These are the prototypical examples of Dedekind domains.

Special Classes of Algebraic Number

- Algebraic Solution
- Gaussian integer
- Eisenstein integer
- Quadratic irrational
- Fundamental unit
- Root of unity
- Gaussian period
- Pisot–Vijayaraghavan number
- Salem number

References

- Lang, Serge (2002), Algebra, Graduate Texts in Mathematics, 211 (Revised third ed.), New York: Springer-Verlag, ISBN 978-0-387-95385-4, MR 1878556
- Ore, Øystein 1948, 1988, Number Theory and Its History, Dover Publications, Inc. New York, ISBN 0-486-65620-9 (pbk.)

Branches of Algebra

Algebra has a number of branches; some of these branches are commutative algebra, elementary algebra, operator algebra, linear algebra, universal algebra and von Neumann algebra. Elementary algebra is the basic part of algebra and one of the main branches of mathematics. This text is a compilation of the various branches of algebra that forms an integral part of the broader subject matter.

Commutative Algebra

Commutative algebra is the branch of algebra that studies commutative rings, their ideals, and modules over such rings. Both algebraic geometry and algebraic number theory build on commutative algebra. Prominent examples of commutative rings include polynomial rings, rings of algebraic integers, including the ordinary integers $\mathbb{Z}$, , and p-adic integers.

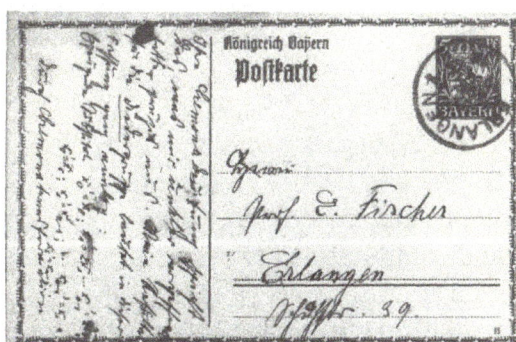

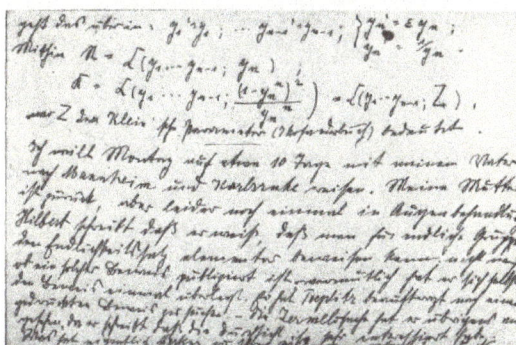

A 1915 postcard from one of the pioneers of commutative algebra, Emmy Noether, to E. Fischer, discussing her work in commutative algebra.

Commutative algebra is the main technical tool in the local study of schemes.

The study of rings which are not necessarily commutative is known as noncommutative algebra; it includes ring theory, representation theory, and the theory of Banach algebras.

Overview

Commutative algebra is essentially the study of the rings occurring in algebraic number theory and algebraic geometry.

In algebraic number theory, the rings of algebraic integers are Dedekind rings, which constitute therefore an important class of commutative rings. Considerations related to modular arithmetic have led to the notion of valuation ring. The restriction of algebraic field extensions to subrings has led to the notions of integral extensions and integrally closed domains as well as the notion of ramification of an extension of valuation rings.

The notion of localization of a ring (in particular the localization with respect to a prime ideal, the localization consisting in inverting a single element and the total quotient ring) is one of the main differences between commutative algebra and the theory of non-commutative rings. It leads to an important class of commutative rings, the local rings that have only one maximal ideal. The set of the prime ideals of a commutative ring is naturally equipped with a topology, the Zariski topology. All these notions are widely used in algebraic geometry and are the basic technical tools for the definition of scheme theory, a generalization of algebraic geometry introduced by Grothendieck.

Many other notions of commutative algebra are counterparts of geometrical notions occurring in algebraic geometry. This is the case of Krull dimension, primary decomposition, regular rings, Cohen–Macaulay rings, Gorenstein rings and many other notions.

History

The subject, first known as ideal theory, began with Richard Dedekind's work on ideals, itself based on the earlier work of Ernst Kummer and Leopold Kronecker. Later, David Hilbert introduced the term *ring* to generalize the earlier term *number ring*. Hilbert introduced a more abstract approach to replace the more concrete and computationally oriented methods grounded in such things as complex analysis and classical invariant theory. In turn, Hilbert strongly influenced Emmy Noether, who recast many earlier results in terms of an ascending chain condition, now known as the Noetherian condition. Another important milestone was the work of Hilbert's student Emanuel Lasker, who introduced primary ideals and proved the first version of the Lasker–Noether theorem.

The main figure responsible for the birth of commutative algebra as a mature subject was Wolfgang Krull, who introduced the fundamental notions of localization and completion of a ring, as well as that of regular local rings. He established the concept of the Krull dimension of a ring, first for Noetherian rings before moving on to expand his theory to cover general valuation rings and Krull rings. To this day, Krull's principal ideal theorem is widely considered the single most important foundational theorem in commutative algebra. These results paved the way for the introduction of commutative algebra into algebraic geometry, an idea which would revolutionize the latter subject.

Much of the modern development of commutative algebra emphasizes modules. Both ideals of a ring R and R-algebras are special cases of R-modules, so module theory encompasses both ideal theory and the theory of ring extensions. Though it was already incipient in Kronecker's work, the modern approach to commutative algebra using module theory is usually credited to Krull and Noether.

Main Tools and Results

Noetherian Rings

In mathematics, more specifically in the area of modern algebra known as ring theory, a Noetherian ring, named after Emmy Noether, is a ring in which every non-empty set of ideals has a maximal element. Equivalently, a ring is Noetherian if it satisfies the ascending chain condition on ideals; that is, given any chain:

$$I_1 \subseteq \cdots I_{k-1} \subseteq I_k \subseteq I_{k+1} \subseteq \cdots$$

there exists an n such that:

$$I_n = I_{n+1} = \cdots$$

For a commutative ring to be Noetherian it suffices that every prime ideal of the ring is finitely generated. (The result is due to I. S. Cohen.)

The notion of a Noetherian ring is of fundamental importance in both commutative and non-commutative ring theory, due to the role it plays in simplifying the ideal structure of a ring. For instance, the ring of integers and the polynomial ring over a field are both Noetherian rings, and consequently, such theorems as the Lasker–Noether theorem, the Krull intersection theorem, and the Hilbert's basis theorem hold for them. Furthermore, if a ring is Noetherian, then it satisfies the descending chain condition on *prime ideals*. This property suggests a deep theory of dimension for Noetherian rings beginning with the notion of the Krull dimension.

Hilbert's Basis Theorem

Theorem. If R is a left (resp. right) Noetherian ring, then the polynomial ring $R[X]$ is also a left (resp. right) Noetherian ring.

Hilbert's basis theorem has some immediate corollaries:

- By induction we see that $R[X_0, X_{n-1}]$ will also be Noetherian.

- Since any affine variety over R^n (i.e. a locus-set of a collection of polynomials) may be written as the locus of an ideal $\mathfrak{a} \subset R[X_0, X_{n-1}]$ and further as the locus of its generators, it follows that every affine variety is the locus of finitely many polynomials — i.e. the intersection of finitely many hypersurfaces.

- If A is a finitely-generated R-algebra, then we know that $A \simeq R[X_0, X_{n-1}]/\mathfrak{a}$, where $\mathfrak{a}$ is an ideal. The basis theorem implies that $\mathfrak{a}$ must be finitely generated, say $\mathfrak{a} = (p_0, p_{N-1})$, i.e. A is finitely presented.

Primary Decomposition

An ideal Q of a ring is said to be *primary* if Q is proper and whenever $xy \in Q$, either $x \in Q$ or $y^n \in Q$ for some positive integer n. In Z, the primary ideals are precisely the ideals of the form (p^e) where p is prime and e is a positive integer. Thus, a primary decomposition of (n) corresponds to representing (n) as the intersection of finitely many primary ideals.

The *Lasker–Noether theorem*, given here, may be seen as a certain generalization of the fundamental theorem of arithmetic:

Lasker-Noether Theorem. Let R be a commutative Noetherian ring and let I be an ideal of R. Then I may be written as the intersection of finitely many primary ideals with distinct radicals; that is:

$$I = \bigcap_{i=1}^{t} Q_i$$

with Q_i primary for all i and $\mathrm{Rad}(Q_i) \neq \mathrm{Rad}(Q_j)$ for $i \neq j$. Furthermore, if:

$$I = \bigcap_{i=1}^{k} P_i$$

is decomposition of I with $\mathrm{Rad}(P_i) \neq \mathrm{Rad}(P_j)$ for $i \neq j$, and both decompositions of I are *irredundant* (meaning that no proper subset of either $\{Q_1, ..., Q_t\}$ or $\{P_1, ..., P_k\}$ yields an intersection equal to I), $t = k$ and (after possibly renumbering the Q_i) $\mathrm{Rad}(Q_i) = \mathrm{Rad}(P_i)$ for all i.

For any primary decomposition of I, the set of all radicals, that is, the set $\{\mathrm{Rad}(Q_1), ..., \mathrm{Rad}(Q_t)\}$ remains the same by the Lasker–Noether theorem. In fact, it turns out that (for a Noetherian ring) the set is precisely the assassinator of the module R/I; that is, the set of all annihilators of R/I (viewed as a module over R) that are prime.

Localization

The localization is a formal way to introduce the "denominators" to a given ring or a module. That is, it introduces a new ring/module out of an existing one so that it consists of fractions

$$\frac{m}{s}.$$

where the denominators s range in a given subset S of R. The archetypal example is the construction of the ring Q of rational numbers from the ring Z of integers.

Completion

A completion is any of several related functors on rings and modules that result in complete topological rings and modules. Completion is similar to localization, and together they are among the most basic tools in analysing commutative rings. Complete commutative rings have simpler structure than the general ones and Hensel's lemma applies to them.

Zariski Topology on Prime Ideals

The Zariski topology defines a topology on the spectrum of a ring (the set of prime ideals). In this formulation, the Zariski-closed sets are taken to be the sets

$$V(I) = \{P \in \mathrm{Spec}(A) \mid I \subseteq P\}$$

where A is a fixed commutative ring and I is an ideal. This is defined in analogy with the classical Zariski topology, where closed sets in affine space are those defined by polynomial equations . To see the connection with the classical picture, note that for any set S of polynomials (over an algebraically closed field), it follows from Hilbert's Nullstellensatz that the points of $V(S)$ (in the old sense) are exactly the tuples $(a_1, ..., a_n)$ such that $(x_1 - a_1, ..., x_n - a_n)$ contains S; moreover, these are maximal ideals and by the "weak" Nullstellensatz, an ideal of any affine coordinate ring is maximal if and only if it is of this form. Thus, $V(S)$ is "the same as" the maximal ideals containing S. Grothendieck's innovation in defining Spec was to replace maximal ideals with all prime ideals; in this formulation it is natural to simply generalize this observation to the definition of a closed set in the spectrum of a ring.

Examples

The fundamental example in commutative algebra is the ring of integers $\mathbb{Z}$. The existence of primes and the unique factorization theorem laid the foundations for concepts such as Noetherian rings and the primary decomposition.

Other important examples are:

- Polynomial rings $R[x_1, ..., x_n]$

- The p-adic integers

- Rings of algebraic integers.

Connections with Algebraic Geometry

Commutative algebra (in the form of polynomial rings and their quotients, used in the definition of algebraic varieties) has always been a part of algebraic geometry. However, in late 1950s, algebraic varieties were subsumed into Alexander Grothendieck's concept of a scheme. Their local objects are affine schemes or prime spectra which are locally ringed spaces which form a category which is antiequivalent (dual) to the category of commutative unital rings, extending the duality between the category of affine algebraic varieties over a field k, and the category of finitely generated reduced k-algebras. The gluing is along Zariski topology; one can glue within the category of locally ringed spaces, but also, using the Yoneda embedding, within the more abstract category of presheaves of sets over the category of affine schemes. The Zariski topology in the set theoretic sense is then replaced by a Zariski topology in the sense of Grothendieck topology. Grothendieck introduced Grothendieck topologies having in mind more exotic but geometrically finer and more sensitive examples than the crude Zariski topology, namely the étale topology, and the two flat Grothendieck topologies: fppf and fpqc; nowadays some other examples became prominent including Nisnevich topology. Sheaves can be furthermore generalized to stacks in the sense of Grothendieck, usually with some additional representability conditions leading to Artin stacks and, even finer, Deligne-Mumford stacks, both often called algebraic stacks.

Elementary Algebra

$$x = \frac{-b \pm \sqrt{b^2 - 4ac}}{2a}$$

The quadratic formula, which is the solution to the quadratic equation $ax^2 + bx + c = 0$ where $a \neq 0$. Here the symbols a, b, c, represent arbitrary numbers, and x is a variable which represents the solution of the equation

Two-dimensional plot (red curve) of the algebraic equation $y = x^2 - x - 2$

Elementary algebra encompasses some of the basic concepts of algebra, one of the main branches of mathematics. It is typically taught to secondary school students and builds on their understanding of arithmetic. Whereas arithmetic deals with specified numbers, algebra introduces quantities without fixed values, known as variables. This use of variables entails a use of algebraic notation and an understanding of the general rules of the operators introduced in arithmetic. Unlike abstract algebra, elementary algebra is not concerned with algebraic structures outside the realm of real and complex numbers.

The use of variables to denote quantities allows general relationships between quantities to be formally and concisely expressed, and thus enables solving a broader scope of problems. Most quantitative results in science and mathematics are expressed as algebraic equations.

Algebraic Notation

Algebraic notation describes how algebra is written. It follows certain rules and conventions, and has its own terminology. For example, the expression $3x^2 - 2xy + c$ has the following components:

1 : Exponent (power), 2 : Coefficient, 3 : term, 4 : operator, 5 : constant, x, y : variables

A *coefficient* is a numerical value, or letter representing a numerical constant, that multiplies a variable (the operator is omitted). A *term* is an addend or a summand, a group of coefficients, variables, constants and exponents that may be separated from the other terms by the plus and minus operators. Letters represent variables and constants. By convention, letters at the beginning of the alphabet (e.g. a, b, c) are typically used to represent constants, and those toward the end of the alphabet (e.g. x, y and z) are used to represent variables. They are usually written in italics.

Algebraic operations work in the same way as arithmetic operations, such as addition, subtraction, multiplication, division and exponentiation. and are applied to algebraic variables and terms. Multiplication symbols are usually omitted, and implied when there is no space between two variables or terms, or when a coefficient is used. For example, $3 \times x^2$ is written as $3x^2$, and $2 \times x \times y$ may be written $2xy$.

Usually terms with the highest power (exponent), are written on the left, for example, x^2 is written to the left of x. When a coefficient is one, it is usually omitted (e.g. $1x^2$ is written x^2). Likewise when the exponent (power) is one, (e.g. $3x^1$ is written $3x$). When the exponent is zero, the result is always 1 (e.g. x^0 is always rewritten to 1). However 0^0, being undefined, should not appear in an expression, and care should be taken in simplifying expressions in which variables may appear in exponents.

Alternative Notation

Other types of notation are used in algebraic expressions when the required formatting is not available, or can not be implied, such as where only letters and symbols are available. For example, exponents are usually formatted using superscripts, e.g. x^2. In plain text, and in the TeX mark-up language, the caret symbol "^" represents exponents, so x^2 is written as "x^2". In programming languages such as Ada, Fortran, Perl, Python and Ruby, a double asterisk is used, so x^2 is written as "x**2". Many programming languages and calculators use a single asterisk to represent the multiplication symbol, and it must be explicitly used, for example, $3x$ is written "3*x".

Concepts

Variables

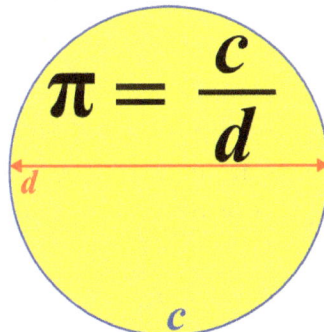

Example of variables showing the relationship between a circle's diameter and its circumference. For any circle, its circumference c, divided by its diameter d, is equal to the constant pi, π (approximately 3.14).

Elementary algebra builds on and extends arithmetic by introducing letters called variables to represent general (non-specified) numbers. This is useful for several reasons.

1. Variables may represent numbers whose values are not yet known. For example, if the temperature of the current day, C, is 20 degrees higher than the temperature of the previous day, P, then the problem can be described algebraically as $C = P + 20$.

2. Variables allow one to describe *general* problems, without specifying the values of the quantities that are involved. For example, it can be stated specifically that 5 minutes is equivalent to $60 \times 5 = 300$ seconds. A more general (algebraic) description may state that the number of seconds, $s = 60 \times m$, where m is the number of minutes.

3. Variables allow one to describe mathematical relationships between quantities that may vary. For example, the relationship between the circumference, c, and diameter, d, of a circle is described by $\pi = c / d$.

4. Variables allow one to describe some mathematical properties. For example, a basic property of addition is commutativity which states that the order of numbers being added together does not matter. Commutativity is stated algebraically as $(a+b)=(b+a)$..

Evaluating Expressions

Algebraic expressions may be evaluated and simplified, based on the basic properties of arithmetic operations (addition, subtraction, multiplication, division and exponentiation). For example,

- Added terms are simplified using coefficients. For example, $x+x+x$ can be simplified as $3x$ (where 3 is a numerical coefficient).

- Multiplied terms are simplified using exponents. For example, $x \times x \times x$ is represented as x^3

- Like terms are added together, for example, $2x^2 + 3ab - x^2 + ab$ is written as $x^2 + 4ab$, because the terms containing x^2 are added together, and, the terms containing ab are added together.

- Brackets can be "multiplied out", using the distributive property. For example, $x(2x+3)$ can be written as $(x \times 2x)+(x \times 3)$ which can be written as $2x^2 + 3x$

- Expressions can be factored. For example, $6x^5 + 3x^2$, by dividing both terms by $3x^2$ can be written as $3x^2(2x^3 +1)$

Equations

An equation states that two expressions are equal using the symbol for equality, $=$ (the equals sign). One of the most well-known equations describes Pythagoras' law relating the length of the sides of a right angle triangle:

$$c^2 = a^2 + b^2$$

This equation states that c^2, representing the square of the length of the side that is the hypotenuse (the side opposite the right angle), is equal to the sum (addition) of the squares of the other two sides whose lengths are represented by a and b.

An equation is the claim that two expressions have the same value and are equal. Some equations are true for all values of the involved variables (such as $a+b=b+a$); such equations are called identities. Conditional equations are true for only some values of the involved variables, e.g. $x^2 -1 = 8$ is true only for $x=3$ and $x=-3$. The values of the variables which make the equation true are the solutions of the equation and can be found through equation solving.

Another type of equation is an inequality. Inequalities are used to show that one side of the equation is greater, or less, than the other. The symbols used for this are: $a > b$ where $>$ represents 'greater than', and $a < b$ where $<$ represents 'less than'. Just like standard equality equations, numbers can be added, subtracted, multiplied or divided. The only exception is that when multiplying or dividing by a negative number, the inequality symbol must be flipped.

Properties of Equality

By definition, equality is an equivalence relation, meaning it has the properties (a) reflexive (i.e. $b = b$), (b) symmetric (i.e. if $a = b$ then $b = a$) (c) transitive (i.e. if $a = b$ and $b = c$ then $a = c$). It also satisfies the important property that if two symbols are used for equal things, then one symbol can be substituted for the other in any true statement about the first and the statement will remain true. This implies the following properties:

- if $a = b$ and $c = d$ then $a + c = b + d$ and $ac = bd$;

- if $a = b$ then $a + c = b + c$;

- more generally, for any function f, if $a = b$ then $f(a) = f(b)$.

Properties of Inequality

The relations *less than* $<$ and greater than $>$ have the property of transitivity:

- If a b and $b < c$ then $a < c$;

- If $a < b$ and $c < d$ then $a + c < b + d$;

- If $a < b$ and $c > 0$ then $ac < bc$;

- If $a < b$ and $c < 0$ then $bc < ac$.

By reversing the inequation, $<$ and $>$ can be swapped, for example:

- $a < b$ is equivalent to $b > a$

Substitution

Substitution is replacing the terms in an expression to create a new expression. Substituting 3 for a in the expression a*5 makes a new expression 3*5 with meaning 15. Substituting the terms of a statement makes a new statement. When the original statement is true independent of the values of the terms, the statement created by substitutions is also true. Hence definitions can be made in symbolic terms and interpreted through substitution: if $a^2 := a * a$, where := means "is defined to equal", substituting 3 for a informs the reader of this statement that 3^2 means 3*3=9. Often it's not known whether the statement is true independent of the values of the terms, and substitution allows one to derive restrictions on the possible values, or show what conditions the statement holds under. For example, taking the statement x+1=0, if x is substituted with 1, this imples 1+1=2=0, which is false, which implies that if x+1=0 then x can't be 1.

If x and y are integers, rationals, or real numbers, then xy=0 implies x=0 or y=0. Suppose abc=0. Then, substituting a for x and bc for y, we learn a=0 or bc=0. Then we can substitute again, letting x=b and y=c, to show that if bc=0 then b=0 or c=0. Therefore, if abc=0, then a=0 or (b=0 or c=0), so abc=0 implies a=0 or b=0 or c=0.

Consider if the original fact were stated as "$ab=0$ implies $a=0$ or $b=0$." Then when we say "suppose $abc=0$," we have a conflict of terms when we substitute. Yet the above logic is still valid to show that if $abc=0$ then $a=0$ or $b=0$ or $c=0$ if instead of letting $a=a$ and $b=bc$ we substitute a for a and b for bc (and with $bc=0$, substituting b for a and c for b). This shows that substituting for the terms in a statement isn't always the same as letting the terms from the statement equal the substituted terms. In this situation it's clear that if we substitute an expression a into the a term of the original equation, the a substituted does not refer to the a in the statement "$ab=0$ implies $a=0$ or $b=0$."

Solving Algebraic Equations

A typical algebra problem.

The following sections lay out examples of some of the types of algebraic equations that may be encountered.

Linear Equations with one Variable

Linear equations are so-called, because when they are plotted, they describe a straight line. The simplest equations to solve are linear equations that have only one variable. They contain only constant numbers and a single variable without an exponent. As an example, consider:

Problem in words: If you double the age of a child and add 4, the resulting answer is 12. How old is the child?

Equivalent equation: $2x + 4 = 12$ where x represent the child's age

To solve this kind of equation, the technique is add, subtract, multiply, or divide both sides of the equation by the same number in order to isolate the variable on one side of the equation. Once the variable is isolated, the other side of the equation is the value of the variable. This problem and its solution are as follows:

1. Equation to solve: $\qquad$ $2x + 4 = 12$

2. Subtract 4 from both sides: $\qquad$ $2x + 4 - 4 = 12 - 4$

3. This simplifies to: $\qquad$ $2x = 8$

4. Divide both sides by 2: $\qquad$ $\dfrac{2x}{2} = \dfrac{8}{2}$

5. This simplifies to the solution: $\quad$ $x = 4$

In words: the child is 4 years old.

The general form of a linear equation with one variable, can be written as: $ax + b = c$

Following the same procedure (i.e. subtract b from both sides, and then divide by a), the general solution is given by $x = \dfrac{c-b}{a}$

Linear Equations with two Variables

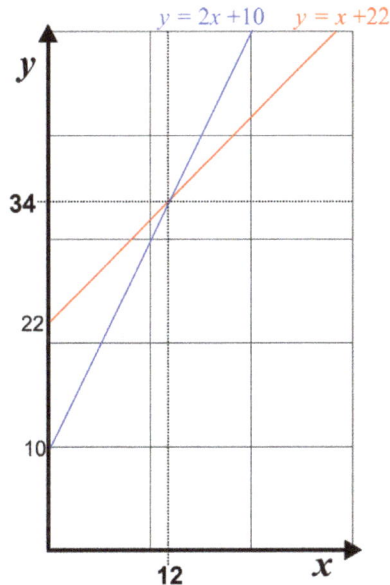

Solving two linear equations with a unique solution at the point that they intersect.

A linear equation with two variables has many (i.e. an infinite number of) solutions. For example:

Problem in words: A father is 22 years older than his son. How old are they?

Equivalent equation: $y = x + 22$ where y is the father's age, x is the son's age.

This cannot be worked out by itself. If the son's age was made known, then there would no longer be two unknowns (variables), and the problem becomes a linear equation with just one variable, that can be solved as described above.

To solve a linear equation with two variables (unknowns), requires two related equations. For example, if it was also revealed that:

Problem in words:	In 10 years, the father will be twice as old as his son.
Equivalent equation:	$y + 10 = 2 \times (x + 10)$
SUBTRACT 10 FROM BOTH SIDES:	$y = 2 \times (x + 10) - 10$
Multiple out brackets:	$y = 2x + 20 - 10$
Simplify:	$y = 2x + 10$

Now there are two related linear equations, each with two unknowns, which enables the production of a linear equation with just one variable, by subtracting one from the other (called the elimination method):

Second equation	$y = 2x + 10$
First equation	$y = x + 22$
Subtract the first equation from the second in order to remove	$(y - y) = (2x - x) + 10 - 22$
Simplify	$0 = x - 12$
Add 12 to both sides	$12 = x$
Rearrange	$x = 12$

In other words, the son is aged 12, and since the father 22 years older, he must be 34. In 10 years time, the son will be 22, and the father will be twice his age, 44. This problem is illustrated on the associated plot of the equations.

Quadratic Equations

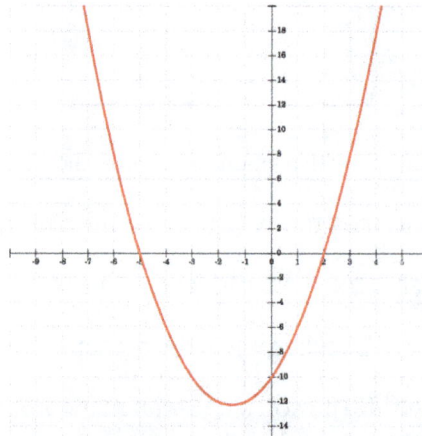

Quadratic equation plot of $y = x^2 + 3x - 10$ showing its roots at $x = -5$ and $x = 2$, , and that the quadratic can be rewritten as $y = (x + 5)(x - 2)$

A quadratic equation is one which includes a term with an exponent of 2, for example, x^2, and no term with higher exponent. The name derives from the Latin *quadrus*, meaning square. In general, a quadratic equation can be expressed in the form $ax^2 + bx + c = 0$, where a is not zero (if it were zero, then the equation would not be quadratic but linear). Because of this a quadratic equation must contain the term ax^2, which is known as the quadratic term. Hence $a \neq 0$, and so we may divide by a and rearrange the equation into the standard form

$$x^2 + px + q = 0$$

where $p = b/a$ and $q = c/a$. Solving this, by a process known as completing the square, leads to the quadratic formula

$$x = \frac{-b \pm \sqrt{b^2 - 4ac}}{2a},$$

where the symbol "±" indicates that both

$$x = \frac{-b + \sqrt{b^2 - 4ac}}{2a} \quad \text{and} \quad x = \frac{-b - \sqrt{b^2 - 4ac}}{2a}$$

are solutions of the quadratic equation.

Quadratic equations can also be solved using factorization (the reverse process of which is expansion, but for two linear terms is sometimes denoted foiling). As an example of factoring:

$$x^2 + 3x - 10 = 0,$$

which is the same thing as

$$(x + 5)(x - 2) = 0.$$

It follows from the zero-product property that either $x = 2$ or $x = -5$ are the solutions, since precisely one of the factors must be equal to zero. All quadratic equations will have two solutions in the complex number system, but need not have any in the real number system. For example,

$$x^2 + 1 = 0$$

has no real number solution since no real number squared equals –1. Sometimes a quadratic equation has a root of multiplicity 2, such as:

$$(x + 1)^2 = 0.$$

For this equation, –1 is a root of multiplicity 2. This means –1 appears two times, since the equation can be rewritten in factored form as

$$[x - (-1)][x - (-1)] = 0.$$

Complex Numbers

All quadratic equations have two solutions in complex numbers, a category that includes real numbers, imaginary numbers, and sums of real and imaginary numbers. Complex numbers first arise in the teaching of quadratic equations and the quadratic formula. For example, the quadratic equation

$$x^2 + x + 1 = 0$$

has solutions

$$x = \frac{-1 + \sqrt{-3}}{2} \quad \text{and} \quad x = \frac{-1 - \sqrt{-3}}{2}.$$

Since $\sqrt{-3}$ is not any real number, both of these solutions for x are complex numbers.

Exponential and Logarithmic Equations

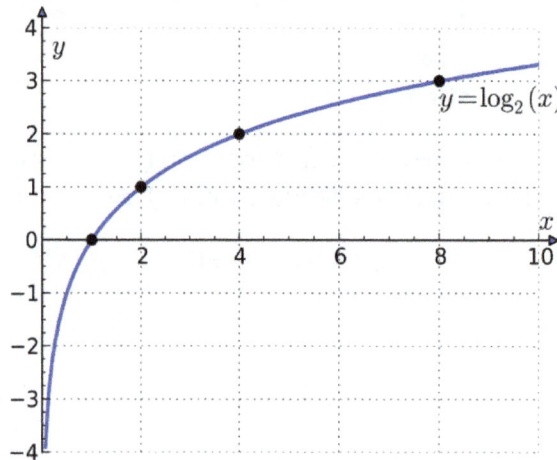

The graph of the logarithm to base 2 crosses the x axis (horizontal axis) at 1 and passes through the points with coordinates (2, 1), (4, 2), and (8, 3). For example, $\log_2(8) = 3$, because $2^3 = 8$. The graph gets arbitrarily close to the y axis, but does not meet or intersect it.

An exponential equation is one which has the form $a^x = b$ for $a > 0$, which has solution

$$X = \log_a b = \frac{\ln b}{\ln a}$$

when $b > 0$. Elementary algebraic techniques are used to rewrite a given equation in the above way before arriving at the solution. For example, if

$$3 \cdot 2^{x-1} + 1 = 10$$

then, by subtracting 1 from both sides of the equation, and then dividing both sides by 3 we obtain

$$2^{x-1} = 3$$

whence

$$x - 1 = \log_2 3$$

or

$$x = \log_2 3 + 1.$$

A logarithmic equation is an equation of the form $\log{(x)} \quad b$ for $a > 0$, which has solution

$$X = a^b.$$

For example, if

$$4\log_5(x-3) - 2 = 6$$

then, by adding 2 to both sides of the equation, followed by dividing both sides by 4, we get

$$\log_5(x-3) = 2$$

whence

$$x - 3 = 5^2 = 25$$

from which we obtain

$$x = 28.$$

Radical Equations

$$\sqrt[2]{x^3} \equiv x^{\frac{3}{2}}$$

Radical equation showing two ways to represent the same expression.
The triple bar means the equation is true for all values of x

A radical equation is one that includes a radical sign, which includes square roots, $\sqrt{x}$, cube roots, $\sqrt[3]{x}$, and nth roots, $\sqrt[n]{x}$.. Recall that an nth root can be rewritten in exponential format, so that $\sqrt[n]{x}$ is equivalent to $x^{\frac{1}{n}}$. Combined with regular exponents (powers), then (the square root of x cubed), can be rewritten as $x^{\frac{3}{2}}$. So a common form of a radical equation is $\sqrt[n]{x^m} = a$ (equivalent to $x^{\frac{m}{n}} = a$) where m and n are integers. It has real solution(s):

m Is odd	m is even and $a \geq 0$	m and n are even and $a < 0$	m is even, n is odd, and $a < 0$
$x = \sqrt[m]{a^n}$ equivalently $x = \left(\sqrt[m]{a}\right)^n$	$x = \pm\sqrt[m]{a^n}$ equivalently $x = \pm\left(\sqrt[m]{a}\right)^n$	$x = \pm\sqrt[m]{a^n}$	No real solution

For example, if:

$$(x+5)^{2/3} = 4,$$

then

$$x + 5 = \pm(\sqrt{4})^3$$
$$x + 5 = \pm 8$$
$$x = -5 \pm 8$$
$$x = 3, -13$$

System of Linear Equations

There are different methods to solve a system of linear equations with two variables.

Elimination Method

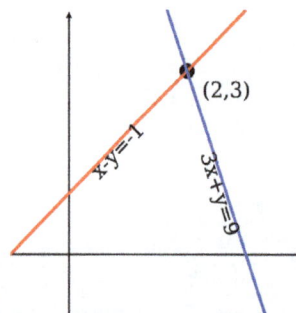

The solution set for the equations $x - y = -1$ and $3x + y = 9$ is the single point (2, 3).

An example of solving a system of linear equations is by using the elimination method:

$$\begin{cases} 4x + 2y = 14 \\ 2x - y = 1. \end{cases}$$

Multiplying the terms in the second equation by 2:

$$4x + 2y = 14$$

$$4x - 2y = 2.$$

Adding the two equations together to get:

$$8x = 16$$

which simplifies to

$$x = 2.$$

Since the fact that $x = 2$ is known, it is then possible to deduce that $y = 3$ by either of the original two equations (by using 2 instead of x) The full solution to this problem is then

$$\begin{cases} x = 2 \\ y = 3. \end{cases}$$

Note that this is not the only way to solve this specific system; y could have been solved before x.

Substitution Method

Another way of solving the same system of linear equations is by substitution.

$$\begin{cases} 4x + 2y = 14 \\ 2x - y = 1. \end{cases}$$

An equivalent for y can be deduced by using one of the two equations. Using the second equation:

$$2x - y = 1$$

Subtracting $2x$ from each side of the equation:

$$\begin{aligned} 2x - 2x - y &= 1 - 2x \\ -y &= 1 - 2x \end{aligned}$$

and multiplying by -1:

$$y = 2x - 1.$$

Using this y value in the first equation in the original system:

$$\begin{aligned} 4x + 2(2x - 1) &= 14 \\ 4x + 4x - 2 &= 14 \\ 8x - 2 &= 14 \end{aligned}$$

Adding 2 on each side of the equation:

$$\begin{aligned} 8 \quad -2 + 2 &= 14 + 2 \\ 8 \qquad\qquad &\quad 16 \end{aligned}$$

which simplifies to

$$x = 2$$

Using this value in one of the equations, the same solution as in the previous method is obtained.

$$\begin{cases} x = 2 \\ y = 3. \end{cases}$$

Note that this is not the only way to solve this specific system; in this case as well, y could have been solved before .

Other Types of Systems of Linear Equations

Inconsistent Systems

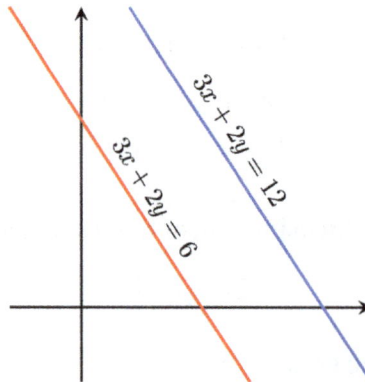

The equations $3x + 2y = 6$ and $3x + 2y = 12$ are parallel and cannot intersect, and is unsolvable.

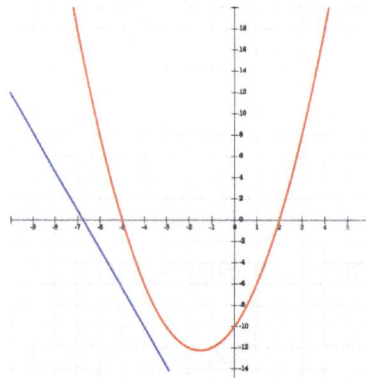

Plot of a quadratic equation (red) and a linear equation (blue) that do not intersect, and consequently for which there is no common solution.

In the above example, a solution exists. However, there are also systems of equations which do not have any solution. Such a system is called inconsistent. An obvious example is

$$\begin{cases} x + y = 1 \\ 0x + 0y = 2. \end{cases}$$

As $0 \neq 2$, the second equation in the system has no solution. Therefore, the system has no solution. However, not all inconsistent systems are recognized at first sight. As an example, let us consider the system

$$\begin{cases} 4x + 2y = 12 \\ -2x - y = -4. \end{cases}$$

Multiplying by 2 both sides of the second equation, and adding it to the first one results in

$$0x + 0y = 4,$$

which has clearly no solution.

Undetermined Systems

There are also systems which have infinitely many solutions, in contrast to a system with a unique solution (meaning, a unique pair of values for x and y) For example:

$$\begin{cases} 4x+2y & =12 \\ -2x-y & =-6 \end{cases}$$

Isolating y in the second equation:

$$y = -2x+6$$

And using this value in the first equation in the system:

$$4x+2(-2x+6)=12$$
$$4x-4x+12 \quad =12$$
$$12=12$$

The equality is true, but it does not provide a value for x. Indeed, one can easily verify (by just filling in some values of x) that for any x there is a solution as long as $y=-2x+6$. There is an infinite number of solutions for this system.

Over- and Underdetermined Systems

Systems with more variables than the number of linear equations are called underdetermined. Such a system, if it has any solutions, does not have a unique one but rather an infinitude of them. An example of such a system is

$$x+2y=10$$

$$y-z=2.$$

When trying to solve it, one is led to express some variables as functions of the other ones if any solutions exist, but cannot express *all* solutions numerically because there are an infinite number of them if there are any.

A system with a greater number of equations than variables is called overdetermined. If an overdetermined system has any solutions, necessarily some equations are linear combinations of the others.

Operator Algebra

In functional analysis, an operator algebra is an algebra of continuous linear operators on a topological vector space with the multiplication given by the composition of mappings. Although it is usually classified as a branch of functional analysis, it has direct applications to representation theory, differential geometry, quantum statistical mechanics, quantum information, and quantum field theory.

Overview

Operator algebras can be used to study arbitrary sets of operators with little algebraic relation *simultaneously*. From this point of view, operator algebras can be regarded as a generalization of spectral theory of a single operator. In general operator algebras are non-commutative rings.

An operator algebra is typically required to be closed in a specified operator topology inside the algebra of the whole continuous linear operators. In particular, it is a set of operators with both algebraic and topological closure properties. In some disciplines such properties are axiomized and algebras with certain topological structure become the subject of the research.

Though algebras of operators are studied in various contexts (for example, algebras of pseudo-differential operators acting on spaces of distributions), the term *operator algebra* is usually used in reference to algebras of bounded operators on a Banach space or, even more specially in reference to algebras of operators on a separable Hilbert space, endowed with the operator norm topology.

In the case of operators on a Hilbert space, the Hermitian adjoint map on operators gives a natural involution which provides an additional algebraic structure which can be imposed on the algebra. In this context, the best studied examples are self-adjoint operator algebras, meaning that they are closed under taking adjoints. These include C*-algebras and von Neumann algebras. C*-algebras can be easily characterized abstractly by a condition relating the norm, involution and multiplication. Such abstractly defined C*-algebras can be identified to a certain closed subalgebra of the algebra of the continuous linear operators on a suitable Hilbert space. A similar result holds for von Neumann algebras.

Commutative self-adjoint operator algebras can be regarded as the algebra of complex valued continuous functions on a locally compact space, or that of measurable functions on a standard measurable space. Thus, general operator algebras are often regarded as a noncommutative generalizations of these algebras, or the structure of the *base space* on which the functions are defined. This point of view is elaborated as the philosophy of noncommutative geometry, which tries to study various non-classical and/or pathological objects by noncommutative operator algebras.

Examples of operator algebras which are not self-adjoint include:

- nest algebras
- many commutative subspace lattice algebras
- many limit algebras

Abstract Algebra

In algebra, which is a broad division of mathematics, abstract algebra (occasionally called modern algebra) is the study of algebraic structures. Algebraic structures include groups, rings, fields, modules, vector spaces, lattices, and algebras. The term *abstract algebra* was coined in the early 20th century to distinguish this area of study from the other parts of algebra.

Algebraic structures, with their associated homomorphisms, form mathematical categories. Category theory is a powerful formalism for analyzing and comparing different algebraic structures.

The permutations of Rubik's Cube form a group, a fundamental concept within abstract algebra.

Universal algebra is a related subject that studies the nature and theories of various types of algebraic structures as a whole. For example, universal algebra studies the overall theory of groups, as distinguished from studying particular groups.

History

As in other parts of mathematics, concrete problems and examples have played important roles in the development of abstract algebra. Through the end of the nineteenth century, many -- perhaps most -- of these problems were in some way related to the theory of algebraic equations. Major themes include:

- Solving of systems of linear equations, which led to linear algebra

- Attempts to find formulae for solutions of general polynomial equations of higher degree that resulted in discovery of groups as abstract manifestations of symmetry

- Arithmetical investigations of quadratic and higher degree forms and diophantine equations, that directly produced the notions of a ring and ideal.

Numerous textbooks in abstract algebra start with axiomatic definitions of various algebraic structures and then proceed to establish their properties. This creates a false impression that in algebra axioms had come first and then served as a motivation and as a basis of further study. The true order of historical development was almost exactly the opposite. For example, the hypercomplex numbers of the nineteenth century had kinematic and physical motivations but challenged comprehension. Most theories that are now recognized as parts of algebra started as collections of disparate facts from various branches of mathematics, acquired a common theme that served as a core around which various results were grouped, and finally became unified on a basis of a common set of concepts. An archetypical example of this progressive synthesis can be seen in the history of group theory.

Early Group Theory

There were several threads in the early development of group theory, in modern language loosely corresponding to *number theory*, *theory of equations*, and *geometry*.

Leonhard Euler considered algebraic operations on numbers modulo an integer, modular arithmetic, in his generalization of Fermat's little theorem. These investigations were taken much further by Carl Friedrich Gauss, who considered the structure of multiplicative groups of residues mod n and established many properties of cyclic and more general abelian groups that arise in this way. In his investigations of composition of binary quadratic forms, Gauss explicitly stated the associative law for the composition of forms, but like Euler before him, he seems to have been more interested in concrete results than in general theory. In 1870, Leopold Kronecker gave a definition of an abelian group in the context of ideal class groups of a number field, generalizing Gauss's work; but it appears he did not tie his definition with previous work on groups, particularly permutation groups. In 1882, considering the same question, Heinrich M. Weber realized the connection and gave a similar definition that involved the cancellation property but omitted the existence of the inverse element, which was sufficient in his context (finite groups).

Permutations were studied by Joseph-Louis Lagrange in his 1770 paper *Réflexions sur la résolution algébrique des équations (Thoughts on the algebraic solution of equations)* devoted to solutions of algebraic equations, in which he introduced Lagrange resolvents. Lagrange's goal was to understand why equations of third and fourth degree admit formulae for solutions, and he identified as key objects permutations of the roots. An important novel step taken by Lagrange in this paper was the abstract view of the roots, i.e. as symbols and not as numbers. However, he did not consider composition of permutations. Serendipitously, the first edition of Edward Waring's *Meditationes Algebraicae (Meditations on Algebra)* appeared in the same year, with an expanded version published in 1782. Waring proved the main theorem on symmetric functions, and specially considered the relation between the roots of a quartic equation and its resolvent cubic. *Mémoire sur la résolution des équations (Memoire on the Solving of Equations)* of Alexandre Vandermonde (1771) developed the theory of symmetric functions from a slightly different angle, but like Lagrange, with the goal of understanding solvability of algebraic equations.

> Kronecker claimed in 1888 that the study of modern algebra began with this first paper of Vandermonde. Cauchy states quite clearly that Vandermonde had priority over Lagrange for this remarkable idea, which eventually led to the study of group theory.

Paolo Ruffini was the first person to develop the theory of permutation groups, and like his predecessors, also in the context of solving algebraic equations. His goal was to establish the impossibility of an algebraic solution to a general algebraic equation of degree greater than four. En route to this goal he introduced the notion of the order of an element of a group, conjugacy, the cycle decomposition of elements of permutation groups and the notions of primitive and imprimitive and proved some important theorems relating these concepts, such as

> if G is a subgroup of S_5 whose order is divisible by 5 then G contains an element of order 5.

Note, however, that he got by without formalizing the concept of a group, or even of a permutation group. The next step was taken by Évariste Galois in 1832, although his work remained unpublished until 1846, when he considered for the first time what is now called the *closure property* of a group of permutations, which he expressed as

> ... if in such a group one has the substitutions S and T then one has the substitution ST.

The theory of permutation groups received further far-reaching development in the hands of Augustin Cauchy and Camille Jordan, both through introduction of new concepts and, primarily, a great wealth of results about special classes of permutation groups and even some general theorems. Among other things, Jordan defined a notion of isomorphism, still in the context of permutation groups and, incidentally, it was he who put the term *group* in wide use.

The abstract notion of a group appeared for the first time in Arthur Cayley's papers in 1854. Cayley realized that a group need not be a permutation group (or even *finite*), and may instead consist of matrices, whose algebraic properties, such as multiplication and inverses, he systematically investigated in succeeding years. Much later Cayley would revisit the question whether abstract groups were more general than permutation groups, and establish that, in fact, any group is isomorphic to a group of permutations.

Modern Algebra

The end of the 19th and the beginning of the 20th century saw a tremendous shift in the methodology of mathematics. Abstract algebra emerged around the start of the 20th century, under the name *modern algebra*. Its study was part of the drive for more intellectual rigor in mathematics. Initially, the assumptions in classical algebra, on which the whole of mathematics (and major parts of the natural sciences) depend, took the form of axiomatic systems. No longer satisfied with establishing properties of concrete objects, mathematicians started to turn their attention to general theory. Formal definitions of certain algebraic structures began to emerge in the 19th century. For example, results about various groups of permutations came to be seen as instances of general theorems that concern a general notion of an *abstract group*. Questions of structure and classification of various mathematical objects came to forefront.

These processes were occurring throughout all of mathematics, but became especially pronounced in algebra. Formal definition through primitive operations and axioms were proposed for many basic algebraic structures, such as groups, rings, and fields. Hence such things as group theory and ring theory took their places in pure mathematics. The algebraic investigations of general fields by Ernst Steinitz and of commutative and then general rings by David Hilbert, Emil Artin and Emmy Noether, building up on the work of Ernst Kummer, Leopold Kronecker and Richard Dedekind, who had considered ideals in commutative rings, and of Georg Frobenius and Issai Schur, concerning representation theory of groups, came to define abstract algebra. These developments of the last quarter of the 19th century and the first quarter of 20th century were systematically exposed in Bartel van der Waerden's *Moderne algebra*, the two-volume monograph published in 1930–1931 that forever changed for the mathematical world the meaning of the word *algebra* from *the theory of equations* to the *theory of algebraic structures*.

Basic Concepts

By abstracting away various amounts of detail, mathematicians have created theories of various algebraic structures that apply to many objects. For instance, almost all systems studied are sets, to which the theorems of set theory apply. Those sets that have a certain binary operation defined on them form magmas, to which the concepts concerning magmas, as well those concerning sets, apply. We can add additional constraints on the algebraic structure, such as associativity (to form semigroups); identity, and inverses (to form groups); and other more complex structures. With additional structure, more theorems could be proved, but the generality is reduced. The "hierarchy" of algebraic objects (in terms of generality) creates a hierarchy of the corresponding theories: for instance, the theorems of group theory apply to rings (algebraic objects that have two binary operations with certain axioms) since a ring is a group over one of its operations. Mathematicians choose a balance between the amount of generality and the richness of the theory.

Examples of algebraic structures with a single binary operation are:

- Magmas
- Quasigroups
- Monoids
- Semigroups
- Groups

More complicated examples include:

- Rings
- Fields
- Modules
- Vector spaces
- Algebras over fields
- Associative algebras
- Lie algebras
- Lattices
- Boolean algebras

Applications

Because of its generality, abstract algebra is used in many fields of mathematics and science. For instance, algebraic topology uses algebraic objects to study topologies. The Poincaré conjecture, proved in 2003, asserts that the fundamental group of a manifold, which encodes information about connectedness, can be used to determine whether a manifold is a sphere or not. Algebraic number theory studies various number rings that generalize the set of integers. Using tools of algebraic number theory, Andrew Wiles proved Fermat's Last Theorem.

In physics, groups are used to represent symmetry operations, and the usage of group theory could simplify differential equations. In gauge theory, the requirement of local symmetry can be used to deduce the equations describing a system. The groups that describe those symmetries are Lie groups, and the study of Lie groups and Lie algebras reveals much about the physical system; for instance, the number of force carriers in a theory is equal to dimension of the Lie algebra, and these bosons interact with the force they mediate if the Lie algebra is nonabelian.

Linear Algebra

Linear algebra is the branch of mathematics concerning vector spaces and linear mappings between such spaces. It includes the study of lines, planes, and subspaces, but is also concerned with properties common to all vector spaces.

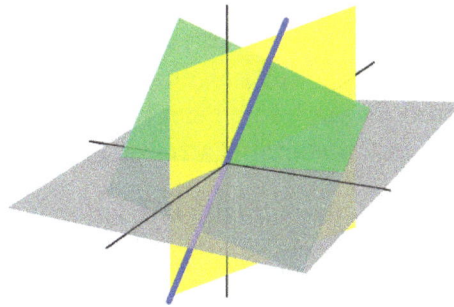

The three-dimensional Euclidean space $\mathbf{R}^3$ is a vector space, and lines and planes passing through the origin are vector subspaces in $\mathbf{R}^3$.

The set of points with coordinates that satisfy a linear equation forms a hyperplane in an n-dimensional space. The conditions under which a set of n hyperplanes intersect in a single point is an important focus of study in linear algebra. Such an investigation is initially motivated by a system of linear equations containing several unknowns. Such equations are naturally represented using the formalism of matrices and vectors.

Linear algebra is central to both pure and applied mathematics. For instance, abstract algebra arises by relaxing the axioms of a vector space, leading to a number of generalizations. Functional analysis studies the infinite-dimensional version of the theory of vector spaces. Combined with calculus, linear algebra facilitates the solution of linear systems of differential equations.

Techniques from linear algebra are also used in analytic geometry, engineering, physics, natural sciences, computer science, computer animation, advanced facial recognition algorithms and the social sciences (particularly in economics). Because linear algebra is such a well-developed theory, nonlinear mathematical models are sometimes approximated by linear models.

History

The study of linear algebra first emerged from the study of determinants, which were used to solve systems of linear equations. Determinants were used by Leibniz in 1693, and subsequently, Gabriel Cramer devised Cramer's Rule for solving linear systems in 1750. Later, Gauss further developed the theory of solving linear systems by using Gaussian elimination, which was initially listed as an advancement in geodesy.

The study of matrix algebra first emerged in England in the mid-1800s. In 1844 Hermann Grassmann published his "Theory of Extension" which included foundational new topics of what is today called linear algebra. In 1848, James Joseph Sylvester introduced the term matrix, which is Latin for "womb". While studying compositions of linear transformations, Arthur Cayley was led to define matrix multiplication and inverses. Crucially, Cayley used a single letter to denote a matrix, thus treating a matrix as an aggregate object. He also realized the connection between matrices and determinants, and wrote "There would be many things to say about this theory of matrices which should, it seems to me, precede the theory of determinants".

In 1882, Hüseyin Tevfik Pasha wrote the book titled "Linear Algebra". The first modern and more precise definition of a vector space was introduced by Peano in 1888; by 1900, a theory of linear transformations of finite-dimensional vector spaces had emerged. Linear algebra took its modern form in the first half of the twentieth century, when many ideas and methods of previous centuries were generalized as abstract algebra. The use of matrices in quantum mechanics, special relativity, and statistics helped spread the subject of linear algebra beyond pure mathematics. The development of computers led to increased research in efficient algorithms for Gaussian elimination and matrix decompositions, and linear algebra became an essential tool for modelling and simulations.

The origin of many of these ideas is discussed in the articles on determinants and Gaussian elimination.

Educational History

Linear algebra first appeared in graduate textbooks in the 1940s and in undergraduate textbooks in the 1950s. Following work by the School Mathematics Study Group, U.S. high schools asked 12th grade students to do "matrix algebra, formerly reserved for college" in the 1960s. In France during the 1960s, educators attempted to teach linear algebra through affine dimensional vector spaces in the first year of secondary school. This was met with a backlash in the 1980s that removed linear algebra from the curriculum. In 1993, the U.S.-based Linear Algebra Curriculum Study Group recommended that undergraduate linear algebra courses be given an application-based "matrix orientation" as opposed to a theoretical orientation.

Scope of Study

Vector Spaces

The main structures of linear algebra are vector spaces. A vector space over a field F is a set V together with two binary operations. Elements of V are called *vectors* and elements of F are called *scalars*. The first operation, *vector addition*, takes any two vectors v and w and outputs a third vector $v + w$. The second operation, *scalar multiplication*, takes any scalar a and any vector v and outputs a new vector av. The operations of addition and multiplication in a vector space must satisfy the following axioms. In the list below, let u, v and w be arbitrary vectors in V, and a and b scalars in F.

Axiom	Signification
Associativity of addition	$u + (v + w) = (u + v) + w$
Commutativity of addition	$u + v = v + u$
Identity element of addition	There exists an element $o \in V$, called the zero vector, such that $v + o = v$ for all $v \in V$.
Inverse elements of addition	For every $v \in V$, there exists an element $-v \in V$, called the additive inverse of v, such that $v + (-v) = o$
Distributivity of scalar multiplication with respect to vector addition	$a(u + v) = au + av$

Distributivity of scalar multiplication with respect to field addition	$(a + b)v = av + bv$
Compatibility of scalar multiplication with field multiplication	$a(bv) = (ab)v$
Identity element of scalar multiplication	$1v = v$, where 1 denotes the multiplicative identity in F.

The first four axioms are those of V being an abelian group under vector addition. Vector spaces may be diverse in nature, for example, containing functions, polynomials or matrices. Linear algebra is concerned with properties common to all vector spaces.

Linear Transformations

Similarly as in the theory of other algebraic structures, linear algebra studies mappings between vector spaces that preserve the vector-space structure. Given two vector spaces V and W over a field F, a linear transformation (also called linear map, linear mapping or linear operator) is a map

$$T : V \rightarrow W$$

that is compatible with addition and scalar multiplication:

$$T(u + v) = T(u) + T(v), \quad T(av) = aT(v)$$

for any vectors $u, v \in V$ and a scalar $a \in$ F.

Additionally for any vectors $u, v \in V$ and scalars $a, b \in$ F:

$$T(au + bv) = T(au) + T(bv) = aT(u) + bT(v)$$

When a bijective linear mapping exists between two vector spaces (that is, every vector from the second space is associated with exactly one in the first), we say that the two spaces are isomorphic. Because an isomorphism preserves linear structure, two isomorphic vector spaces are "essentially the same" from the linear algebra point of view. One essential question in linear algebra is whether a mapping is an isomorphism or not, and this question can be answered by checking if the determinant is nonzero. If a mapping is not an isomorphism, linear algebra is interested in finding its range (or image) and the set of elements that get mapped to zero, called the kernel of the mapping.

Linear transformations have geometric significance. For example, 2 × 2 real matrices denote standard planar mappings that preserve the origin.

Subspaces, Span, and Basis

Again, in analogue with theories of other algebraic objects, linear algebra is interested in subsets of vector spaces that are themselves vector spaces; these subsets are called linear subspaces. For example, both the range and kernel of a linear mapping are subspaces, and are thus often called the range space and the nullspace; these are important examples of subspaces. Another important way of forming a subspace is to take a linear combination of a set of vectors $v_1, v_2, ..., v_k$:

$$a_1 v_1 + a_2 v_2 + \cdots + a_k v_k,$$

where $a_1, a_2, ..., a_k$ are scalars. The set of all linear combinations of vectors $v_1, v_2, ..., v_k$ is called their span, which forms a subspace.

A linear combination of any system of vectors with all zero coefficients is the zero vector of V. If this is the only way to express the zero vector as a linear combination of $v_1, v_2, ..., v_k$ then these vectors are linearly independent. Given a set of vectors that span a space, if any vector w is a linear combination of other vectors (and so the set is not linearly independent), then the span would remain the same if we remove w from the set. Thus, a set of linearly dependent vectors is redundant in the sense that there will be a linearly independent subset which will span the same subspace. Therefore, we are mostly interested in a linearly independent set of vectors that spans a vector space V, which we call a basis of V. Any set of vectors that spans V contains a basis, and any linearly independent set of vectors in V can be extended to a basis. It turns out that if we accept the axiom of choice, every vector space has a basis; nevertheless, this basis may be unnatural, and indeed, may not even be constructible. For instance, there exists a basis for the real numbers, considered as a vector space over the rationals, but no explicit basis has been constructed.

Any two bases of a vector space V have the same cardinality, which is called the dimension of V. The dimension of a vector space is well-defined by the dimension theorem for vector spaces. If a basis of V has finite number of elements, V is called a finite-dimensional vector space. If V is finite-dimensional and U is a subspace of V, then $\dim U \le \dim V$. If U_1 and U_2 are subspaces of V, then

$$\dim(U_1 + U_2) = \dim U_1 + \dim U_2 - \dim(U_1 \cap U_2).$$

One often restricts consideration to finite-dimensional vector spaces. A fundamental theorem of linear algebra states that all vector spaces of the same dimension are isomorphic, giving an easy way of characterizing isomorphism.

Matrix Theory

A particular basis $\{v_1, v_2, ..., v_n\}$ of V allows one to construct a coordinate system in V: the vector with coordinates $(a_1, a_2, ..., a_n)$ is the linear combination

$$a_1 v_1 + a_2 v_2 + \cdots + a_n v_n.$$

The condition that $v_1, v_2, ..., v_n$ span V guarantees that each vector v can be assigned coordinates, whereas the linear independence of $v_1, v_2, ..., v_n$ assures that these coordinates are unique (i.e. there is only one linear combination of the basis vectors that is equal to v). In this way, once a basis of a vector space V over F has been chosen, V may be identified with the coordinate n-space F^n. Under this identification, addition and scalar multiplication of vectors in V correspond to addition and scalar multiplication of their coordinate vectors in F^n. Furthermore, if V and W are an n-dimensional and m-dimensional vector space over F, and a basis of V and a basis of W have been fixed, then any linear transformation $T: V \to W$ may be encoded by an $m \times n$ matrix A with entries in the field F, called the matrix of T with respect to these bases. Two matrices that encode the same linear transformation in different bases are called similar. Matrix theory replaces the study of linear transformations, which were defined axiomatically, by the study of matrices, which are concrete objects. This major technique distinguishes linear algebra from theories of other algebraic structures, which usually cannot be parameterized so concretely.

There is an important distinction between the coordinate n-space R^n and a general finite-dimensional vector space V. While R^n has a standard basis $\{e_1, e_2, ..., e_n\}$, a vector space V typically does not come equipped with such a basis and many different bases exist (although they all consist of the same number of elements equal to the dimension of V).

One major application of the matrix theory is calculation of determinants, a central concept in linear algebra. While determinants could be defined in a basis-free manner, they are usually introduced via a specific representation of the mapping; the value of the determinant does not depend on the specific basis. It turns out that a mapping has an inverse if and only if the determinant has an inverse (every non-zero real or complex number has an inverse). If the determinant is zero, then the nullspace is nontrivial. Determinants have other applications, including a systematic way of seeing if a set of vectors is linearly independent (we write the vectors as the columns of a matrix, and if the determinant of that matrix is zero, the vectors are linearly dependent). Determinants could also be used to solve systems of linear equations, but in real applications, Gaussian elimination is a faster method.

Eigenvalues and Eigenvectors

In general, the action of a linear transformation may be quite complex. Attention to low-dimensional examples gives an indication of the variety of their types. One strategy for a general n-dimensional transformation T is to find "characteristic lines" that are invariant sets under T. If v is a non-zero vector such that Tv is a scalar multiple of v, then the line through 0 and v is an invariant set under T and v is called a characteristic vector or eigenvector. The scalar λ such that $Tv = \lambda v$ is called a characteristic value or eigenvalue of T.

To find an eigenvector or an eigenvalue, we note that

$$Tv - \lambda v = (T - \lambda I)v = 0,$$

where I is the identity matrix. For there to be nontrivial solutions to that equation, $\det(T - \lambda I) = 0$. The determinant is a polynomial, and so the eigenvalues are not guaranteed to exist if the field is R. Thus, we often work with an algebraically closed field such as the complex numbers when dealing with eigenvectors and eigenvalues so that an eigenvalue will always exist. It would be particularly nice if given a transformation T taking a vector space V into itself we can find a basis for V consisting of eigenvectors. If such a basis exists, we can easily compute the action of the transformation on any vector: if $v_1, v_2, ..., v_n$ are linearly independent eigenvectors of a mapping of n-dimensional spaces T with (not necessarily distinct) eigenvalues $\lambda_1, \lambda_2, ..., \lambda_n$, and if $v = a_1 v_1 + ... + a_n v_n$, then,

$$T(v) = T(a_1 v_1) + \cdots + T(a_n v_n) = a_1 T(v_1) + \cdots + a_n T(v_n) = a_1 \lambda_1 v_1 + \cdots + a_n \lambda_n v_n.$$

Such a transformation is called a diagonalizable matrix since in the eigenbasis, the transformation is represented by a diagonal matrix. Because operations like matrix multiplication, matrix inversion, and determinant calculation are simple on diagonal matrices, computations involving matrices are much simpler if we can bring the matrix to a diagonal form. Not all matrices are diagonalizable (even over an algebraically closed field).

Inner-product Spaces

Besides these basic concepts, linear algebra also studies vector spaces with additional structure, such as an inner product. The inner product is an example of a bilinear form, and it gives the vector space a geometric structure by allowing for the definition of length and angles. Formally, an *inner product* is a map

$$\langle \cdot, \cdot \rangle : V \times V \to F$$

that satisfies the following three axioms for all vectors u, v, w in V and all scalars a in F:

- Conjugate symmetry:

$$\langle u, v \rangle = \overline{\langle v, u \rangle}.$$

Note that in R, it is symmetric.

- Linearity in the first argument:

$$\langle au, v \rangle = a \langle u, v \rangle.$$

$$\langle u + v, w \rangle = \langle u, w \rangle + \langle v, w \rangle.$$

- Positive-definiteness:

$$\langle v, v \rangle \geq 0 \text{ with equality only for } v = 0.$$

We can define the length of a vector v in V by

$$\| v \|^2 = \langle v, v \rangle,$$

and we can prove the Cauchy–Schwarz inequality:

$$| \langle u, v \rangle | \leq \| u \| \cdot \| v \|.$$

In particular, the quantity

$$\frac{| \langle u, v \rangle |}{\| u \| \cdot \| v \|} \leq 1,$$

and so we can call this quantity the cosine of the angle between the two vectors.

Two vectors are orthogonal if $\langle u, v \rangle = 0$. An orthonormal basis is a basis where all basis vectors have length 1 and are orthogonal to each other. Given any finite-dimensional vector space, an orthonormal basis could be found by the Gram–Schmidt procedure. Orthonormal bases are particularly nice to deal with, since if $v = a_1 v_1 + \ldots + a_n v_n$, then $a_i = \langle v, v_i \rangle$.

The inner product facilitates the construction of many useful concepts. For instance, given a transform T, we can define its Hermitian conjugate T^* as the linear transform satisfying

$$\langle Tu, v \rangle = \langle u, T^*v \rangle.$$

If T satisfies $TT^* = T^*T$, we call T normal. It turns out that normal matrices are precisely the matrices that have an orthonormal system of eigenvectors that span V.

Some Main Useful Theorems

- A matrix is invertible, or non-singular, if and only if the linear map represented by the matrix is an isomorphism.

- Any vector space over a field F of dimension n is isomorphic to F^n as a vector space over F.

- Corollary: Any two vector spaces over F of the same finite dimension are isomorphic to each other.

- A linear map is an isomorphism if and only if the determinant is nonzero.

Applications

Because of the ubiquity of vector spaces, linear algebra is used in many fields of mathematics, natural sciences, computer science, and social science. Below are just some examples of applications of linear algebra.

Solution of Linear Systems

Linear algebra provides the formal setting for the linear combination of equations used in the Gaussian method. Suppose the goal is to find and describe the solution(s), if any, of the following system of linear equations:

$$\begin{aligned} 2x + y - z &= 8 & (L_1) \\ -3x - y + 2z &= -11 & (L_2) \\ -2x + y + 2z &= -3 & (L_3) \end{aligned}$$

The Gaussian-elimination algorithm is as follows: eliminate x from all equations below L_1, and then eliminate y from all equations below L_2. This will put the system into triangular form. Then, using back-substitution, each unknown can be solved for.

In the example, x is eliminated from L_2 by adding $(3/2)L_1$ to L_2. x is then eliminated from L_3 by adding L_1 to L_3. Formally:

$$L_2 + \tfrac{3}{2}L_1 \rightarrow L_2$$

$$L_3 + L_1 \rightarrow L_3$$

The result is:

$$\begin{aligned} 2x + y - z &= 8 \\ \frac{1}{2}y + \frac{1}{2}z &= 1 \\ 2y + z &= 5 \end{aligned}$$

Now y is eliminated from L_3 by adding $-4L_2$ to L_3:

$$L_3 + -4L_2 \to L_3$$

The result is:

$$2x + y - z = 8$$
$$\frac{1}{2}y + \frac{1}{2}z = 1$$
$$-z = 1$$

This result is a system of linear equations in triangular form, and so the first part of the algorithm is complete.

The last part, back-substitution, consists of solving for the known in reverse order. It can thus be seen that

$$z = -1 \quad (L_3)$$

Then, z can be substituted into L_2, which can then be solved to obtain

$$y = 3 \quad (L_2)$$

Next, z and y can be substituted into L_1, which can be solved to obtain

$$x = 2 \quad (L_1)$$

The system is solved.

We can, in general, write any system of linear equations as a matrix equation:

$$Ax = b.$$

The solution of this system is characterized as follows: first, we find a particular solution x_0 of this equation using Gaussian elimination. Then, we compute the solutions of $Ax = 0$; that is, we find the null space N of A. The solution set of this equation is given by $x_0 + N = \{x_0 + n : n \in N\}$. If the number of variables is equal to the number of equations, then we can characterize when the system has a unique solution: since N is trivial if and only if $\det A \neq 0$, the equation has a unique solution if and only if $\det A \neq 0$.

Least-squares Best Fit Line

The least squares method is used to determine the best fit line for a set of data. This line will minimize the sum of the squares of the residuals.

Fourier Series Expansion

Fourier series are a representation of a function $f : [-\pi, \pi] \to R$ as a trigonometric series:

$$f(x) = \frac{a_0}{2} + \sum_{n=1}^{\infty} [a_n \cos(nx) + b_n \sin(nx)].$$

This series expansion is extremely useful in solving partial differential equations. In this article, we will not be concerned with convergence issues; it is nice to note that all Lipschitz-continuous functions have a converging Fourier series expansion, and nice enough discontinuous functions have a Fourier series that converges to the function value at most points.

The space of all functions that can be represented by a Fourier series form a vector space (technically speaking, we call functions that have the same Fourier series expansion the "same" function, since two different discontinuous functions might have the same Fourier series). Moreover, this space is also an inner product space with the inner product

$$\langle f, g \rangle = \frac{1}{\pi} \int_{-\pi}^{\pi} f(x) g(x) dx.$$

The functions $g_n(x) = \sin(nx)$ for $n > 0$ and $h_n(x) = \cos(nx)$ for $n \geq 0$ are an orthonormal basis for the space of Fourier-expandable functions. We can thus use the tools of linear algebra to find the expansion of any function in this space in terms of these basis functions. For instance, to find the coefficient a_k, we take the inner product with h_k:

$$\langle f, h_k \rangle = \frac{a_0}{2} \langle h_0, h_k \rangle + \sum_{n=1}^{\infty} [a_n \langle h_n, h_k \rangle + b_n \langle g_n, h_k \rangle],$$

and by orthonormality, $\langle f, h_k \rangle = a_k$; that is,

$$a_k = \frac{1}{\pi} \int_{-\pi}^{\pi} f(x) \cos(kx) dx.$$

Quantum Mechanics

Quantum mechanics is highly inspired by notions in linear algebra. In quantum mechanics, the physical state of a particle is represented by a vector, and observables (such as momentum, energy, and angular momentum) are represented by linear operators on the underlying vector space. More concretely, the wave function of a particle describes its physical state and lies in the vector space L^2 (the functions $\varphi: R^3 \to C$ such that $\int_{-\infty}^{\infty} \int_{-\infty}^{\infty} \int_{-\infty}^{\infty} |\phi|^2 \, dxdydz$ is finite), and it evolves according to the Schrödinger equation. Energy is represented as the operator $H = -\frac{\hbar^2}{2m} \nabla^2 + V(x, y, z)$, where V is the potential energy. H is also known as the Hamiltonian operator. The eigenvalues of H represents the possible energies that can be observed. Given a particle in some state φ, we can expand φ into a linear combination of eigenstates of H. The component of H in each eigenstate determines the probability of measuring the corresponding eigenvalue, and the measurement forces the particle to assume that eigenstate (wave function collapse).

Geometric Introduction

Many of the principles and techniques of linear algebra can be seen in the geometry of lines in a real two dimensional plane E. When formulated using vectors and matrices the geometry of points and lines in the plane can be extended to the geometry of points and hyperplanes in high-dimensional spaces.

Point coordinates in the plane E are ordered pairs of real numbers, (x,y), and a line is defined as the set of points (x,y) that satisfy the linear equation

$$\lambda : ax + by + c = 0,$$

where a, b and c are not all zero. Then,

$$\lambda : \begin{bmatrix} a & b & c \end{bmatrix} \begin{Bmatrix} x \\ y \\ 1 \end{Bmatrix} = 0,$$

or

$$A\mathbf{x} = 0,$$

where $\mathbf{x} = (x, y, 1)$ is the 3×1 set of homogeneous coordinates associated with the point (x, y).

Homogeneous coordinates identify the plane E with the $z = 1$ plane in three dimensional space. The x–y coordinates in E are obtained from homogeneous coordinates $\mathbf{y} = (y_1, y_2, y_3)$ by dividing by the third component (if it is nonzero) to obtain $\mathbf{y} = (y_1/y_3, y_2/y_3, 1)$.

The linear equation, λ, has the important property, that if $\mathbf{x}_1$ and $\mathbf{x}_2$ are homogeneous coordinates of points on the line, then the point $\alpha \mathbf{x}_1 + \beta \mathbf{x}_2$ is also on the line, for any real α and β.

Now consider the equations of the two lines λ_1 and λ_2,

$$\lambda_1 : a_1 x + b_1 y + c_1 = 0, \quad \lambda_2 : a_2 x + b_2 y + c_2 = 0,$$

which forms a system of linear equations. The intersection of these two lines is defined by $\mathbf{x} = (x, y, 1)$ that satisfy the matrix equation,

$$\lambda_{1,2} : \begin{bmatrix} a_1 & b_1 & c_1 \\ a_2 & b_2 & c_2 \end{bmatrix} \begin{Bmatrix} x \\ y \\ 1 \end{Bmatrix} = \begin{Bmatrix} 0 \\ 0 \end{Bmatrix},$$

or using homogeneous coordinates,

$$B\mathbf{x} = 0.$$

The point of intersection of these two lines is the unique non-zero solution of these equations. In homogeneous coordinates, the solutions are multiples of the following solution:

$$x_1 = \begin{vmatrix} b_1 & c_1 \\ b_2 & c_2 \end{vmatrix}, x_2 = - \begin{vmatrix} a_1 & c_1 \\ a_2 & c_2 \end{vmatrix}, x_3 = \begin{vmatrix} a_1 & b_1 \\ a_2 & b_2 \end{vmatrix}$$

if the rows of B are linearly independent (i.e., λ_1 and λ_2 represent distinct lines). Divide through by x_3 to get Cramer's rule for the solution of a set of two linear equations in two unknowns. Notice that this yields a point in the $z = 1$ plane only when the 2×2 submatrix associated with x_3 has a non-zero determinant.

It is interesting to consider the case of three lines, λ_1, λ_2 and λ_3, which yield the matrix equation,

$$\lambda_{1,2,3}: \begin{bmatrix} a_1 & b_1 & c_1 \\ a_2 & b_2 & c_2 \\ a_3 & b_3 & c_3 \end{bmatrix} \begin{Bmatrix} x \\ y \\ 1 \end{Bmatrix} = \begin{Bmatrix} 0 \\ 0 \\ 0 \end{Bmatrix}.$$

which in homogeneous form yields,

$$C\mathbf{x} = 0.$$

Clearly, this equation has the solution x = (0,0,0), which is not a point on the $z = 1$ plane E. For a solution to exist in the plane E, the coefficient matrix C must have rank 2, which means its determinant must be zero. Another way to say this is that the columns of the matrix must be linearly dependent.

Introduction to Linear Transformations

Another way to approach linear algebra is to consider linear functions on the two dimensional real plane $E=R^2$. Here R denotes the set of real numbers. Let x=(x, y) be an arbitrary vector in E and consider the linear function $\lambda: E \to R$, given by

$$\lambda: \begin{bmatrix} a & b \end{bmatrix} \begin{Bmatrix} x \\ y \end{Bmatrix} = c,$$

or

$$A\mathbf{x} = c.$$

This transformation has the important property that if Ay=d, then

$$A(\alpha\mathbf{x} + \beta\mathbf{y}) = \alpha A\mathbf{x} + \beta A\mathbf{y} = \alpha c + \beta d.$$

This shows that the sum of vectors in E map to the sum of their images in R. This is the defining characteristic of a linear map, or linear transformation. For this case, where the image space is a real number the map is called a linear functional.

Consider the linear functional a little more carefully. Let i=(1,0) and j =(0,1) be the natural basis vectors on E, so that x=xi+yj. It is now possible to see that

$$A\mathbf{x} = A(x\mathbf{i} + y\mathbf{j}) = xA\mathbf{i} + yA\mathbf{j} = \begin{bmatrix} A\mathbf{i} & A\mathbf{j} \end{bmatrix} \begin{Bmatrix} x \\ y \end{Bmatrix} = \begin{bmatrix} a & b \end{bmatrix} \begin{Bmatrix} x \\ y \end{Bmatrix} = c.$$

Thus, the columns of the matrix A are the image of the basis vectors of E in R.

This is true for any pair of vectors used to define coordinates in E. Suppose we select a non-orthogonal non-unit vector basis v and w to define coordinates of vectors in E. This means a vector x has coordinates (α,β), such that x=αv+βw. Then, we have the linear functional

$$\lambda: A\mathbf{x} = \begin{bmatrix} A\mathbf{v} & A\mathbf{w} \end{bmatrix} \begin{Bmatrix} \alpha \\ \beta \end{Bmatrix} = \begin{bmatrix} d & e \end{bmatrix} \begin{Bmatrix} \alpha \\ \beta \end{Bmatrix} = c,$$

where Av=d and Aw=e are the images of the basis vectors v and w. This is written in matrix form as

$$\begin{bmatrix} a & b \end{bmatrix} \begin{bmatrix} v_1 & w_1 \\ v_2 & w_2 \end{bmatrix} = \begin{bmatrix} d & e \end{bmatrix}.$$

Coordinates Relative to a Basis

This leads to the question of how to determine the coordinates of a vector x relative to a general basis v and w in E. Assume that we know the coordinates of the vectors, x, v and w in the natural basis i=(1,0) and j =(0,1). Our goal is two find the real numbers α, β, so that x=αv+βw, that is

$$\begin{Bmatrix} x \\ y \end{Bmatrix} = \begin{bmatrix} v_1 & w_1 \\ v_2 & w_2 \end{bmatrix} \begin{Bmatrix} \alpha \\ \beta \end{Bmatrix}.$$

To solve this equation for α, β, we compute the linear coordinate functionals σ and τ for the basis v, w, which are given by,

$$\sigma = \begin{bmatrix} \sigma_1 & \sigma_2 \end{bmatrix} = \frac{1}{v_1 w_2 - v_2 w_1} \begin{bmatrix} w_2 & -w_1 \end{bmatrix}, \tau = \begin{bmatrix} \tau_1 & \tau_2 \end{bmatrix} = \frac{1}{v_1 w_2 - v_2 w_1} \begin{bmatrix} -v_2 & v_1 \end{bmatrix},$$

The functionals σ and τ compute the components of x along the basis vectors v and w, respectively, that is,

$$\sigma \mathbf{x} = \alpha, \tau \mathbf{x} = \beta,$$

which can be written in matrix form as

$$\begin{bmatrix} \sigma_1 & \sigma_2 \\ \tau_1 & \tau_2 \end{bmatrix} \begin{Bmatrix} x \\ y \end{Bmatrix} = \begin{Bmatrix} \alpha \\ \beta \end{Bmatrix}.$$

These coordinate functionals have the properties,

$$\sigma \mathbf{v} = 1, \sigma \mathbf{w} = 0, \tau \mathbf{w} = 1, \tau \mathbf{v} = 0.$$

These equations can be assembled into the single matrix equation,

$$\begin{bmatrix} \sigma_1 & \sigma_2 \\ \tau_1 & \tau_2 \end{bmatrix} \begin{bmatrix} v_1 & w_1 \\ v_2 & w_2 \end{bmatrix} = \begin{bmatrix} 1 & 0 \\ 0 & 1 \end{bmatrix}.$$

Thus, the matrix formed by the coordinate linear functionals is the inverse of the matrix formed by the basis vectors.

Inverse Image

The set of points in the plane E that map to the same image in R under the linear functional λ define a line in E. This line is the image of the inverse map, $\lambda^{-1}\colon R \to E$. This inverse image is the set of the points $x=(x, y)$ that solve the equation,

$$A\mathbf{x} = \begin{bmatrix} a & b \end{bmatrix} \begin{Bmatrix} x \\ y \end{Bmatrix} = c.$$

Notice that a linear functional operates on known values for $x=(x, y)$ to compute a value c in R, while the inverse image seeks the values for $x=(x, y)$ that yield a specific value c.

In order to solve the equation, we first recognize that only one of the two unknowns (x,y) can be determined, so we select y to be determined, and rearrange the equation

$$by = c - ax.$$

Solve for y and obtain the inverse image as the set of points,

$$\mathbf{x}(t) = \begin{Bmatrix} 0 \\ c/b \end{Bmatrix} + t \begin{Bmatrix} 1 \\ -a/b \end{Bmatrix} = \mathbf{p} + t\mathbf{h}.$$

For convenience the free parameter x has been relabeled t.

The vector p defines the intersection of the line with the y-axis, known as the y-intercept. The vector h satisfies the homogeneous equation,

$$A\mathbf{h} = \begin{bmatrix} a & b \end{bmatrix} \begin{Bmatrix} 1 \\ -a/b \end{Bmatrix} = 0.$$

Notice that if h is a solution to this homogeneous equation, then $t\,h$ is also a solution.

The set of points of a linear functional that map to zero define the *kernel* of the linear functional. The line can be considered to be the set of points h in the kernel translated by the vector p.

Generalizations and Related Topics

Since linear algebra is a successful theory, its methods have been developed and generalized in other parts of mathematics. In module theory, one replaces the field of scalars by a ring. The concepts of linear independence, span, basis, and dimension (which is called rank in module theory) still make sense. Nevertheless, many theorems from linear algebra become false in module theory. For instance, not all modules have a basis (those that do are called free modules), the rank of a free module is not necessarily unique, not every linearly independent subset of a module can be extended to form a basis, and not every subset of a module that spans the space contains a basis.

In multilinear algebra, one considers multivariable linear transformations, that is, mappings that are linear in each of a number of different variables. This line of inquiry naturally leads to the idea of the dual space, the vector space V^* consisting of linear maps $f\colon V \to F$ where F is the field of scalars. Multilinear maps $T\colon V^n \to F$ can be described via tensor products of elements of V^*.

If, in addition to vector addition and scalar multiplication, there is a bilinear vector product $V \times V$

$\rightarrow V$, the vector space is called an algebra; for instance, associative algebras are algebras with an associate vector product (like the algebra of square matrices, or the algebra of polynomials).

Functional analysis mixes the methods of linear algebra with those of mathematical analysis and studies various function spaces, such as L^p spaces.

Representation theory studies the actions of algebraic objects on vector spaces by representing these objects as matrices. It is interested in all the ways that this is possible, and it does so by finding subspaces invariant under all transformations of the algebra. The concept of eigenvalues and eigenvectors is especially important.

Algebraic geometry considers the solutions of systems of polynomial equations.

There are several related topics in the field of Computer Programming that utilizes much of the techniques and theorems Linear Algebra encompasses and refers to.

Universal Algebra

Universal algebra (sometimes called general algebra) is the field of mathematics that studies algebraic structures themselves, not examples ("models") of algebraic structures. For instance, rather than take particular groups as the object of study, in universal algebra one takes "the theory of groups" as an object of study.

Basic Idea

In universal algebra, an algebra (or algebraic structure) is a set A together with a collection of operations on A. An **n**-ary operation on A is a function that takes n elements of A and returns a single element of A. Thus, a 0-ary operation (or *nullary operation*) can be represented simply as an element of A, or a *constant*, often denoted by a letter like a. A 1-ary operation (or *unary operation*) is simply a function from A to A, often denoted by a symbol placed in front of its argument, like $\sim x$. A 2-ary operation (or *binary operation*) is often denoted by a symbol placed between its arguments, like $x * y$. Operations of higher or unspecified *arity* are usually denoted by function symbols, with the arguments placed in parentheses and separated by commas, like $f(x,y,z)$ or $f(x_1,...,x_n)$. Some researchers allow infinitary operations, such as $\bigwedge_{\alpha \in J} x_\alpha$ where J is an infinite index set, thus leading into the algebraic theory of complete lattices. One way of talking about an algebra, then, is by referring to it as an algebra of a certain type Ω, where Ω is an ordered sequence of natural numbers representing the arity of the operations of the algebra.

Equations

After the operations have been specified, the nature of the algebra can be further limited by axioms, which in universal algebra often take the form of identities, or equational laws. An example is the associative axiom for a binary operation, which is given by the equation $x * (y * z) = (x * y) * z$. The axiom is intended to hold for all elements x, y, and z of the set A.

Varieties

An algebraic structure that can be defined by identities is called a variety, and these are sufficiently important that some authors consider varieties the only object of study in universal algebra, while others consider them an object.

Restricting one's study to varieties rules out:

- quantification, including universal quantification ($\forall$), except before an equation, and existential quantification ($\exists$)
    - Predicate logic in particular is ruled out
- All relations except equality, in particular inequalities, both $a \neq b$ and order relations

In this narrower definition, universal algebra can be seen as a special branch of model theory, typically dealing with structures having operations only (i.e. the type can have symbols for functions but not for relations other than equality), and in which the language used to talk about these structures uses equations only.

Not all algebraic structures in a wider sense fall into this scope. For example, ordered groups are not studied in mainstream universal algebra because they involve an ordering relation.

A more fundamental restriction is that universal algebra cannot study the class of fields, because there is no type (a.k.a. signature) in which all field laws can be written as equations (inverses of elements are defined for all *non-zero* elements in a field, so inversion cannot simply be added to the type).

One advantage of this restriction is that the structures studied in universal algebra can be defined in any category that has *finite products*. For example, a topological group is just a group in the category of topological spaces.

Examples

Most of the usual algebraic systems of mathematics are examples of varieties, but not always in an obvious way – the usual definitions often involve quantification or inequalities.

Groups

To see how this works, let's consider the definition of a group. Normally a group is defined in terms of a single binary operation $*$, subject to these axioms:

- Associativity (as in the previous section): $x * (y * z) = (x * y) * z$;   formally: $\forall x,y,z.\ x*(y*z)=(x*y)*z$.
- Identity element: There exists an element e such that for each element x, $e * x = x = x * e$; formally: $\exists e\ \forall x.\ e*x=x=x*e$.
- Inverse element: It can easily be seen that the identity element is unique. If this unique identity element is denoted by e then for each x, there exists an element i such that $x * i = e = i * x$;   formally: $\forall x\ \exists i.\ x*i=e=i*x$.

(Some authors also use an axiom called "closure", stating that $x * y$ belongs to the set A whenever x and y do. But from a universal algebraist's point of view, that is already implied by calling $*$ a binary operation.)

This definition of a group is problematic from the point of view of universal algebra. The reason is that the axioms of the identity element and inversion are not stated purely in terms of equational laws but also have clauses involving the phrase "there exists ... such that ...". This is inconvenient; the list of group properties can be simplified to universally quantified equations by adding a nullary operation e and a unary operation $\sim$ in addition to the binary operation $*$. Then list the axioms for these three operations as follows:

- Associativity: $x * (y * z) = (x * y) * z$.

- Identity element: $e * x = x = x * e$; formally: $\forall x.\ e*x=x=x*e$.

- Inverse element: $x * (\sim x) = e = (\sim x) * x$ formally: $\forall x.\ x*\sim x=e=\sim x*x$.

(Of course, we usually write "x^{-1}" instead of "$\sim x$", which shows that the notation for operations of low arity is not *always* as given in the second paragraph.)

What has changed is that in the usual definition there are:

- a single binary operation (signature (2))

- 1 equational law (associativity)

- 2 quantified laws (identity and inverse)

...while in the universal algebra definition there are

- 3 operations: one binary, one unary, and one nullary (signature (2,1,0))

- 3 equational laws (associativity, identity, and inverse)

- no quantified laws (except for outermost universal quantifiers which are allowed in varieties)

It is important to check that this really does capture the definition of a group. The reason that it might not is that specifying one of these universal groups might give more information than specifying one of the usual kind of group. After all, nothing in the usual definition said that the identity element e was *unique*; if there is another identity element e', then it is ambiguous which one should be the value of the nullary operator e. Proving that it is unique is a common beginning exercise in classical group theory textbooks. The same thing is true of inverse elements. So, the universal algebraist's definition of a group is equivalent to the usual definition.

At first glance this is simply a technical difference, replacing quantified laws with equational laws. However, it has immediate practical consequences – when defining a group object in category theory, where the object in question may not be a set, one must use equational laws (which make sense in general categories), and cannot use quantified laws (which do not make sense, as objects in general categories do not have elements). Further, the perspective of universal algebra insists not only that the inverse and identity exist, but that they be maps in the category. The basic example is of a topological group – not only must the inverse exist element-wise, but the inverse map must be continuous (some authors also require the identity map to be a closed inclusion, hence cofibration, again referring to properties of the map).

Basic Constructions

We assume that the type, Ω, has been fixed. Then there are three basic constructions in universal algebra: homomorphic image, subalgebra, and product.

A homomorphism between two algebras A and B is a function $h: A \to B$ from the set A to the set B such that, for every operation f_A of A and corresponding f_B of B (of arity, say, n), $h(f_A(x_1,...,x_n)) = f_B(h(x_1),...,h(x_n))$. (Sometimes the subscripts on f are taken off when it is clear from context which algebra the function is from.) For example, if e is a constant (nullary operation), then $h(e_A) = e_B$. If $\sim$ is a unary operation, then $h(\sim x) = \sim h(x)$. If $*$ is a binary operation, then $h(x * y) = h(x) * h(y)$. And so on. A few of the things that can be done with homomorphisms, as well as definitions of certain special kinds of homomorphisms, are listed under the entry Homomorphism. In particular, we can take the homomorphic image of an algebra, $h(A)$.

A subalgebra of A is a subset of A that is closed under all the operations of A. A product of some set of algebraic structures is the cartesian product of the sets with the operations defined coordinatewise.

Some Basic Theorems

- The isomorphism theorems, which encompass the isomorphism theorems of groups, rings, modules, etc.

- Birkhoff's HSP Theorem, which states that a class of algebras is a variety if and only if it is closed under homomorphic images, subalgebras, and arbitrary direct products.

Motivations and Applications

In addition to its unifying approach, universal algebra also gives deep theorems and important examples and counterexamples. It provides a useful framework for those who intend to start the study of new classes of algebras. It can enable the use of methods invented for some particular classes of algebras to other classes of algebras, by recasting the methods in terms of universal algebra (if possible), and then interpreting these as applied to other classes. It has also provided conceptual clarification; as J.D.H. Smith puts it, *"What looks messy and complicated in a particular framework may turn out to be simple and obvious in the proper general one."*

In particular, universal algebra can be applied to the study of monoids, rings, and lattices. Before universal algebra came along, many theorems (most notably the isomorphism theorems) were proved separately in all of these fields, but with universal algebra, they can be proven once and for all for every kind of algebraic system.

The 1956 paper by Higgins referenced below has been well followed up for its framework for a range of particular algebraic systems, while his 1963 paper is notable for its discussion of algebras with operations which are only partially defined, typical examples for this being categories and groupoids. This leads on to the subject of higher-dimensional algebra which can be defined as the study of algebraic theories with partial operations whose domains are defined under geometric conditions. Notable examples of these are various forms of higher-dimensional categories and groupoids.

Generalizations

A more generalised programme along these lines is carried out by category theory. Given a list of operations and axioms in universal algebra, the corresponding algebras and homomorphisms are the objects and morphisms of a category. Category theory applies to many situations where universal algebra does not, extending the reach of the theorems. Conversely, many theorems that hold in universal algebra do not generalise all the way to category theory. Thus both fields of study are useful.

A more recent development in category theory that generalizes operations is operad theory – an operad is a set of operations, similar to a universal algebra.

Another development is partial algebra where the operators can be partial functions.

An important generalization of universal algebra theory is model theory, which is sometimes described as "universal algebra + logic".

History

In Alfred North Whitehead's book *A Treatise on Universal Algebra,* published in 1898, the term *universal algebra* had essentially the same meaning that it has today. Whitehead credits William Rowan Hamilton and Augustus De Morgan as originators of the subject matter, and James Joseph Sylvester with coining the term itself.

At the time structures such as Lie algebras and hyperbolic quaternions drew attention to the need to expand algebraic structures beyond the associatively multiplicative class. In a review Alexander Macfarlane wrote: "The main idea of the work is not unification of the several methods, nor generalization of ordinary algebra so as to include them, but rather the comparative study of their several structures." At the time George Boole's algebra of logic made a strong counterpoint to ordinary number algebra, so the term "universal" served to calm strained sensibilities.

Whitehead's early work sought to unify quaternions (due to Hamilton), Grassmann's Ausdehnungslehre, and Boole's algebra of logic. Whitehead wrote in his book:

> *"Such algebras have an intrinsic value for separate detailed study; also they are worthy of comparative study, for the sake of the light thereby thrown on the general theory of symbolic reasoning, and on algebraic symbolism in particular. The comparative study necessarily presupposes some previous separate study, comparison being impossible without knowledge."*

Whitehead, however, had no results of a general nature. Work on the subject was minimal until the early 1930s, when Garrett Birkhoff and Øystein Ore began publishing on universal algebras. Developments in metamathematics and category theory in the 1940s and 1950s furthered the field, particularly the work of Abraham Robinson, Alfred Tarski, Andrzej Mostowski, and their students (Brainerd 1967).

In the period between 1935 and 1950, most papers were written along the lines suggested by Birkhoff's papers, dealing with free algebras, congruence and subalgebra lattices, and homomorphism theorems. Although the development of mathematical logic had made applications to algebra possible, they came about slowly; results published by Anatoly Maltsev in the 1940s went unnoticed because of the war. Tarski's lecture at the 1950 International Congress of Mathematicians in Cambridge ushered in a new period in which model-theoretic aspects were developed, mainly by Tarski himself, as well as C.C. Chang, Leon Henkin, Bjarni Jónsson, Roger Lyndon, and others.

In the late 1950s, Edward Marczewski emphasized the importance of free algebras, leading to the publication of more than 50 papers on the algebraic theory of free algebras by Marczewski himself, together with Jan Mycielski, Władysław Narkiewicz, Witold Nitka, J. Płonka, S. Świerczkowski, K. Urbanik, and others.

Von Neumann Algebra

In mathematics, a von Neumann algebra or W*-algebra is a *-algebra of bounded operators on a Hilbert space that is closed in the weak operator topology and contains the identity operator. They were originally introduced by John von Neumann, motivated by his study of single operators, group representations, ergodic theory and quantum mechanics. His double commutant theorem shows that the analytic definition is equivalent to a purely algebraic definition as an algebra of symmetries.

Two basic examples of von Neumann algebras are as follows. The ring $L^\infty(R)$ of essentially bounded measurable functions on the real line is a commutative von Neumann algebra, which acts by pointwise multiplication on the Hilbert space $L^2(R)$ of square integrable functions. The algebra $B(H)$ of all bounded operators on a Hilbert space H is a von Neumann algebra, non-commutative if the Hilbert space has dimension at least 2.

Von Neumann algebras were first studied by von Neumann (1930) in 1929; he and Francis Murray developed the basic theory, under the original name of rings of operators, in a series of papers written in the 1930s and 1940s (F.J. Murray & J. von Neumann 1936, 1937, 1943; J. von Neumann 1938, 1940, 1943, 1949), reprinted in the collected works of von Neumann (1961).

Introductory accounts of von Neumann algebras are given in the online notes of Jones (2003) and Wassermann (1991) and the books by Dixmier (1981), Schwartz (1967), Blackadar (2005) and Sakai (1971). The three volume work by Takesaki (1979) gives an encyclopedic account of the theory. The book by Connes (1994) discusses more advanced topics.

Definitions

There are three common ways to define von Neumann algebras.

The first and most common way is to define them as weakly closed *-algebras of bounded operators (on a Hilbert space) containing the identity. In this definition the weak (operator) topology can be replaced by many other common topologies including the strong, ultrastrong or ultraweak operator topologies. The *-algebras of bounded operators that are closed in the norm topology are C*-algebras, so in particular any von Neumann algebra is a C*-algebra.

The second definition is that a von Neumann algebra is a subset of the bounded operators closed under * and equal to its double commutant, or equivalently the commutant of some subset closed under *. The von Neumann double commutant theorem (von Neumann 1930) says that the first two definitions are equivalent.

The first two definitions describe a von Neumann algebra concretely as a set of operators acting on some given Hilbert space. Sakai (1971) showed that von Neumann algebras can also be defined abstractly as C*-algebras that have a predual; in other words the von Neumann algebra, considered as a Banach space, is the dual of some other Banach space called the predual. The predual of a von Neumann algebra is in fact unique up to isomorphism. Some authors use "von Neumann algebra" for the algebras together with a Hilbert space action, and "W*-algebra" for the abstract concept, so a von Neumann algebra is a W*-algebra together with a Hilbert space and a suitable faithful unital action on the Hilbert space. The concrete and abstract definitions of a von Neumann algebra are similar to the concrete and abstract definitions of a C*-algebra, which can be defined either as norm-closed *-algebras of operators on a Hilbert space, or as Banach *-algebras such that $||aa^*||=||a||\,||a^*||$.

Terminology

Some of the terminology in von Neumann algebra theory can be confusing, and the terms often have different meanings outside the subject.

- A factor is a von Neumann algebra with trivial center, i.e. a center consisting only of scalar operators.

- A finite von Neumann algebra is one which is the direct integral of finite factors (meaning the von Neumann algebra has a faithful normal tracial state $\tau: M \to C$). Similarly, properly infinite von Neumann algebras are the direct integral of properly infinite factors.

- A von Neumann algebra that acts on a separable Hilbert space is called separable. Note that such algebras are rarely separable in the norm topology.

- The von Neumann algebra generated by a set of bounded operators on a Hilbert space is the smallest von Neumann algebra containing all those operators.

- The tensor product of two von Neumann algebras acting on two Hilbert spaces is defined to be the von Neumann algebra generated by their algebraic tensor product, considered as operators on the Hilbert space tensor product of the Hilbert spaces.

By forgetting about the topology on a von Neumann algebra, we can consider it a (unital) *-algebra, or just a ring. Von Neumann algebras are semihereditary: every finitely generated submodule of a projective module is itself projective. There have been several attempts to axiomatize the underlying rings of von Neumann algebras, including Baer *-rings and AW* algebras. The *-algebra of affiliated operators of a finite von Neumann algebra is a von Neumann regular ring. (The von Neumann algebra itself is in general not von Neumann regular.)

Commutative Von Neumann Algebras

The relationship between commutative von Neumann algebras and measure spaces is analogous to that between commutative C*-algebras and locally compact Hausdorff spaces. Every commutative von Neumann algebra is isomorphic to $L^\infty(X)$ for some measure space (X, μ) and conversely, for every σ-finite measure space X, the *-algebra $L^\infty(X)$ is a von Neumann algebra.

Due to this analogy, the theory of von Neumann algebras has been called noncommutative measure theory, while the theory of C*-algebras is sometimes called noncommutative topology (Connes 1994).

Projections

Operators E in a von Neumann algebra for which $E = EE = E^*$ are called projections; they are exactly the operators which give an orthogonal projection of H onto some closed subspace. A subspace of the Hilbert space H is said to belong to the von Neumann algebra M if it is the image of some projection in M. This establishes a 1:1 correspondence between projections of M and subspaces that belong to M. Informally these are the closed subspaces that can be described using elements of M, or that M "knows" about.

It can be shown that the closure of the image of any operator in M and the kernel of any operator in M belongs to M. Also, the closure of the image under an operator of M of any subspace belonging to M also belongs to M. (These results are a consequence of the polar decomposition).

Comparison Theory of Projections

The basic theory of projections was worked out by Murray & von Neumann (1936). Two subspaces belonging to M are called (Murray–von Neumann) equivalent if there is a partial isometry mapping the first isomorphically onto the other that is an element of the von Neumann algebra (informally, if M "knows" that the subspaces are isomorphic). This induces a natural equivalence relation on projections by defining E to be equivalent to F if the corresponding subspaces are equivalent, or in other words if there is a partial isometry of H that maps the image of E isometrically to the image of F and is an element of the von Neumann algebra. Another way of stating this is that E is equivalent to F if $E=uu^*$ and $F=u^*u$ for some partial isometry u in M.

The equivalence relation ~ thus defined is additive in the following sense: Suppose $E_1 \sim F_1$ and $E_2 \sim F_2$. If $E_1 \perp E_2$ and $F_1 \perp F_2$, then $E_1 + E_2 \sim F_1 + F_2$. Additivity would *not* generally hold if one were to require unitary equivalence in the definition of ~, i.e. if we say E is equivalent to F if $u^*Eu = F$ for some unitary u.

The subspaces belonging to M are partially ordered by inclusion, and this induces a partial order $\leq$ of projections. There is also a natural partial order on the set of *equivalence classes* of projections, induced by the partial order $\leq$ of projections. If M is a factor, $\leq$ is a total order on equivalence classes of projections, described in the section on traces below.

A projection (or subspace belonging to M) E is said to be a finite projection if there is no projection $F < E$ (meaning $F \leq E$ and $F \neq E$) that is equivalent to E. For example, all finite-dimensional projections (or subspaces) are finite (since isometries between Hilbert spaces leave the dimension fixed), but the identity operator on an infinite-dimensional Hilbert space is not finite in the von Neumann algebra of all bounded operators on it, since it is isometrically isomorphic to a proper subset of itself. However it is possible for infinite dimensional subspaces to be finite.

Orthogonal projections are noncommutative analogues of indicator functions in $L^\infty(R)$. $L^\infty(R)$ is the $||\cdot||_\infty$-closure of the subspace generated by the indicator functions. Similarly, a von Neumann algebra is generated by its projections; this is a consequence of the spectral theorem for self-adjoint operators.

The projections of a finite factor form a continuous geometry.

Factors

A von Neumann algebra N whose center consists only of multiples of the identity operator is called a factor. von Neumann (1949) showed that every von Neumann algebra on a separable Hilbert space is isomorphic to a direct integral of factors. This decomposition is essentially unique. Thus, the problem of classifying isomorphism classes of von Neumann algebras on separable Hilbert spaces can be reduced to that of classifying isomorphism classes of factors.

Murray & von Neumann (1936) showed that every factor has one of 3 types as described below. The type classification can be extended to von Neumann algebras that are not factors, and a von Neumann algebra is of type X if it can be decomposed as a direct integral of type X factors; for example, every commutative von Neumann algebra has type I_1. Every von Neumann algebra can be written uniquely as a sum of von Neumann algebras of types I, II, and III.

There are several other ways to divide factors into classes that are sometimes used:

- A factor is called discrete (or occasionally tame) if it has type I, and continuous (or occasionally wild) if it has type II or III.

- A factor is called semifinite if it has type I or II, and purely infinite if it has type III.

- A factor is called finite if the projection 1 is finite and properly infinite otherwise. Factors of types I and II may be either finite or properly infinite, but factors of type III are always properly infinite.

Type I Factors

A factor is said to be of type I if there is a minimal projection $E \neq 0$, i.e. a projection E such that there is no other projection F with $0 < F < E$. Any factor of type I is isomorphic to the von Neumann algebra of *all* bounded operators on some Hilbert space; since there is one Hilbert space for every cardinal number, isomorphism classes of factors of type I correspond exactly to the cardinal numbers. Since many authors consider von Neumann algebras only on separable Hilbert spaces, it is customary to call the bounded operators on a Hilbert space of finite dimension n a factor of type I_n, and the bounded operators on a separable infinite-dimensional Hilbert space, a factor of type I_∞.

Type Ii Factors

A factor is said to be of type II if there are no minimal projections but there are non-zero finite projections. This implies that every projection E can be halved in the sense that there are equivalent projections F and G such that $E = F + G$. If the identity operator in a type II factor is finite, the factor is said to be of type II_1; otherwise, it is said to be of type II_∞. The best understood factors of type II are the hyperfinite type II_1 factor and the hyperfinite type II_∞ factor, found by Murray & von Neumann (1936). These are the unique hyperfinite factors of types II_1 and II_∞; there are an uncountable number of other factors of these types that are the subject of intensive study. Murray & von Neumann (1937) proved the fundamental result that a factor of type II_1 has a unique finite tracial state, and the set of traces of projections is [0,1].

A factor of type II_∞ has a semifinite trace, unique up to rescaling, and the set of traces of projections is [0,∞]. The set of real numbers λ such that there is an automorphism rescaling the trace by a factor of λ is called the fundamental group of the type II_∞ factor.

The tensor product of a factor of type II_1 and an infinite type I factor has type II_∞, and conversely any factor of type II_∞ can be constructed like this. The fundamental group of a type II_1 factor is defined to be the fundamental group of its tensor product with the infinite (separable) factor of type I. For many years it was an open problem to find a type II factor whose fundamental group was not the group of positive reals, but Connes then showed that the von Neumann group algebra of a countable discrete group with Kazhdan's property T (the trivial representation is isolated in the dual space), such as SL(3,Z), has a countable fundamental group. Subsequently Sorin Popa showed that the fundamental group can be trivial for certain groups, including the semidirect product of Z^2 by SL(2,Z).

An example of a type II_1 factor is the von Neumann group algebra of a countable infinite discrete group such that every non-trivial conjugacy class is infinite. McDuff (1969) found an uncountable family of such groups with non-isomorphic von Neumann group algebras, thus showing the existence of uncountably many different separable type II_1 factors.

Type III factors

Lastly, type III factors are factors that do not contain any nonzero finite projections at all. In their first paper Murray & von Neumann (1936) were unable to decide whether or not they existed; the first examples were later found by von Neumann (1940). Since the identity operator is always infinite in those factors, they were sometimes called type III_∞ in the past, but recently that notation has been superseded by the notation III_λ, where λ is a real number in the interval [0,1]. More precisely, if the Connes spectrum (of its modular group) is 1 then the factor is of type III_0, if the Connes spectrum is all integral powers of λ for $0 < \lambda < 1$, then the type is III_λ, and if the Connes spectrum is all positive reals then the type is III_1. (The Connes spectrum is a closed subgroup of the positive reals, so these are the only possibilities.) The only trace on type III factors takes value ∞ on all non-zero positive elements, and any two non-zero projections are equivalent. At one time type III factors were considered to be intractable objects, but Tomita–Takesaki theory has led to a good structure theory. In particular, any type III factor can be written in a canonical way as the crossed product of a type II_∞ factor and the real numbers.

The Predual

Any von Neumann algebra M has a predual M_*, which is the Banach space of all ultraweakly continuous linear functionals on M. As the name suggests, M is (as a Banach space) the dual of its predual. The predual is unique in the sense that any other Banach space whose dual is M is canonically isomorphic to M_*. Sakai (1971) showed that the existence of a predual characterizes von Neumann algebras among C* algebras.

The definition of the predual given above seems to depend on the choice of Hilbert space that M acts on, as this determines the ultraweak topology. However the predual can also be defined without using the Hilbert space that M acts on, by defining it to be the space generated by all positive normal linear functionals on M. (Here "normal" means that it preserves suprema when applied to increasing nets of self adjoint operators; or equivalently to increasing sequences of projections.)

The predual M_* is a closed subspace of the dual M^* (which consists of all norm-continuous linear functionals on M) but is generally smaller. The proof that M_* is (usually) not the same as M^* is nonconstructive and uses the axiom of choice in an essential way; it is very hard to exhibit explicit elements of M^* that are not in M_*. For example, exotic positive linear forms on the von Neumann algebra $l^\infty(Z)$ are given by free ultrafilters; they correspond to exotic *-homomorphisms into C and describe the Stone–Čech compactification of Z.

Examples:

- The predual of the von Neumann algebra $L^\infty(R)$ of essentially bounded functions on R is the Banach space $L^1(R)$ of integrable functions. The dual of $L^\infty(R)$ is strictly larger than $L^1(R)$ For example, a functional on $L^\infty(R)$ that extends the Dirac measure δ_0 on the closed subspace of bounded continuous functions $C^0_b(R)$ cannot be represented as a function in $L^1(R)$.

- The predual of the von Neumann algebra $B(H)$ of bounded operators on a Hilbert space H is the Banach space of all trace class operators with the trace norm $||A|| = \text{Tr}(|A|)$. The Banach space of trace class operators is itself the dual of the C*-algebra of compact operators (which is not a von Neumann algebra).

Weights, States, and Traces

Weights and their special cases states and traces are discussed in detail in (Takesaki 1979).

- A weight ω on a von Neumann algebra is a linear map from the set of positive elements (those of the form a^*a) to $[0,\infty]$.

- A positive linear functional is a weight with $\omega(1)$ finite (or rather the extension of ω to the whole algebra by linearity).

- A state is a weight with $\omega(1) = 1$.

- A trace is a weight with $\omega(aa^*) = \omega(a^*a)$ for all a.

- A tracial state is a trace with $\omega(1) = 1$.

Any factor has a trace such that the trace of a non-zero projection is non-zero and the trace of a projection is infinite if and only if the projection is infinite. Such a trace is unique up to rescaling. For factors that are separable or finite, two projections are equivalent if and only if they have the same trace. The type of a factor can be read off from the possible values of this trace as follows:

- Type I_n: 0, x, 2x, ...,nx for some positive x (usually normalized to be $1/n$ or 1).

- Type I_∞: 0, x, 2x, ...,∞ for some positive x (usually normalized to be 1).

- Type II_1: $[0,x]$ for some positive x (usually normalized to be 1).

- Type II$_\infty$: [0,∞].

- Type III: 0,∞.

If a von Neumann algebra acts on a Hilbert space containing a norm 1 vector v, then the functional $a \rightarrow (av,v)$ is a normal state. This construction can be reversed to give an action on a Hilbert space from a normal state: this is the GNS construction for normal states.

Modules Over a Factor

Given an abstract separable factor, one can ask for a classification of its modules, meaning the separable Hilbert spaces that it acts on. The answer is given as follows: every such module H can be given an M-dimension $\dim_M(H)$ (not its dimension as a complex vector space) such that modules are isomorphic if and only if they have the same M-dimension. The M-dimension is additive, and a module is isomorphic to a subspace of another module if and only if it has smaller or equal M-dimension.

A module is called standard if it has a cyclic separating vector. Each factor has a standard representation, which is unique up to isomorphism. The standard representation has an antilinear involution J such that $JMJ = M'$. For finite factors the standard module is given by the GNS construction applied to the unique normal tracial state and the M-dimension is normalized so that the standard module has M-dimension 1, while for infinite factors the standard module is the module with M-dimension equal to ∞.

The possible M-dimensions of modules are given as follows:

- Type I$_n$ (n finite): The M-dimension can be any of $0/n$, $1/n$, $2/n$, $3/n$, ..., ∞. The standard module has M-dimension 1 (and complex dimension n^2.)

- Type I$_\infty$ The M-dimension can be any of 0, 1, 2, 3, ..., ∞. The standard representation of $B(H)$ is $H \otimes H$; its M-dimension is ∞.

- Type II$_1$: The M-dimension can be anything in [0, ∞]. It is normalized so that the standard module has M-dimension 1. The M-dimension is also called the coupling constant of the module H.

- Type II$_\infty$: The M-dimension can be anything in [0, ∞]. There is in general no canonical way to normalize it; the factor may have outer automorphisms multiplying the M-dimension by constants. The standard representation is the one with M-dimension ∞.

- Type III: The M-dimension can be 0 or ∞. Any two non-zero modules are isomorphic, and all non-zero modules are standard.

Amenable Von Neumann Algebras

Connes (1976) and others proved that the following conditions on a von Neumann algebra M on a separable Hilbert space H are all equivalent:

- M is hyperfinite or AFD or approximately finite dimensional or approximately finite: this means the algebra contains an ascending sequence of finite dimensional subalgebras with dense union. (Warning: some authors use "hyperfinite" to mean "AFD and finite".)

- M is amenable: this means that the derivations of M with values in a normal dual Banach bimodule are all inner.

- M has Schwartz's property P: for any bounded operator T on H the weak operator closed convex hull of the elements uTu^* contains an element commuting with M.

- M is semidiscrete: this means the identity map from M to M is a weak pointwise limit of completely positive maps of finite rank.

- M has property E or the Hakeda–Tomiyama extension property: this means that there is a projection of norm 1 from bounded operators on H to M '.

- M is injective: any completely positive linear map from any self adjoint closed subspace containing 1 of any unital C*-algebra A to M can be extended to a completely positive map from A to M.

There is no generally accepted term for the class of algebras above; Connes has suggested that amenable should be the standard term.

The amenable factors have been classified: there is a unique one of each of the types I_n, I_∞, II_1, II_∞, III_λ, for $0 < \lambda \le 1$, and the ones of type III_0 correspond to certain ergodic flows. (For type III_0 calling this a classification is a little misleading, as it is known that there is no easy way to classify the corresponding ergodic flows.) The ones of type I and II_1 were classified by Murray & von Neumann (1943), and the remaining ones were classified by Connes (1976), except for the type III_1 case which was completed by Haagerup.

All amenable factors can be constructed using the group-measure space construction of Murray and von Neumann for a single ergodic transformation. In fact they are precisely the factors arising as crossed products by free ergodic actions of Z or Z/nZ on abelian von Neumann algebras $L^\infty(X)$. Type I factors occur when the measure space X is atomic and the action transitive. When X is diffuse or non-atomic, it is equivalent to [0,1] as a measure space. Type II factors occur when X admits an equivalent finite (II_1) or infinite (II_∞) measure, invariant under an action of Z. Type III factors occur in the remaining cases where there is no invariant measure, but only an invariant measure class: these factors are called Krieger factors.

Tensor Products of Von Neumann Algebras

The Hilbert space tensor product of two Hilbert spaces is the completion of their algebraic tensor product. One can define a tensor product of von Neumann algebras (a completion of the algebraic tensor product of the algebras considered as rings), which is again a von Neumann algebra, and act on the tensor product of the corresponding Hilbert spaces. The tensor product of two finite algebras is finite, and the tensor product of an infinite algebra and a non-zero algebra is infinite. The type of the tensor product of two von Neumann algebras (I, II, or III) is the maximum of their types. The commutation theorem for tensor products states that

$$(M \otimes N)' = M' \otimes N',$$

where M' denotes the commutant of M.

The tensor product of an infinite number of von Neumann algebras, if done naively, is usually a ridiculously large non-separable algebra. Instead von Neumann (1938) showed that one should choose a state on each of the von Neumann algebras, use this to define a state on the algebraic tensor product, which can be used to product a Hilbert space and a (reasonably small) von Neumann algebra. Araki & Woods (1968) studied the case where all the factors are finite matrix algebras; these factors are called Araki-Woods factors or ITPFI factors (ITPFI stands for "infinite tensor product of finite type I factors"). The type of the infinite tensor product can vary dramatically as the states are changed; for example, the infinite tensor product of an infinite number of type I_2 factors can have any type depending on the choice of states. In particular Powers (1967) found an uncountable family of non-isomorphic hyperfinite type III_λ factors for $0 < \lambda < 1$, called Powers factors, by taking an infinite tensor product of type I_2 factors, each with the state given by:

$$x \mapsto \mathrm{Tr} \begin{pmatrix} \dfrac{1}{\lambda+1} & 0 \\ 0 & \dfrac{\lambda}{\lambda+1} \end{pmatrix} x.$$

All hyperfinite von Neumann algebras not of type III_0 are isomorphic to Araki-Woods factors, but there are uncountably many of type III_0 that are not.

Bimodules and Subfactors

A bimodule (or correspondence) is a Hilbert space H with module actions of two commuting von Neumann algebras. Bimodules have a much richer structure than that of modules. Any bimodule over two factors always gives a subfactor since one of the factors is always contained in the commutant of the other. There is also a subtle relative tensor product operation due to Connes on bimodules. The theory of subfactors, initiated by Vaughan Jones, reconciles these two seemingly different points of view.

Bimodules are also important for the von Neumann group algebra M of a discrete group Γ. Indeed if V is any unitary representation of Γ, then, regarding Γ as the diagonal subgroup of $\Gamma \times \Gamma$, the corresponding induced representation on $l^2(\Gamma, V)$ is naturally a bimodule for two commuting copies of M. Important representation theoretic properties of Γ can be formulated entirely in terms of bimodules and therefore make sense for the von Neumann algebra itself. For example Connes and Jones gave a definition of an analogue of Kazhdan's Property T for von Neumann algebras in this way.

Non-amenable Factors

Von Neumann algebras of type I are always amenable, but for the other types there are an uncountable number of different non-amenable factors, which seem very hard to classify, or even distinguish from each other. Nevertheless Voiculescu has shown that the class of non-amenable factors coming from the group-measure space construction is disjoint from the class coming from group von Neumann algebras of free groups. Later Narutaka Ozawa proved that group von Neumann algebras of hyperbolic groups yield prime type II_1 factors, i.e. ones that cannot be factored as tensor products of type II_1 factors, a result first proved by Leeming Ge for free group factors using

Voiculescu's free entropy. Popa's work on fundamental groups of non-amenable factors represents another significant advance. The theory of factors "beyond the hyperfinite" is rapidly expanding at present, with many new and surprising results; it has close links with rigidity phenomena in geometric group theory and ergodic theory.

Examples

- The essentially bounded functions on a σ-finite measure space form a commutative (type I_1) von Neumann algebra acting on the L^2 functions. For certain non-σ-finite measure spaces, usually considered pathological, $L^\infty(X)$ is not a von Neumann algebra; for example, the σ-algebra of measurable sets might be the countable-cocountable algebra on an uncountable set.

- The bounded operators on any Hilbert space form a von Neumann algebra, indeed a factor, of type I.

- If we have any unitary representation of a group G on a Hilbert space H then the bounded operators commuting with G form a von Neumann algebra G', whose projections correspond exactly to the closed subspaces of H invariant under G. Equivalent subrepresentations correspond to equivalent projections in G'. The double commutant G'' of G is also a von Neumann algebra.

- The von Neumann group algebra of a discrete group G is the algebra of all bounded operators on $H = l^2(G)$ commuting with the action of G on H through right multiplication. One can show that this is the von Neumann algebra generated by the operators corresponding to multiplication from the left with an element $g \in G$. It is a factor (of type II_1) if every non-trivial conjugacy class of G is infinite (for example, a non-abelian free group), and is the hyperfinite factor of type II_1 if in addition G is a union of finite subgroups (for example, the group of all permutations of the integers fixing all but a finite number of elements).

- The tensor product of two von Neumann algebras, or of a countable number with states, is a von Neumann algebra as described in the section above.

- The crossed product of a von Neumann algebra by a discrete (or more generally locally compact) group can be defined, and is a von Neumann algebra. Special cases are the group-measure space construction of Murray and von Neumann and Krieger factors.

- The von Neumann algebras of a measurable equivalence relation and a measurable groupoid can be defined. These examples generalise von Neumann group algebras and the group-measure space construction.

Applications

Von Neumann algebras have found applications in diverse areas of mathematics like knot theory, statistical mechanics, Quantum field theory, Local quantum physics, Free probability, Noncommutative geometry, representation theory, geometry, and probability.

For instance, C-star algebra provides an alternative axiomatization to probability theory. In this case the method goes by the name of Gelfand–Naimark–Segal construction. This is analogous to the two approaches to measure and integration, where one has the choice to construct measures of sets first and define integrals later, or construct integrals first and define set measures as integrals of characteristic functions.

References

- Lewis Hirsch, Arthur Goodman, Understanding Elementary Algebra With Geometry: A Course for College Students, Publisher: Cengage Learning, 2005, ISBN 0534999727, 9780534999728, 654.

- Richard N. Aufmann, Joanne Lockwood, Introductory Algebra: An Applied Approach, Publisher Cengage Learning, 2010, ISBN 1439046042, 9781439046043.

- William L. Hosch (editor), The Britannica Guide to Algebra and Trigonometry, Britannica Educational Publishing, The Rosen Publishing Group, 2010, ISBN 1615302190, 9781615302192.

- James E. Gentle, Numerical Linear Algebra for Applications in Statistics, Publisher: Springer, 1998, ISBN 0387985425, 9780387985428.

- Matsumura, Hideyuki, Commutative algebra. Second edition. Mathematics Lecture Note Series, 56. Benjamin/Cummings Publishing Co., Inc., Reading, Mass., 1980. xv+313 pp. ISBN 0-8053-7026-9

- Ron Larson, Robert Hostetler, Bruce H. Edwards, Algebra And Trigonometry: A Graphing Approach, Publisher: Cengage Learning, 2007, ISBN 061885195X, 9780618851959.

- Sin Kwai Meng, Chip Wai Lung, Ng Song Beng, "Algebraic notation", in Mathematics Matters Secondary 1 Express Textbook, Publisher Panpac Education Pte Ltd, ISBN 9812738827, 9789812738820.

- Jerome E. Kaufmann, Karen L. Schwitters, Algebra for College Students, Publisher Cengage Learning, 2010, ISBN 0538733543, 9780538733540, 803 pages, page 222

- Ramesh Bangia, Dictionary of Information Technology, Publisher Laxmi Publications, Ltd., 2010, ISBN 9380298153, 9789380298153.

- Matthew A. Telles, Python Power!: The Comprehensive Guide, Publisher Course Technology PTR, 2008, ISBN 1598631586, 9781598631586.

- William P. Berlinghoff, Fernando Q. Gouvêa, Math through the Ages: A Gentle History for Teachers and Others, Publisher MAA, 2004, ISBN 0883857367, 9780883857366.

- Thomas Sonnabend, Mathematics for Teachers: An Interactive Approach for Grades K-8, Publisher: Cengage Learning, 2009, ISBN 0495561665, 9780495561668.

- Lewis Hirsch, Arthur Goodman, Understanding Elementary Algebra With Geometry: A Course for College Students, Publisher: Cengage Learning, 2005, ISBN 0534999727, 9780534999728.

- Lawrence S. Leff, College Algebra: Barron's Ez-101 Study Keys, Publisher: Barron's Educational Series, 2005, ISBN 0764129147, 9780764129148.

- Andrew Marx, Shortcut Algebra I: A Quick and Easy Way to Increase Your Algebra I Knowledge and Test Scores, Publisher Kaplan Publishing, 2007, ISBN 1419552880, 9781419552885.

- Mark Clark, Cynthia Anfinson, Beginning Algebra: Connecting Concepts Through Applications, Publisher Cengage Learning, 2011, ISBN 0534419380, 9780534419387.

- Alan S. Tussy, R. David Gustafson, Elementary and Intermediate Algebra, Publisher Cengage Learning, 2012, ISBN 1111567689, 9781111567682.

- Chris Carter, Physics: Facts and Practice for A Level, Publisher Oxford University Press, 2001, ISBN 019914768X, 9780199147687.

- Sinha, The Pearson Guide to Quantitative Aptitude for CAT 2/ePublisher: Pearson Education India, 2010, ISBN 8131723666, 9788131723661.

Key Concepts of Algebra

The major concepts of algebra are discussed in this text. Identity element, inverse element, commutative property, associative property and isomorphism are some of the key concepts of algebra. The section strategically encompasses and incorporates the major components and key concepts of algebra, providing a complete understanding.

Identity Element

In mathematics, an identity element or neutral element is a special type of element of a set with respect to a binary operation on that set, which leaves other elements unchanged when combined with them. This concept is used in algebraic structures such as groups. The term *identity element* is often shortened to *identity* when there is no possibility of confusion.

Let $(S, *)$ be a set S with a binary operation $*$ on it. Then an element e of S is called a left identity if $e * a = a$ for all a in S, and a right identity if $a * e = a$ for all a in S. If e is both a left identity and a right identity, then it is called a two-sided identity, or simply an identity.

An identity with respect to addition is called an additive identity (often denoted as 0) and an identity with respect to multiplication is called a multiplicative identity (often denoted as 1). The distinction is used most often for sets that support both binary operations, such as rings. The multiplicative identity is often called the unit in the latter context, where, though, a unit is often used in a broader sense, to mean an element with a multiplicative inverse.

Examples

Set	Operation	Identity
Real numbers	+ (addition)	0
Real numbers	· (multiplication)	1
Positive integers	Least common multiple	1
Non-negative integers	Greatest common divisor	0 (under most definitions of GCD)
m-by-n Matrices	+ (addition)	Zero matrix
n-by-n square matrices	Matrix multiplication	I_n (identity matrix)
m-by-n matrices	∘ (Hadamard product)	$J_{m,n}$ (Matrix of ones)
All functions from a set, M, to itself	∘ (function composition)	Identity function
All distributions on a group, G	* (convolution)	δ (Dirac delta)
Extended real numbers	Minimum/infimum	$+\infty$

Extended real numbers	Maximum/supremum	$-\infty$
Subsets of a set M	∩ (intersection)	M
Sets	∪ (union)	∅ (empty set)
Strings, lists	Concatenation	Empty string, empty list
A Boolean algebra	∧ (logical and)	⊤ (truth)
A Boolean algebra	∨ (logical or)	⊥ (falsity)
A Boolean algebra	⊕ (exclusive or)	⊥ (falsity)
Knots	Knot sum	Unknot
Compact surfaces	# (connected sum)	S^2
Two elements, {e, f}	∗ defined by $e \ast e = f \ast e = e$ and $f \ast f = e \ast f = f$	Both e and f are left identities, but there is no right identity and no two-sided identity

Properties

As the last example (a semigroup) shows, it is possible for $(S, \ast)$ to have several left identities. In fact, every element can be a left identity. Similarly, there can be several right identities. But if there is both a right identity and a left identity, then they are equal and there is just a single two-sided identity. To see this, note that if l is a left identity and r is a right identity then $l = l \ast r = r$. In particular, there can never be more than one two-sided identity. If there were two, e and f, then $e \ast f$ would have to be equal to both e and f.

It is also quite possible for $(S, \ast)$ to have *no* identity element. A common example of this is the cross product of vectors; in this case, the absence of an identity element is related to the fact that the direction of any nonzero cross product is always orthogonal to any element multiplied – so that it is not possible to obtain a non-zero vector in the same direction as the original. Another example would be the additive semigroup of positive natural numbers.

Inverse Element

In abstract algebra, the idea of an inverse element generalises concepts of a negation (sign reversal) in relation to addition, and a reciprocal in relation to multiplication. The intuition is of an element that can 'undo' the effect of combination with another given element. While the precise definition of an inverse element varies depending on the algebraic structure involved, these definitions coincide in a group.

The word 'inverse' is derived from Latin: *inversus* that means 'turned upside down', 'overturned'.

Formal Definitions

In a Unital Magma

Let S be a set closed under a binary operation $*$ (i.e., a magma). If e is an identity element of $(S, *)$ (i.e., S is a unital magma) and y, then x is called a left inverse of b and b is called a right inverse of a. If an element x is both a left inverse and a right inverse of y, then x is called a two-sided inverse, or simply an inverse, of y. An element with a two-sided inverse in S is called invertible in S. An element with an inverse element only on one side is left invertible, resp. right invertible. A unital magma in which all elements are invertible is called a loop. A loop whose binary operation satisfies the associative law is a group.

Just like $(S, *)$ can have several left identities or several right identities, it is possible for an element to have several left inverses or several right inverses (but note that their definition above uses a *two-sided* identity e). It can even have several left inverses *and* several right inverses.

If the operation $*$ is associative then if an element has both a left inverse and a right inverse, they are equal. In other words, in a monoid (an associative unital magma) every element has at most one inverse. In a monoid, the set of (left and right) invertible elements is a group, called the group of units of S, and denoted by $U(S)$ or H_1.

A left-invertible element is left-cancellative, and analogously for right and two-sided.

In a Semigroup

The definition in the previous section generalizes the notion of inverse in group relative to the notion of identity. It's also possible, albeit less obvious, to generalize the notion of an inverse by dropping the identity element but keeping associativity, i.e. in a semigroup.

In a semigroup S an element x is called (von Neumann) regular if there exists some element z in S such that $xzx = x$; z is sometimes called a pseudoinverse. An element y is called (simply) an inverse of x if $xyx = x$ and $y = yxy$. Every regular element has at least one inverse: if $x = xzx$ then it is easy to verify that $y = zxz$ is an inverse of x as defined in this section. Another easy to prove fact: if y is an inverse of x then $e = xy$ and $f = yx$ are idempotents, that is $ee = e$ and $ff = f$. Thus, every pair of (mutually) inverse elements gives rise to two idempotents, and $ex = xf = x$, $ye = fy = y$, and e acts as a left identity on x, while f acts a right identity, and the left/right roles are reversed for y. This simple observation can be generalized using Green's relations: every idempotent e in an arbitrary semigroup is a left identity for R_e and right identity for L_e. An intuitive description of this fact is that every pair of mutually inverse elements produces a local left identity, and respectively, a local right identity.

In a monoid, the notion of inverse as defined in the previous section is strictly narrower than the definition given in this section. Only elements in the Green class H_1 have an inverse from the unital magma perspective, whereas for any idempotent e, the elements of H_e have an inverse as defined in this section. Under this more general definition, inverses need not be unique (or exist) in an arbitrary semigroup or monoid. If all elements are regular, then the semigroup (or monoid) is called regular, and every element has at least one inverse. If every element has exactly one inverse as defined in this section, then the semigroup is called an inverse semigroup. Finally, an inverse semigroup with only one idempotent is a group. An inverse semigroup may have an absorbing element 0 because 000 = 0, whereas a group may not.

Outside semigroup theory, a unique inverse as defined in this section is sometimes called a quasi-inverse. This is generally justified because in most applications (e.g. all examples in this article) associativity holds, which makes this notion a generalization of the left/right inverse relative to an identity.

U-semigroups

A natural generalization of the inverse semigroup is to define an (arbitrary) unary operation $°$ such that $(a°)° = a$ for all a in S; this endows S with a type $\langle 2,1 \rangle$ algebra. A semigroup endowed with such an operation is called a **U-semigroup**. Although it may seem that $a°$ will be the inverse of a, this is not necessarily the case. In order to obtain interesting notion(s), the unary operation must somehow interact with the semigroup operation. Two classes of U-semigroups have been studied:

- **I-semigroups**, in which the interaction axiom is $aa°a = a$

- *-semigroups, in which the interaction axiom is $(ab)° = b°a°$. Such an operation is called an involution, and typically denoted by a^*

Clearly a group is both an I-semigroup and a *-semigroup. A class of semigroups important in semigroup theory are completely regular semigroups; these are I-semigroups in which one additionally has $aa° = a°a$; in other words every element has commuting pseudoinverse $a°$. There are few concrete examples of such semigroups however; most are completely simple semigroups. In contrast, a subclass of *-semigroups, the *-regular semigroups (in the sense of Drazin), yield one of best known examples of a (unique) pseudoinverse, the Moore–Penrose inverse. In this case however the involution a^* is not the pseudoinverse. Rather, the pseudoinverse of x is the unique element y such that $xyx = x, yxy = y, (xy)^* = xy, (yx)^* = yx$. Since *-regular semigroups generalize inverse semigroups, the unique element defined this way in a *-regular semigroup is called the generalized inverse or Penrose–Moore inverse.

Rings and Semirings

Examples

All examples in this section involve associative operators, thus we shall use the terms left/right inverse for the unital magma-based definition, and quasi-inverse for its more general version.

Real Numbers

Every real number x has an additive inverse (i.e. an inverse with respect to addition) given by $-x$. Every nonzero real number x has a multiplicative inverse (i.e. an inverse with respect to multiplication) given by $\frac{1}{x}$ (or x^{-1}). By contrast, zero has no multiplicative inverse, but it has a unique quasi-inverse, "0(ZERO)" itself.

Functions and Partial Functions

A function g is the left (resp. right) inverse of a function f (for function composition), if and only if $g \circ f$ (resp. $f \circ g$) is the identity function on the domain (resp. codomain) of f. The inverse of a function f is often written f^{-1}, but this notation is sometimes ambiguous. Only bijections have two-sided inverses, but *any* function has a quasi-inverse, i.e. the full transformation monoid is regular. The monoid of partial functions is also regular, whereas the monoid of injective partial transformations is the prototypical inverse semigroup.

Galois Connections

The lower and upper adjoints in a (monotone) Galois connection, L and G are quasi-inverses of each other, i.e. $LGL = L$ and $GLG = G$ and one uniquely determines the other. They are not left or right inverses of each other however.

Matrices

A square matrix M with entries in a field K is invertible (in the set of all square matrices of the same size, under matrix multiplication) if and only if its determinant is different from zero. If the determinant of M is zero, it is impossible for it to have a one-sided inverse; therefore a left inverse or right inverse implies the existence of the other one.

More generally, a square matrix over a commutative ring R is invertible if and only if its determinant is invertible in R.

Non-square matrices of full rank have several one-sided inverses:

- For $A : m \times n \mid m > n$ we have a left inverse: $\underbrace{(A^T A)^{-1} A^T}_{A_{\text{left}}^{-1}} A = I_n$

- For $A : m \times n \mid m < n$ we have a right inverse: $A \underbrace{A^T (AA^T)^{-1}}_{A_{\text{right}}^{-1}} = I_m$

The left inverse can be used to determine the least norm solution of $Ax = b$, which is also the least squares formula for regression and is given by $x = (A^T A)^{-1} A^T b$.

No rank deficient matrix has any (even one-sided) inverse. However, the Moore–Penrose pseudo-inverse exists for all matrices, and coincides with the left or right (or true) inverse when it exists.

As an example of matrix inverses, consider:

$$A : 2 \times 3 = \begin{bmatrix} 1 & 2 & 3 \\ 4 & 5 & 6 \end{bmatrix}$$

So, as $m < n$, we have a right inverse, $A_{\text{right}}^{-1} = A^T (AA^T)^{-1}$. By components it is computed as

$$AA^T = \begin{bmatrix} 1 & 2 & 3 \\ 4 & 5 & 6 \end{bmatrix} \cdot \begin{bmatrix} 1 & 4 \\ 2 & 5 \\ 3 & 6 \end{bmatrix} = \begin{bmatrix} 14 & 32 \\ 32 & 77 \end{bmatrix}$$

$$(AA^T)^{-1} = \begin{bmatrix} 14 & 32 \\ 32 & 77 \end{bmatrix}^{-1} = \frac{1}{54}\begin{bmatrix} 77 & -32 \\ -32 & 14 \end{bmatrix}$$

$$A^T(AA^T)^{-1} = \frac{1}{54}\begin{bmatrix} 1 & 4 \\ 2 & 5 \\ 3 & 6 \end{bmatrix} \cdot \begin{bmatrix} 77 & -32 \\ -32 & 14 \end{bmatrix} = \frac{1}{18}\begin{bmatrix} -17 & 8 \\ -2 & 2 \\ 13 & -4 \end{bmatrix} = A_{\text{right}}^{-1}$$

The left inverse doesn't exist, because

$$A^T A = \begin{bmatrix} 1 & 4 \\ 2 & 5 \\ 3 & 6 \end{bmatrix} \cdot \begin{bmatrix} 1 & 2 & 3 \\ 4 & 5 & 6 \end{bmatrix} = \begin{bmatrix} 17 & 22 & 27 \\ 22 & 29 & 36 \\ 27 & 36 & 45 \end{bmatrix}$$

which is a singular matrix, and cannot be inverted.

Quasiregular Element

This chapter addresses the notion of quasiregularity in the context of ring theory, a branch of modern algebra. For other notions of quasiregularity in mathematics.

In mathematics, specifically ring theory, the notion of quasiregularity provides a computationally convenient way to work with the Jacobson radical of a ring. Intuitively, quasiregularity captures what it means for an element of a ring to be "bad"; that is, have undesirable properties. Although a "bad element" is necessarily quasiregular, quasiregular elements need not be "bad," in a rather vague sense. In this article, we primarily concern ourselves with the notion of quasiregularity for unital rings. However, one section is devoted to the theory of quasiregularity in non-unital rings, which constitutes an important aspect of noncommutative ring theory.

Definition

Let R be a ring (with unity) and let r be an element of R. Then r is said to be quasiregular, if $1 - r$ is a unit in R; that is, invertible under multiplication. The notions of right or left quasiregularity correspond to the situations where $1 - r$ has a right or left inverse, respectively.

An element x of a non-unital ring is said to be right quasiregular if there is y such that $x + y - xy = 0$. The notion of a left quasiregular element is defined in an analogous manner. The element y is sometimes referred to as a right quasi-inverse of x. If the ring is unital, this definition quasiregularity coincides with that given above. If one writes $x \cdot y = x + y - xy$, then this binary operation $\cdot$ is associative. In fact, the map $(R, \cdot) \to (R, \times); x \mapsto 1 - x$ (where $\times$ denotes the multiplication of the ring R) is a monoid isomorphism. Therefore, if an element possesses both a left and right quasi-inverse, they are equal.

Note that some authors use different definitions. They call an element x right quasiregular if there exists y such that $x + y + xy = 0$, which is equivalent to saying that $1 + x$ has a right inverse when the ring is unital. If we write $x \circ y = x + y + xy$, then $(-x)^{\circ}(-y) = -(xy)$, so we can easily go from one set-up to the other by changing signs. For example, x is right quasiregular in one set-up iff $-x$ is right quasiregular in the other set-up.

Examples

- If R is a ring, then the additive identity of R is always quasiregular.

- If x^2 is right (resp. left) quasiregular, then x is right (resp. left) quasiregular.

- If R is a rng, every nilpotent element of R is quasiregular. This fact is supported by an elementary computation:

  If $x^{n+1} = 0$, then

  $(1-x)(1+x+x^2+\cdots+x^n) = 1$ (or $(1+x)(1-x+x^2-\cdots+(-x)^n) = 1$ if we follow the second convention).

  From this we see easily that the quasi-inverse of x is $-x-x^2-\cdots-x^n$ (or $-x+x^2-\cdots+(-x)^n$).

- In the second convention, a matrix is quasiregular in a matrix ring if it does not possess -1 as an eigenvalue. More generally, a bounded operator is quasiregular if -1 is not in its spectrum.

- In a unital Banach algebra, if $\|x\| < 1$, then the geometric series $\sum_{0}^{\infty} x^n$ converges. Consequently, every such x is quasiregular.

- If R is a ring and $S = R[[X_1,...,X_n]]$ denotes the ring of formal power series in n intederminants over R, an element of S is quasiregular if and only its constant term is quasiregular as an element of R.

Properties

- Every element of the Jacobson radical of a (not necessarily commutative) ring is quasiregular. In fact, the Jacobson radical of a ring can be characterized as the unique right ideal of the ring, maximal with respect to the property that every element is right quasiregular. However, a right quasiregular element need not necessarily be a member of the Jacobson radical. This justifies the remark in the beginning of the article - "bad elements" are quasiregular, although quasiregular elements are not necessarily "bad." Elements of the Jacobson radical of a ring, are often deemed to be "bad."

- If an element of a ring is nilpotent and central, then it is a member of the ring's Jacobson radical. This is because the principal right ideal generated by that element consists of quasiregular (in fact, nilpotent) elements only.

- If an element, r, of a ring is idempotent, it cannot be a member of the ring's Jacobson radical. This is because idempotent elements cannot be quasiregular. This property, as well as the one above, justify the remark given at the top of the article that the notion of quasiregularity is computationally convenient when working with the Jacobson radical.

Generalization to Semirings

The notion of quasiregular element readily generalizes to semirings. If a is an element of a semiring S, then an affine map from S to itself is $\mu_a(r) = ra + 1$.. An element a of S is said to be right quasiregular if μ_a has a fixed point, which need not be unique. Each such fixed point is called a left quasi-inverse of a. If b is a left quasi-inverse of a and additionally $b = ab + 1$, then b it is called a quasi-inverse of a; any element of the semiring that has a quasi-inverse is said to be quasiregular. It is possible that some but not all elements of a semiring be quasiregular; for example, in the semiring of nonnegative reals with the usual addition and multiplication of reals, μ_a has the fixed point $\dfrac{1}{1-a}$ for all $a < 1$, but has no fixed point for $a \geq 1$. If every element of a semiring is quasiregular then the semiring is called a quasi-regular semiring, closed semiring, or occasionally a Lehmann semiring (the latter honoring the paper of Daniel J. Lehmann.)

Examples of quasi-regular semirings are provided by the Kleene algebras (prominently among them, the algebra of regular expressions), in which the quasi-inverse is lifted to the role of a unary operation (denoted by a^*) defined as the least fixedpoint solution. Kleene algebras are additively idempotent but not all quasi-regular semirings are so. We can extend the example of nonnegative reals to include infinity and it becomes a quasi-regular semiring with the quasi-inverse of any element $a \geq 1$ being the infinity. This quasi-regular semiring is not additively idempotent however, so it is not a Kleene algebra. It is however a complete semiring. More generally, all complete semirings are quasiregular. The term *closed semiring* is actually used by some authors to mean complete semiring rather than just quasiregular.

Conway semirings are also quasiregular; the two Conway axioms are actually independent, i.e. there are semirings satisfying only the product-star [Conway] axiom, $(ab)^* = 1 + a(ba)^*b$, but not the sum-star axiom, $(a+b)^* = (a^*b)^*a^*$ and vice versa; it is the product-star [Conway] axiom that implies that a semiring is quasiregular. Additionally, a commutative semiring is quasiregular if and only if it satisfies the product-star Conway axiom.

Quasiregular semirings appear in algebraic path problems, a generalization of the shortest path problem.

Commutative Property

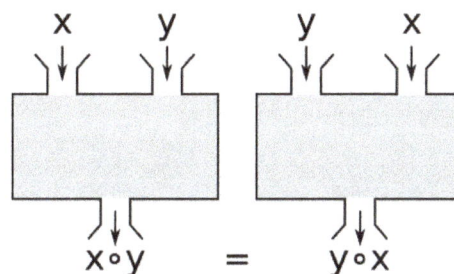

An operation $\circ$ is commutative iff $x \circ y = y \circ x$ for each x and y. This image illustrates this property with the concept of an operation as a "calculation machine". It doesn't matter for the output $x \circ y$ or $y \circ x$ respectively which order the arguments x and y have – the final outcome is the same.

In mathematics, a binary operation is commutative if changing the order of the operands does not change the result. It is a fundamental property of many binary operations, and many mathematical proofs depend on it. Most familiar as the name of the property that says "3 + 4 = 4 + 3" or "2 × 5 = 5 × 2", the property can also be used in more advanced settings. The name is needed because there are operations, such as division and subtraction, that do not have it (for example, "3 − 5 ≠ 5 − 3"), such operations are *not* commutative, or *noncommutative operations*. The idea that simple operations, such as multiplication and addition of numbers, are commutative was for many years implicitly assumed and the property was not named until the 19th century when mathematics started to become formalized. A corresponding property exists for binary relations; a binary relation is said to be symmetric if the relation applies regardless of the order over some set; for example, equality is symmetric as two mathematical objects are equal regardless of the order of the two.

Common Uses

The *commutative property* (or *commutative law*) is a property generally associated with binary operations and functions. If the commutative property holds for a pair of elements under a certain binary operation then the two elements are said to *commute* under that operation.

Mathematical Definitions

The term "commutative" is used in several related senses.

- A binary operation $*$ on a set S is called *commutative* if:

$$x * y = y * x \qquad \text{for all } x, y \in S$$

An operation that does not satisfy the above property is called *non-commutative*.

- One says that x *commutes* with y under $*$ if:

$$x * y = y * x$$

- A binary function $f : A \times A \to B$ is called *commutative* if:

$$f(x, y) = f(y, x) \qquad \text{for all } x, y \in A$$

Examples

Commutative Operations in Everyday Life

The cumulation of apples, which can be seen as an addition of natural numbers, is commutative.

- Putting on socks resembles a commutative operation since which sock is put on first is unimportant. Either way, the result (having both socks on), is the same. In contrast, putting on underwear and trousers is not commutative.

- The commutativity of addition is observed when paying for an item with cash. Regardless of the order the bills are handed over in, they always give the same total.

Commutative Operations in Mathematics

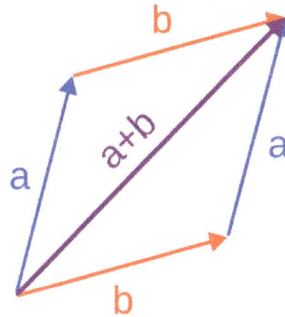

The addition of vectors is commutative, because $\vec{a} + \vec{b} = \vec{b} + \vec{a}$.

Two well-known examples of commutative binary operations:

- The addition of real numbers is commutative, since

  $$y + z = z + y \qquad \text{for all } y, z \in \mathbb{R}$$

  For example $4 + 5 = 5 + 4$, since both expressions equal 9.

- The multiplication of real numbers is commutative, since

  $$yz = zy \qquad \text{for all } y, z \in \mathbb{R}$$

  For example, $3 \times 5 = 5 \times 3$, since both expressions equal 15.

- Some binary truth functions are also commutative, since the truth tables for the functions are the same when one changes the order of the operands.

  For example, the logical biconditional function p ↔ q is equivalent to q ↔ p. This function is also written as p IFF q, or as p ≡ q, or as Epq.

  The last form is an example of the most concise notation in the article on truth functions, which lists the sixteen possible binary truth functions of which eight are commutative: Vpq = Vqp; Apq (OR) = Aqp; Dpq (NAND) = Dqp; Epq (IFF) = Eqp; Jpq = Jqp; Kpq (AND) = Kqp; Xpq (NOR) = Xqp; Opq = Oqp.

- Further examples of commutative binary operations include addition and multiplication of complex numbers, addition and scalar multiplication of vectors, and intersection and union of sets.

Noncommutative Operations in Everyday Life

- Concatenation, the act of joining character strings together, is a noncommutative operation. For example,

$$EA + T = EAT \neq TEA = T + EA$$

- Washing and drying clothes resembles a noncommutative operation; washing and then drying produces a markedly different result to drying and then washing.

- Rotating a book 90° around a vertical axis then 90° around a horizontal axis produces a different orientation than when the rotations are performed in the opposite order.

- The twists of the Rubik's Cube are noncommutative. This can be studied using group theory.

- Also thought processes are noncommutative: A person asked a question (A) and then a question (B) may give different answers to each question than a person asked first (B) and then (A), because asking a question may change the person's state of mind.

Noncommutative Operations in Mathematics

Some non-commutative binary operations:

- Subtraction is non-commutative, since $0 - 1 \neq 1 - 0$

- Division is non-commutative, since $1 \div 2 \neq 2 \div 1$

- Some truth functions are non-commutative, since the truth tables for the functions are different when one changes the order of the operands.

For example, the truth tables for $f(A, B) = A \wedge \neg B$ (A AND NOT B) and $f(B, A) = B \wedge \neg A$ are

A	B	**f** (A, B)	**f** (B, A)
F	F	F	F
F	T	F	T
T	F	T	F
T	T	F	F

- Matrix multiplication is non-commutative since

$$\begin{bmatrix} 0 & 2 \\ 0 & 1 \end{bmatrix} = \begin{bmatrix} 1 & 1 \\ 0 & 1 \end{bmatrix} \cdot \begin{bmatrix} 0 & 1 \\ 0 & 1 \end{bmatrix} \neq \begin{bmatrix} 0 & 1 \\ 0 & 1 \end{bmatrix} \cdot \begin{bmatrix} 1 & 1 \\ 0 & 1 \end{bmatrix} = \begin{bmatrix} 0 & 1 \\ 0 & 1 \end{bmatrix}$$

History and Etymology

Records of the implicit use of the commutative property go back to ancient times. The Egyptians used the commutative property of multiplication to simplify computing products. Euclid is known to have assumed the commutative property of multiplication in his book *Elements*. Formal uses of the commutative property arose in the late 18th and early 19th centuries, when mathematicians began to work on a theory of functions. Today the commutative property is a well known and basic property used in most branches of mathematics.

(14)

f et f, sont telles qu'elles donnent
el que soit l'ordre dans lequel on les
ppelées *commutatives entre elles.*

; $aEz = Eaz$;....

The first known use of the term was in a French Journal published in 1814

The first recorded use of the term *commutative* was in a memoir by François Servois in 1814, which used the word *commutatives* when describing functions that have what is now called the commutative property. The word is a combination of the French word *commuter* meaning "to substitute or switch" and the suffix *-ative* meaning "tending to" so the word literally means "tending to substitute or switch." The term then appeared in English in *Philosophical Transactions of the Royal Society* in 1844.

Propositional Logic

Rule of Replacement

In truth-functional propositional logic, *commutation*, or *commutativity* refer to two valid rules of replacement. The rules allow one to transpose propositional variables within logical expressions in logical proofs. The rules are:

$$(P \lor Q) \Leftrightarrow (Q \lor P)$$

and

$$(P \land Q) \Leftrightarrow (Q \land P)$$

where " $\Leftrightarrow$ " is a metalogical symbol representing "can be replaced in a proof with."

Truth Functional Connectives

Commutativity is a property of some logical connectives of truth functional propositional logic. The following logical equivalences demonstrate that commutativity is a property of particular connectives. The following are truth-functional tautologies.

Commutativity of conjunction

$$(P \land Q) \leftrightarrow (Q \land P)$$

Commutativity of disjunction

$$(P \lor Q) \leftrightarrow (Q \lor P)$$

Commutativity of implication (also called the law of permutation)

$$(P \rightarrow (Q \rightarrow R)) \leftrightarrow (Q \rightarrow (P \rightarrow R))$$

Commutativity of equivalence (also called the complete commutative law of equivalence)

$$(P \leftrightarrow Q) \leftrightarrow (Q \leftrightarrow P)$$

Set Theory

In group and set theory, many algebraic structures are called commutative when certain operands satisfy the commutative property. In higher branches of mathematics, such as analysis and linear algebra the commutativity of well-known operations (such as addition and multiplication on real and complex numbers) is often used (or implicitly assumed) in proofs.

Mathematical Structures and Commutativity

- A commutative semigroup is a set endowed with a total, associative and commutative operation.

- If the operation additionally has an identity element, we have a commutative monoid

- An abelian group, or *commutative group* is a group whose group operation is commutative.

- A commutative ring is a ring whose multiplication is commutative. (Addition in a ring is always commutative.)

- In a field both addition and multiplication are commutative.

Related Properties

Associativity

The associative property is closely related to the commutative property. The associative property of an expression containing two or more occurrences of the same operator states that the order operations are performed in does not affect the final result, as long as the order of terms doesn't change. In contrast, the commutative property states that the order of the terms does not affect the final result.

Most commutative operations encountered in practice are also associative. However, commutativity does not imply associativity. A counterexample is the function

$$f(x, y) = \frac{x + y}{2},$$

which is clearly commutative (interchanging x and y does not affect the result), but it is not associative (since, for example, $f(-4, f(0, +4)) = -1$ but $f(f(-4, 0), +4) = +1$). More such examples may be found in Commutative non-associative magmas.

Symmetry

Some forms of symmetry can be directly linked to commutativity. When a commutative operator is written as a binary function then the resulting function is symmetric across the line $y = x$. As an example, if we let a function f represent addition (a commutative operation) so that $f(x,y) = x + y$ then f is a symmetric function, which can be seen in the image on the right.

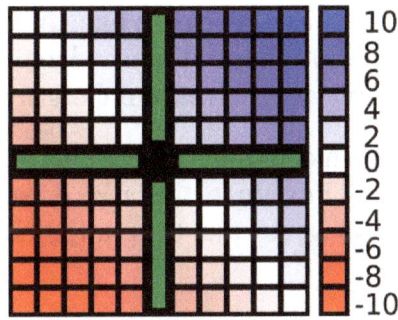

Graph showing the symmetry of the addition function

For relations, a symmetric relation is analogous to a commutative operation, in that if a relation R is symmetric, then $aRb \Leftrightarrow bRa$.

Non-commuting Operators in Quantum Mechanics

In quantum mechanics as formulated by Schrödinger, physical variables are represented by linear operators such as x (meaning multiply by x), and $\dfrac{d}{dx}$. These two operators do not commute as may be seen by considering the effect of their compositions $x\dfrac{d}{dx}$ and $\dfrac{d}{dx}x$ (also called products of operators) on a one-dimensional wave function $\psi(x)$:

$$x\frac{d}{dx}\psi = x\psi' \neq \frac{d}{dx}x\psi = \psi + x\psi'$$

According to the uncertainty principle of Heisenberg, if the two operators representing a pair of variables do not commute, then that pair of variables are mutually complementary, which means they cannot be simultaneously measured or known precisely. For example, the position and the linear momentum in the x-direction of a particle are represented respectively by the operators x and $-i\hbar\dfrac{\partial}{\partial x}$ (where $\hbar$ is the reduced Planck constant). This is the same example except for the constant $-i\hbar$, so again the operators do not commute and the physical meaning is that the position and linear momentum in a given direction are complementary.

Associative Property

In mathematics, the associative property is a property of some binary operations. In propositional logic, associativity is a valid rule of replacement for expressions in logical proofs.

Within an expression containing two or more occurrences in a row of the same associative operator, the order in which the operations are performed does not matter as long as the sequence of the operands is not changed. That is, rearranging the parentheses in such an expression will not change its value. Consider the following equations:

$$(2+3)+4 = 2+(3+4) = 9$$

$$2\times(3\times4) = (2\times3)\times4 = 24.$$

Even though the parentheses were rearranged on each line, the values of the expressions were not altered. Since this holds true when performing addition and multiplication on any real numbers, it can be said that "addition and multiplication of real numbers are associative operations".

Associativity is not to be confused with commutativity, which addresses whether or not the order of two operands changes the result. For example, the order doesn't matter in the multiplication of real numbers, that is, $a \times b = b \times a$, so we say that the multiplication of real numbers is a commutative operation.

Associative operations are abundant in mathematics; in fact, many algebraic structures (such as semigroups and categories) explicitly require their binary operations to be associative.

However, many important and interesting operations are non-associative; some examples include subtraction, exponentiation and the vector cross product. In contrast to the theoretical counterpart, the addition of floating point numbers in computer science is not associative, and is an important source of rounding error.

Definition

$$S \times S \times S \xrightarrow{\ *\times 1\ } S \times S$$

$$\begin{array}{ccc} & & \\ {\scriptstyle 1\times *}\downarrow & & \downarrow{\scriptstyle *} \\ S \times S & \xrightarrow{\quad * \quad} & S \end{array}$$

A binary operation $*$ on the set S is associative when this diagram commutes. That is, when the two paths from $S \times S \times S$ to S compose to the same function from $S \times S \times S$ to S.

Formally, a binary operation $*$ on a set S is called associative if it satisfies the associative law:

$(x * y) * z = x * (y * z)$ for all x, y, z in S.

Here, $*$ is used to replace the symbol of the operation, which may be any symbol, and even the absence of symbol (juxtaposition) as for multiplication.

$(xy)z = x(yz) = xyz$ for all x, y, z in S.

The associative law can also be expressed in functional notation thus: $f(f(x, y), z) = f(x, f(y, z))$.

Generalized Associative Law

If a binary operation is associative, repeated application of the operation produces the same result regardless how valid pairs of parenthesis are inserted in the expression. This is called the generalized associative law. For instance, a product of four elements may be written in five possible ways:

1. ((ab)c)d

2. (ab)(cd)

3. (a(bc))d

4. a((bc)d)

5. a(b(cd))

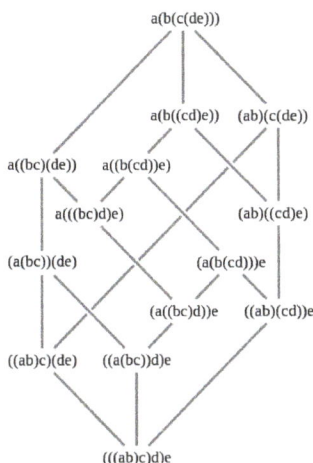

a(b(c(de)))

a(b((cd)e))   (ab)(c(de))

a((bc)(de))   a((b(cd))e)

a(((bc)d)e)   (ab)((cd)e)

(a(bc))(de)   (a(b(cd))e

(a((bc)d))e   ((ab)(cd))e

((ab)c)(de)   ((a(bc))d)e

(((ab)c)d)e

In the absence of the associative property, five factors *a, b, c, d, e* result in a Tamari lattice of order four, possibly different products.

If the product operation is associative, the generalized associative law says that all these formulas will yield the same result, making the parenthesis unnecessary. Thus "the" product can be written unambiguously as

abcd.

As the number of elements increases, the number of possible ways to insert parentheses grows quickly, but they remain unnecessary for disambiguation.

Examples

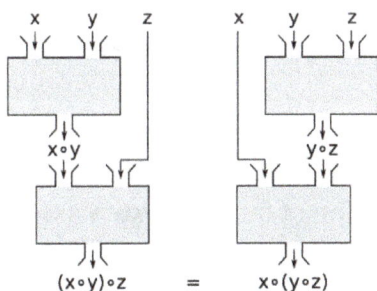

In associative operations is . $(x^{\circ}y)^{\circ}z = x^{\circ}(y^{\circ}z)$.

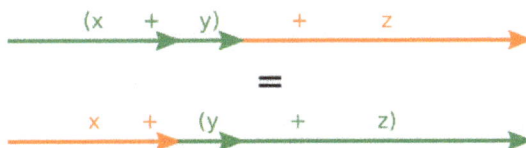

The addition of real numbers is associative.

Some examples of associative operations include the following.

- The concatenation of the three strings "hello", " ", "world" can be computed by concatenating the first two strings (giving "hello ") and appending the third string ("world"), or by joining the second and third string (giving "world") and concatenating the first string ("hello") with the result. The two methods produce the same result; string concatenation is associative (but not commutative).

- In arithmetic, addition and multiplication of real numbers are associative; i.e.,

$$(x+y)+z = x+(y+z) = x+y+z$$
$$(xy)z = x(yz) = xyz$$

for all $x, y, z \in \mathbb{R}$.

Because of associativity, the grouping parentheses can be omitted without ambiguity.

- Addition and multiplication of complex numbers and quaternions are associative. Addition of octonions is also associative, but multiplication of octonions is non-associative.

- The greatest common divisor and least common multiple functions act associatively.

$$\gcd(\gcd(x, y), z) = \gcd(x, \gcd(y, z)) = \gcd(x, y, z)$$
$$\operatorname{lcm}(\operatorname{lcm}(x, y), z) = \operatorname{lcm}(x, \operatorname{lcm}(y, z)) = \operatorname{lcm}(x, y, z)$$

for all $x, y, z \in \mathbb{Z}$.

- Taking the intersection or the union of sets:

$$(A \cap B) \cap C = A \cap (B \cap C) = A \cap B \cap C$$
$$(A \cup B) \cup C = A \cup (B \cup C) = A \cup B \cup C$$

for all sets A, B, C.

- If M is some set and S denotes the set of all functions from M to M, then the operation of functional composition on S is associative:

$$(f \circ g) \circ h = f \circ (g \circ h) = f \circ g \circ h \qquad \text{for all } f, g, h \in S.$$

- Slightly more generally, given four sets M, N, P and Q, with h: M to N, g: N to P, and f: P to Q, then

$$(f \circ g) \circ h = f \circ (g \circ h) = f \circ g \circ h$$

as before. In short, composition of maps is always associative.

- Consider a set with three elements, A, B, and C. The following operation:

×	A	B	C
A	A	A	A
B	A	B	C
C	A	A	A

is associative. Thus, for example, A(BC)=(AB)C = A. This operation is not commutative.

- Because matrices represent linear transformation functions, with matrix multiplication representing functional composition, one can immediately conclude that matrix multiplication is associative.

Propositional Logic

Rule of Replacement

In standard truth-functional propositional logic, *association*, or *associativity* are two valid rules of replacement. The rules allow one to move parentheses in logical expressions in logical proofs. The rules are:

$$(P \vee (Q \vee R)) \Leftrightarrow ((P \vee Q) \vee R)$$

and

$$(P \wedge (Q \wedge R)) \Leftrightarrow ((P \wedge Q) \wedge R),$$

where "$\Leftrightarrow$" is a metalogical symbol representing "can be replaced in a proof with."

Truth Functional Connectives

Associativity is a property of some logical connectives of truth-functional propositional logic. The following logical equivalences demonstrate that associativity is a property of particular connectives. The following are truth-functional tautologies.

Associativity of Disjunction:

$$((P \vee Q) \vee R) \leftrightarrow (P \vee (Q \vee R))$$

$$(P \vee (Q \vee R)) \leftrightarrow ((P \vee Q) \vee R)$$

Associativity of conjunction:

$$((P \wedge Q) \wedge R) \leftrightarrow (P \wedge (Q \wedge R))$$

$$(P \wedge (Q \wedge R)) \leftrightarrow ((P \wedge Q) \wedge R)$$

Associativity of equivalence:

$$((P \leftrightarrow Q) \leftrightarrow R) \leftrightarrow (P \leftrightarrow (Q \leftrightarrow R))$$

$$(P \leftrightarrow (Q \leftrightarrow R)) \leftrightarrow ((P \leftrightarrow Q) \leftrightarrow R)$$

Non-associativity

A binary operation on a set S that does not satisfy the associative law is called non-associative. Symbolically,

$$(x * y) * z \neq x * (y * z) \qquad \text{for some } x, y, z \in S.$$

For such an operation the order of evaluation *does* matter. For example:

- Subtraction

$$(5-3)-2 \neq 5-(3-2)$$

- Division

$$(4/2)/2 \neq 4/(2/2)$$

- Exponentiation

$$2^{(1^2)} \neq (2^1)^2$$

Also note that infinite sums are not generally associative, for example:

$$(1-1)+(1-1)+(1-1)+(1-1)+(1-1)+(1-1)+\ldots=0$$

whereas

$$1+(-1+1)+(-1+1)+(-1+1)+(-1+1)+(-1+1)+(-1+\ldots=1$$

The study of non-associative structures arises from reasons somewhat different from the mainstream of classical algebra. One area within non-associative algebra that has grown very large is that of Lie algebras. There the associative law is replaced by the Jacobi identity. Lie algebras abstract the essential nature of infinitesimal transformations, and have become ubiquitous in mathematics.

There are other specific types of non-associative structures that have been studied in depth; these tend to come from some specific applications or areas such as combinatorial mathematics. Other examples are Quasigroup, Quasifield, Non-associative ring, Non-associative algebra and Commutative non-associative magmas.

Nonassociativity of Floating Point Calculation

In mathematics, addition and multiplication of real numbers is associative. By contrast, in computer science, the addition and multiplication of floating point numbers is *not* associative, as rounding errors are introduced when dissimilar-sized values are joined together.

To illustrate this, consider a floating point representation with a 4-bit mantissa:
$(1.000_2 \times 2^0 + 1.000_2 \times 2^0) + 1.000_2 \times 2^4 = 1.000_2 \times 2^1 + 1.000_2 \times 2^4 = 1.001_2 \times 2^4$
$1.000_2 \times 2^0 + (1.000_2 \times 2^0 + 1.000_2 \times 2^4) = 1.000_2 \times 2^0 + 1.000_2 \times 2^4 = 1.000_2 \times 2^4$

Even though most computers compute with a 24 or 53 bits of mantissa, this is an important source of rounding error, and approaches such as the Kahan Summation Algorithm are ways to minimise the errors. It can be especially problematic in parallel computing.

Notation for Non-associative Operations

In general, parentheses must be used to indicate the order of evaluation if a non-associative operation appears more than once in an expression. However, mathematicians agree on a particular order of evaluation for several common non-associative operations. This is simply a notational convention to avoid parentheses.

A left-associative operation is a non-associative operation that is conventionally evaluated from left to right, i.e.,

$$
\left.
\begin{aligned}
x * y * z &= (x * y) * z \\
w * x * y * z &= ((w * x) * y) * z \\
\text{etc.}
\end{aligned}
\right\} \text{for all } w, x, y, z \in S
$$

while a right-associative operation is conventionally evaluated from right to left:

$$
\left.
\begin{aligned}
x * y * z &= x * (y * z) \\
w * x * y * z &= w * (x * (y * z)) \\
\text{etc.}
\end{aligned}
\right\} \text{for all } w, x, y, z \in S
$$

Both left-associative and right-associative operations occur. Left-associative operations include the following:

- Subtraction and division of real numbers:

$$x - y - z = (x - y) - z \qquad \text{for all } x, y, z \in \mathbb{R};$$

$$x / y / z = (x / y) / z \qquad \text{for all } x, y, z \in \mathbb{R} \text{ with } y \neq 0, z \neq 0.$$

- Function application:

$$(f \, x \, y) = ((f \, x) \, y)$$

This notation can be motivated by the currying isomorphism.

Right-associative operations include the following:

- Exponentiation of real numbers:

$$x^{y^z} = x^{(y^z)}$$

- One reason exponentiation is right-associative is that a repeated left-associative exponentiation operation would be less useful. Multiple appearances could (and would) be rewritten with multiplication:

$$(x^y)^z = x^{(yz)}$$

An additional argument for exponentiation being right-associative is that the superscript inherently behaves as a set of parentheses; e.g. in the expression 2^{x+3} the addition is performed before the exponentiation despite there being no explicit parentheses $2^{(x+3)}$ wrapped around it. Thus given an expression such as x^{y^z}, it makes sense to require evaluating the full exponent y^z of the base x first.

- Tetration via the up-arrow operator:

$$a \uparrow\uparrow b = {}^b a$$

- Function definition

$$\mathbb{Z} \to \mathbb{Z} \to \mathbb{Z} = \mathbb{Z} \to (\mathbb{Z} \to \mathbb{Z})$$

$$x \mapsto y \mapsto x - y = x \mapsto (y \mapsto x - y)$$

Using right-associative notation for these operations can be motivated by the Curry-Howard correspondence and by the currying isomorphism.

Non-associative operations for which no conventional evaluation order is defined include the following.

- Taking the Cross product of three vectors:

$$\vec{a} \times (\vec{b} \times \vec{c}) \neq (\vec{a} \times \vec{b}) \times \vec{c} \qquad \text{for some } \vec{a}, \vec{b}, \vec{c} \in \mathbb{R}^3$$

- Taking the pairwise average of real numbers:

$$\frac{(x+y)/2 + z}{2} \neq \frac{x + (y+z)/2}{2} \qquad \text{for all } x, y, z \in \mathbb{R} \text{ with } x \neq z.$$

- Taking the relative complement of sets $(A \setminus B) \setminus C$ is not the same as $A \setminus (B \setminus C)$. (Compare material nonimplication in logic.)

Distributive Property

$$ab + ac = a(b+c)$$

Visualization of distributive law for positive numbers

In abstract algebra and formal logic, the distributive property of binary operations generalizes the distributive law from elementary algebra. In propositional logic, distribution refers to two valid rules of replacement. The rules allow one to reformulate conjunctions and disjunctions within logical proofs.

For example, in arithmetic:

$$2 \cdot (1 + 3) = (2 \cdot 1) + (2 \cdot 3), \text{ but } 2 / (1 + 3) \neq (2 / 1) + (2 / 3).$$

In the left-hand side of the first equation, the 2 multiplies the sum of 1 and 3; on the right-hand side, it multiplies the 1 and the 3 individually, with the products added afterwards. Because these give the same final answer (8), it is said that multiplication by 2 *distributes* over addition of 1 and 3. Since one could have put any real numbers in place of 2, 1, and 3 above, and still have obtained a true equation, we say that multiplication of real numbers *distributes* over addition of real numbers.

Definition

Given a set S and two binary operators $*$ and $+$ on S, we say that the operation:

$*$ is *left-distributive* over $+$ if, given any elements x, y, and z of S,

$$x * (y + z) = (x * y) + (x * z),$$

$*$ is *right-distributive* over $+$ if, given any elements x, y, and z of S,

$$(y + z) * x = (y * x) + (z * x), \text{ and}$$

$*$ is *distributive* over $+$ if it is left- and right-distributive.

Notice that when $*$ is commutative, the three conditions above are logically equivalent.

Meaning

The operators used for examples in this section are the binary operations of addition ( ) and multiplication (+) of numbers.

There is a distinction between left-distributivity and right-distributivity:

$$a \cdot (b \pm c) = a \cdot b \pm a \cdot c \quad \text{(left-distributive)}$$

$$(a \pm b) \cdot c = a \cdot c \pm b \cdot c \quad \text{(right-distributive)}$$

In either case, the distributive property can be described in words as:

To multiply a sum (or difference) by a factor, each summand (or minuend and subtrahend) is multiplied by this factor and the resulting products are added (or subtracted).

If the operation outside the parentheses (in this case, the multiplication) is commutative, then left-distributivity implies right-distributivity and vice versa.

One example of an operation that is "only" right-distributive is division, which is not commutative:

$$(a \pm b) \div c = a \div c \pm b \div c$$

In this case, left-distributivity does not apply:

$$a \div (b \pm c) \neq a \div b \pm a \div c$$

The distributive laws are among the axioms for rings and fields. Examples of structures in which two operations are mutually related to each other by the distributive law are Boolean algebras such as the algebra of sets or the switching algebra. There are also combinations of operations that are not mutually distributive over each other; for example, addition is not distributive over multiplication.

Multiplying sums can be put into words as follows: When a sum is multiplied by a sum, multiply each summand of a sum with each summand of the other sums (keeping track of signs), and then adding up all of the resulting products.

Examples

Real Numbers

In the following examples, the use of the distributive law on the set of real numbers $\mathbb{R}$ is illustrated. When multiplication is mentioned in elementary mathematics, it usually refers to this kind of multiplication. From the point of view of algebra, the real numbers form a field, which ensures the validity of the distributive law.

First example (mental and written multiplication)

During mental arithmetic, distributivity is often used unconsciously:

$$6 \cdot 16 = 6 \cdot (10 + 6) = 6 \cdot 10 + 6 \cdot 6 = 60 + 36 = 96$$

Thus, to calculate $6 \cdot 16$ in your head, you first multiply $6 \cdot 10$ and $6 \cdot 6$ and add the intermediate results. Written multiplication is also based on the distributive law.

Second example (with variables)

$$3a^2b \cdot (4a - 5b) = 3a^2b \cdot 4a - 3a^2b \cdot 5b = 12a^3b - 15a^2b^2$$

Third example (with two sums)

$$(a + b) \cdot (a - b) = a \cdot (a - b) + b \cdot (a - b) = a^2 - ab + ba - b^2 :$$

$$= (a + b) \cdot a - (a + b) \cdot b = a^2 + ba - ab - b^2 = a^2 - b^2$$

Here the distributive law was applied twice, and it does not matter which bracket is first multiplied out.

Fourth Example

Here the distributive law is applied the other way around compared to the previous examples. Consider

$$12a^3b^2 - 30a^4bc + 18a^2b^3c^2.$$

Since the factor $6a^2b$ occurs in all summand, it can be factored out. That is, due to the distributive law one obtains

$$12a^3b^2 - 30a^4bc + 18a^2b^3c^2 = 6a^2b(2ab - 5a^2c + 3b^2c^2).$$

Matrices

The distributive law is valid for matrix multiplication. More precisely,

$$(A+B)\cdot C = A\cdot C + B\cdot C$$

for all $l \times m$-matrices A, B and $m \times n$-matrices C, as well as

$$A\cdot(B+C) = A\cdot B + A\cdot C$$

for all $l \times m$-matrices A and $m \times n$-matrices B, C. Because the commutative property does not hold for matrix multiplication, the second law does not follow from the first law. In this case, they are two different laws.

Other Examples

1. Multiplication of ordinal numbers, in contrast, is only left-distributive, not right-distributive.

2. The cross product is left- and right-distributive over vector addition, though not commutative.

3. The union of sets is distributive over intersection, and intersection is distributive over union.

4. Logical disjunction ("or") is distributive over logical conjunction ("and"), and conjunction is distributive over disjunction.

5. For real numbers (and for any totally ordered set), the maximum operation is distributive over the minimum operation, and vice versa: $\max(a, \min(b, c)) = \min(\max(a, b), \max(a, c))$ and $\min(a, \max(b, c)) = \max(\min(a, b), \min(a, c))$.

6. For integers, the greatest common divisor is distributive over the least common multiple, and vice versa: $\gcd(a, \mathrm{lcm}(b, c)) = \mathrm{lcm}(\gcd(a, b), \gcd(a, c))$ and $\mathrm{lcm}(a, \gcd(b, c)) = \gcd(\mathrm{lcm}(a, b), \mathrm{lcm}(a, c))$.

7. For real numbers, addition distributes over the maximum operation, and also over the minimum operation: $a + \max(b, c) = \max(a + b, a + c)$ and $a + \min(b, c) = \min(a + b, a + c)$.

Propositional Logic

Rule of Replacement

In standard truth-functional propositional logic, *distribution* in logical proofs uses two valid rules of replacement to expand individual occurrences of certain logical connectives, within some formula, into separate applications of those connectives across subformulas of the given formula. The rules are:

$$(P \wedge (Q \vee R)) \Leftrightarrow ((P \wedge Q) \vee (P \wedge R))$$

and

$$(P \vee (Q \wedge R)) \Leftrightarrow ((P \vee Q) \wedge (P \vee R))$$

where "$\Leftrightarrow$", also written $\equiv$, is a metalogical symbol representing "can be replaced in a proof with" or "is logically equivalent to".

Truth Functional Connectives

Distributivity is a property of some logical connectives of truth-functional propositional logic. The following logical equivalences demonstrate that distributivity is a property of particular connectives. The following are truth-functional tautologies.

Distribution of conjunction over conjunction

$$(P \wedge (Q \wedge R)) \leftrightarrow ((P \wedge Q) \wedge (P \wedge R))$$

Distribution of conjunction over disjunction

$$(P \wedge (Q \vee R)) \leftrightarrow ((P \wedge Q) \vee (P \wedge R))$$

Distribution of disjunction over conjunction

$$(P \vee (Q \wedge R)) \leftrightarrow ((P \vee Q) \wedge (P \vee R))$$

Distribution of disjunction over disjunction

$$(P \vee (Q \vee R)) \leftrightarrow ((P \vee Q) \vee (P \vee R))$$

Distribution of implication

$$(P \rightarrow (Q \rightarrow R)) \leftrightarrow ((P \rightarrow Q) \rightarrow (P \rightarrow R))$$

Distribution of implication over equivalence

$$(P \rightarrow (Q \leftrightarrow R)) \leftrightarrow ((P \rightarrow Q) \leftrightarrow (P \rightarrow R))$$

Distribution of disjunction over equivalence

$$(P \vee (Q \leftrightarrow R)) \leftrightarrow ((P \vee Q) \leftrightarrow (P \vee R))$$

Double distribution

$$((P \wedge Q) \vee (R \wedge S)) \leftrightarrow (((P \vee R) \wedge (P \vee S)) \wedge ((Q \vee R) \wedge (Q \vee S)))$$

$$((P \vee Q) \wedge (R \vee S)) \leftrightarrow (((P \wedge R) \vee (P \wedge S)) \vee ((Q \wedge R) \vee (Q \wedge S)))$$

Distributivity and Rounding

In practice, the distributive property of multiplication (and division) over addition may appear to be compromised or lost because of the limitations of arithmetic precision. For example, the identity $\frac{1}{3}+\frac{1}{3}+\frac{1}{3} = (1 + 1 + 1) / 3$ appears to fail if the addition is conducted in decimal arithmetic; however, if many significant digits are used, the calculation will result in a closer approximation to the correct results. For example, if the arithmetical calculation takes the form: 0.33333 + 0.33333 + 0.33333 = 0.99999 ≠ 1, this result is a closer approximation than if fewer significant digits had been used. Even when fractional numbers can be represented exactly in arithmetical form, errors will be introduced if those arithmetical values are rounded or truncated. For example, buying two books, each priced at £14.99 before a tax of 17.5%, in two separate transactions will actually save £0.01, over buying them together: £14.99 × 1.175 = £17.61 to the nearest £0.01, giving a total expenditure of £35.22, but £29.98 × 1.175 = £35.23. Methods such as banker's rounding may help in some cases, as may increasing the precision used, but ultimately some calculation errors are inevitable.

Distributivity in Rings

Distributivity is most commonly found in rings and distributive lattices.

A ring has two binary operations (commonly called "+" and "*"), and one of the requirements of a ring is that * must distribute over +. Most kinds of numbers (example 1) and matrices (example 4) form rings. A lattice is another kind of algebraic structure with two binary operations, ∧ and ∨. If either of these operations (say ∧) distributes over the other (∨), then ∨ must also distribute over ∧, and the lattice is called distributive.

Examples 4 and 5 are Boolean algebras, which can be interpreted either as a special kind of ring (a Boolean ring) or a special kind of distributive lattice (a Boolean lattice). Each interpretation is responsible for different distributive laws in the Boolean algebra. Examples 6 and 7 are distributive lattices which are not Boolean algebras.

Failure of one of the two distributive laws brings about near-rings and near-fields instead of rings and division rings respectively. The operations are usually configured to have the near-ring or near-field distributive on the right but not on the left.

Rings and distributive lattices are both special kinds of rigs, certain generalizations of rings. Those numbers in example 1 that don't form rings at least form rigs. Near-rigs are a further generalization of rigs that are left-distributive but not right-distributive; example 2 is a near-rig.

Generalizations of Distributivity

In several mathematical areas, generalized distributivity laws are considered. This may involve the weakening of the above conditions or the extension to infinitary operations. Especially in order theory one finds numerous important variants of distributivity, some of which include infinitary operations, such as the infinite distributive law; others being defined in the presence of only *one* binary operation, such as the according definitions and their relations are given in the article distributivity (order theory). This also includes the notion of a completely distributive lattice.

In the presence of an ordering relation, one can also weaken the above equalities by replacing = by either ≤ or ≥. Naturally, this will lead to meaningful concepts only in some situations. An application of this principle is the notion of sub-distributivity as explained in the article on interval arithmetic.

In category theory, if (S, μ, η) and (S', μ', η') are monads on a category C, a distributive law $S.S' \to S'.S$ is a natural transformation $\lambda : S.S' \to S'.S$ such that (S', λ) is a lax map of monads $S \to S$ and (S, λ) is a colax map of monads $S' \to S'$. This is exactly the data needed to define a monad structure on $S'.S$: the multiplication map is $S'\mu.\mu'S^2.S'\lambda S$ and the unit map is $\eta'S.\eta$.

A generalized distributive law has also been proposed in the area of information theory.

Notions of Antidistributivity

The ubiquitous identity that relates inverses to the binary operation in any group, namely $(xy)^{-1} = y^{-1}x^{-1}$, which is taken as an axiom in the more general context of a semigroup with involution, has sometimes been called an antidistributive property (of inversion as a unary operation).

In the context of a near-ring, which removes the commutativity of the additively written group and assumes only one-sided distributivity, one can speak of (two-sided) distributive elements but also of antidistributive elements. The latter reverse the order of (the non-commutative) addition; assuming a left-nearring (i.e. one which all elements distribute when multiplied on the left), then an antidistributive element a reverses the order of addition when multiplied to the right: $(x + y)a = ya + xa$.

In the study of propositional logic and Boolean algebra, the term antidistributive law is sometimes used to denote the interchange between conjunction and disjunction when implication factors over them:

- $(a \lor b) \Rightarrow c \equiv (a \Rightarrow c) \land (b \Rightarrow c)$

- $(a \land b) \Rightarrow c \equiv (a \Rightarrow c) \lor (b \Rightarrow c)$

These two tautologies are a direct consequence of the duality in De Morgan's laws.

Isomorphism

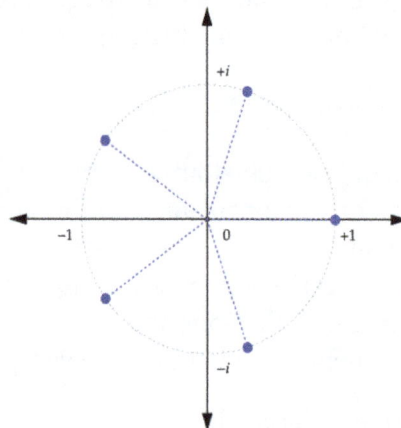

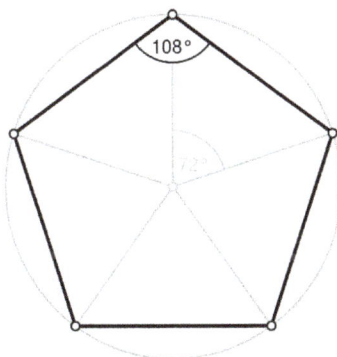

The group of fifth roots of unity under multiplication is isomorphic to the group of rotations of the regular pentagon under composition.

In mathematics, an isomorphism is a homomorphism or morphism (i.e. a mathematical mapping) that admits an inverse. Two mathematical objects are isomorphic if an isomorphism exists between them. An *automorphism* is an isomorphism whose source and target coincide. The interest of isomorphisms lies in the fact that two isomorphic objects cannot be distinguished by using only the properties used to define morphisms; thus isomorphic objects may be considered the same as long as one considers only these properties and their consequences.

For most algebraic structures, including groups and rings, a homomorphism is an isomorphism if and only if it is bijective.

In topology, where the morphisms are continuous functions, isomorphisms are also called *homeomorphisms* or *bicontinuous functions*. In mathematical analysis, where the morphisms are differentiable functions, isomorphisms are also called *diffeomorphisms*.

A canonical isomorphism is a canonical map that is an isomorphism. Two objects are said to be canonically isomorphic if there is a canonical isomorphism between them. For example, the canonical map from a finite-dimensional vector space V to its second dual space is a canonical isomorphism; on the other hand, V is isomorphic to its dual space but not canonically in general.

Isomorphisms are formalized using category theory. A morphism $f : X \to Y$ in a category is an isomorphism if it admits a two-sided inverse, meaning that there is another morphism $g : Y \to X$ in that category such that $gf = 1_X$ and $fg = 1_Y$, where 1_X and 1_Y are the identity morphisms of X and Y, respectively.

Examples

Logarithm and Exponential

Let $\mathbb{R}^+$ be the multiplicative group of positive real numbers, and let $\mathbb{R}$ be the additive group of real numbers.

The logarithm function $\log : \mathbb{R}^+ \to \mathbb{R}$ satisfies $\log(xy) = \log x + \log y$ for all $x, y \in \mathbb{R}^+$, so it is a group homomorphism. The exponential function $\exp : \mathbb{R} \to \mathbb{R}^+$ satisfies $\exp(x + y) = (\exp x)(\exp y)$ for all $x, y \in \mathbb{R}$, so it too is a homomorphism.

The identities $\log \exp x = x$ and $\exp \log y = y$ show that $\log$ and $\exp$ are inverses of each other.

Since log is a homomorphism that has an inverse that is also a homomorphism, log is an isomorphism of groups.

Because log is an isomorphism, it translates multiplication of positive real numbers into addition of real numbers. This facility makes it possible to multiply real numbers using a ruler and a table of logarithms, or using a slide rule with a logarithmic scale.

Integers Modulo 6

Consider the group $(\mathbb{Z}_6, +)$, the integers from 0 to 5 with addition modulo 6. Also consider the group $(\mathbb{Z}_2 \times \mathbb{Z}_3, +)$, the ordered pairs where the x coordinates can be 0 or 1, and the y coordinates can be 0, 1, or 2, where addition in the x-coordinate is modulo 2 and addition in the y-coordinate is modulo 3.

These structures are isomorphic under addition, if you identify them using the following scheme:

$(0,0) \rightarrow 0$

$(1,1) \rightarrow 1$

$(0,2) \rightarrow 2$

$(1,0) \rightarrow 3$

$(0,1) \rightarrow 4$

$(1,2) \rightarrow 5$

or in general $(a,b) \rightarrow (3a + 4b) \bmod 6$.

For example, note that $(1,1) + (1,0) = (0,1)$, which translates in the other system as $1 + 3 = 4$.

Even though these two groups "look" different in that the sets contain different elements, they are indeed isomorphic: their structures are exactly the same. More generally, the direct product of two cyclic groups $\mathbb{Z}_n$ and $\mathbb{Z}_n$ is isomorphic to $(\mathbb{Z}_{mn}, +)$ if and only if m and n are coprime.

Relation-preserving Isomorphism

If one object consists of a set X with a binary relation R and the other object consists of a set Y with a binary relation S then an isomorphism from X to Y is a bijective function $f : X \rightarrow Y$ such that:

$$S(f(u), f(v)) \Leftrightarrow R(u,v)$$

S is reflexive, irreflexive, symmetric, antisymmetric, asymmetric, transitive, total, trichotomous, a partial order, total order, strict weak order, total preorder (weak order), an equivalence relation, or a relation with any other special properties, if and only if R is.

For example, R is an ordering $\leq$ and S an ordering $\sqsubseteq$, then an isomorphism from X to Y is a bijective function $f : X \rightarrow Y$ such that

$$f(u) \sqsubseteq f(v) \Leftrightarrow u \leq v.$$

Such an isomorphism is called an *order isomorphism* or (less commonly) an *isotone isomorphism*.

If $X = Y$, then this is a relation-preserving automorphism.

Isomorphism vs. Bijective Morphism

In a concrete category (that is, roughly speaking, a category whose objects are sets and morphisms are mappings between sets), such as the category of topological spaces or categories of algebraic objects like groups, rings, and modules, an isomorphism must be bijective on the underlying sets. In algebraic categories (specifically, categories of varieties in the sense of universal algebra), an isomorphism is the same as a homomorphism which is bijective on underlying sets. However, there are concrete categories in which bijective morphisms are not necessarily isomorphisms (such as the category of topological spaces), and there are categories in which each object admits an underlying set but in which isomorphisms need not be bijective (such as the homotopy category of CW-complexes).

Applications

In abstract algebra, two basic isomorphisms are defined:

- Group isomorphism, an isomorphism between groups
- Ring isomorphism, an isomorphism between rings. (Note that isomorphisms between fields are actually ring isomorphisms)

Just as the automorphisms of an algebraic structure form a group, the isomorphisms between two algebras sharing a common structure form a heap. Letting a particular isomorphism identify the two structures turns this heap into a group.

In mathematical analysis, the Laplace transform is an isomorphism mapping hard differential equations into easier algebraic equations.

In category theory, let the category C consist of two classes, one of *objects* and the other of *morphisms*. Then a general definition of isomorphism that covers the previous and many other cases is: an isomorphism is a morphism $f: a \to b$ that has an inverse, i.e. there exists a morphism $g: b \to a$ with $fg = 1_b$ and $gf = 1_a$. For example, a bijective linear map is an isomorphism between vector spaces, and a bijective continuous function whose inverse is also continuous is an isomorphism between topological spaces, called a homeomorphism.

In graph theory, an isomorphism between two graphs G and H is a bijective map f from the vertices of G to the vertices of H that preserves the "edge structure" in the sense that there is an edge from vertex u to vertex v in G if and only if there is an edge from $f(u)$ to $f(v)$ in H.

In mathematical analysis, an isomorphism between two Hilbert spaces is a bijection preserving addition, scalar multiplication, and inner product.

In early theories of logical atomism, the formal relationship between facts and true propositions was theorized by Bertrand Russell and Ludwig Wittgenstein to be isomorphic. An example of this line of thinking can be found in Russell's Introduction to Mathematical Philosophy.

In cybernetics, the Good Regulator or Conant-Ashby theorem is stated "Every Good Regulator of a system must be a model of that system". Whether regulated or self-regulating an isomorphism is required between regulator part and the processing part of the system.

Relation with Equality

In certain areas of mathematics, notably category theory, it is valuable to distinguish between *equality* on the one hand and *isomorphism* on the other. Equality is when two objects are exactly the same, and everything that's true about one object is true about the other, while an isomorphism implies everything that's true about a designated part of one object's structure is true about the other's. For example, the sets

$$A = \{x \in \mathbb{Z} \mid x^2 < 2\} \text{ and } B = \{-1, 0, 1\}$$

are *equal*; they are merely different presentations—the first an intensional one (in set builder notation), and the second extensional (by explicit enumeration)—of the same subset of the integers. By contrast, the sets $\{A,B,C\}$ and $\{1,2,3\}$ are not *equal*—the first has elements that are letters, while the second has elements that are numbers. These are isomorphic as sets, since finite sets are determined up to isomorphism by their cardinality (number of elements) and these both have three elements, but there are many choices of isomorphism—one isomorphism is

$$A \mapsto 1, B \mapsto 2, C \mapsto 3, \text{ while another is } A \mapsto 3, B \mapsto 2, C \mapsto 1,$$

and no one isomorphism is intrinsically better than any other. On this view and in this sense, these two sets are not equal because one cannot consider them *identical*: one can choose an isomorphism between them, but that is a weaker claim than identity—and valid only in the context of the chosen isomorphism.

Sometimes the isomorphisms can seem obvious and compelling, but are still not equalities. As a simple example, the genealogical relationships among Joe, John, and Bobby Kennedy are, in a real sense, the same as those among the American football quarterbacks in the Manning family: Archie, Peyton, and Eli. The father-son pairings and the elder-brother-younger-brother pairings correspond perfectly. That similarity between the two family structures illustrates the origin of the word *isomorphism*. But because the Kennedys are not the same people as the Mannings, the two genealogical structures are merely isomorphic and not equal.

Another example is more formal and more directly illustrates the motivation for distinguishing equality from isomorphism: the distinction between a finite-dimensional vector space V and its dual space $V^* = \{\varphi: V \to K\}$ of linear maps from V to its field of scalars K. These spaces have the same dimension, and thus are isomorphic as abstract vector spaces (since algebraically, vector spaces are classified by dimension, just as sets are classified by cardinality), but there is no "natural" choice of isomorphism $V \to V^*$. If one chooses a basis for V, then this yields an isomorphism: For all $u.\ v \in V$,

$$v \mapsto \tilde{\phi}_v \in V^* \quad \text{such that} \quad \phi_v(u) = v^\mathrm{T} u.$$

This corresponds to transforming a column vector (element of V) to a row vector (element of V^*) by transpose, but a different choice of basis gives a different isomorphism: the isomorphism "depends on the choice of basis". More subtly, there *is* a map from a vector space V to its double dual $V^{**} = \{ x \colon V^* \to K \}$ that does not depend on the choice of basis: For all $v \in V$ and $\varphi \in V^*$,

$$v \overset{\sim}{\mapsto} x_v \in V^{**} \quad \text{such that} \quad x_v(\phi) = \phi(v).$$

This leads to a third notion, that of a natural isomorphism: while V and V^{**} are different sets, there is a "natural" choice of isomorphism between them. This intuitive notion of "an isomorphism that does not depend on an arbitrary choice" is formalized in the notion of a natural transformation; briefly, that one may *consistently* identify, or more generally map from, a vector space to its double dual, $V \to V^{**}$, for *any* vector space in a consistent way. Formalizing this intuition is a motivation for the development of category theory.

However, there is a case where the distinction between natural isomorphism and equality is usually not made. That is for the objects that may be characterized by a universal property. In fact, there is a unique isomorphism, necessarily natural, between two objects sharing the same universal property. A typical example is the set of real numbers, which may be defined through infinite decimal expansion, infinite binary expansion, Cauchy sequences, Dedekind cuts and many other ways. Formally these constructions define different objects, which all are solutions of the same universal property. As these objects have exactly the same properties, one may forget the method of construction and considering them as equal. This is what everybody does when talking of "*the* set of the real numbers". The same occurs with quotient spaces: they are commonly constructed as sets of equivalence classes. However, talking of set of sets may be counterintuitive, and quotient spaces are commonly considered as a pair of a set of undetermined objects, often called "points", and a surjective map onto this set.

If one wishes to draw a distinction between an arbitrary isomorphism (one that depends on a choice) and a natural isomorphism (one that can be done consistently), one may write $\approx$ for an unnatural isomorphism and $\cong$ for a natural isomorphism, as in $V \approx V^*$ and $V \cong V^{**}$. This convention is not universally followed, and authors who wish to distinguish between unnatural isomorphisms and natural isomorphisms will generally explicitly state the distinction.

Generally, saying that two objects are *equal* is reserved for when there is a notion of a larger (ambient) space that these objects live in. Most often, one speaks of equality of two subsets of a given set (as in the integer set example above), but not of two objects abstractly presented. For example, the 2-dimensional unit sphere in 3-dimensional space

$$S^2 := \{(x, y, z) \in \mathbb{R}^3 \mid x^2 + y^2 + z^2 = 1\} \text{ and the Riemann sphere } \widehat{\mathbb{C}}$$

which can be presented as the one-point compactification of the complex plane $\mathbb{C} \cup \{\infty\}$ *or* as the complex projective line (a quotient space)

$$\mathbf{P}^1_{\mathbb{C}} := (\mathbb{C}^2 \setminus \{(0,0)\}) / (\mathbb{C}^*)$$

are three different descriptions for a mathematical object, all of which are isomorphic, but not *equal* because they are not all subsets of a single space: the first is a subset of $\mathbb{R}^3$, the second is $\mathbb{C} \cong \mathbb{R}^2$ plus an additional point, and the third is a subquotient of $\mathbb{C}^2$

In the context of category theory, objects are usually at most isomorphic—indeed, a motivation for the development of category theory was showing that different constructions in homology theory yielded equivalent (isomorphic) groups. Given maps between two objects X and Y, however, one asks if they are equal or not (they are both elements of the set $\text{Hom}(X, Y)$, hence equality is the proper relationship), particularly in commutative diagrams.

References

- M. Kilp, U. Knauer, A.V. Mikhalev, Monoids, Acts and Categories with Applications to Wreath Products and Graphs, De Gruyter Expositions in Mathematics vol. 29, Walter de Gruyter, 2000, ISBN 3-11-015248-7, p. 14–15

- Howie, John M. (1995). Fundamentals of Semigroup Theory. Clarendon Press. ISBN 0-19-851194-9. contains all of the semigroup material herein except *-regular semigroups.

- Lam, Tsit-Yuen (2003). Exercises in Classical Ring Theory. Problem Books in Mathematics (2nd ed.). Springer-Verlag. ISBN 978-0387005003.

- Milies, César Polcino; Sehgal, Sudarshan K. (2002). An introduction to group rings. Springer. ISBN 978-1-4020-0238-0.

- Robins, R. Gay, and Charles C. D. Shute. 1987. The Rhind Mathematical Papyrus: An Ancient Egyptian Text. London: British Museum Publications Limited. ISBN 0-7141-0944-4

- Ayres, Frank, Schaum's Outline of Modern Abstract Algebra, McGraw-Hill; 1st edition (June 1, 1965). ISBN 0-07-002655-6.

- Vinberg, Ėrnest Borisovich (2003). A Course in Algebra. American Mathematical Society. p. 3. ISBN 9780821834138.

Major Theorems in Algebra

This text provides a plethora of the major theorems of algebra. The main theorems in algebra are fundamental theorem of algebra, Abel-Ruffini theorem, binomial theorem, Chevalley-Warning theorem and Boolean prime ideal theorem. The topics discussed in the chapter are of great importance to broaden the existing knowledge on the theorems of algebra.

Fundamental Theorem of Algebra

The fundamental theorem of algebra states that every non-constant single-variable polynomial with complex coefficients has at least one complex root. This includes polynomials with real coefficients, since every real number is a complex number with an imaginary part equal to zero.

Equivalently (by definition), the theorem states that the field of complex numbers is algebraically closed.

The theorem is also stated as follows: every non-zero, single-variable, degree n polynomial with complex coefficients has, counted with multiplicity, exactly n roots. The equivalence of the two statements can be proven through the use of successive polynomial division.

In spite of its name, there is no purely algebraic proof of the theorem, since any proof must use the completeness of the reals (or some other equivalent formulation of completeness), which is not an algebraic concept. Additionally, it is not fundamental for modern algebra; its name was given at a time when the study of algebra was mainly concerned with the solutions of polynomial equations with real or complex coefficients.

History

Peter Roth, in his book *Arithmetica Philosophica* (published in 1608, at Nürnberg, by Johann Lantzenberger), wrote that a polynomial equation of degree n (with real coefficients) *may* have n solutions. Albert Girard, in his book *L'invention nouvelle en l'Algèbre* (published in 1629), asserted that a polynomial equation of degree n has n solutions, but he did not state that they had to be real numbers. Furthermore, he added that his assertion holds "unless the equation is incomplete", by which he meant that no coefficient is equal to 0. However, when he explains in detail what he means, it is clear that he actually believes that his assertion is always true; for instance, he shows that the equation $x^4 = 4x - 3$, although incomplete, has four solutions (counting multiplicities): 1 (twice), $-1 + i\sqrt{2}$, and $-1 - i\sqrt{2}$.

As will be mentioned again below, it follows from the fundamental theorem of algebra that every non-constant polynomial with real coefficients can be written as a product of polynomials with real coefficients whose degree is either 1 or 2. However, in 1702 Leibniz said that no polynomial of the type $x^4 + a^4$ (with a real and distinct from 0) can be written in such a way. Later, Nikolaus Bernoulli made the same assertion concerning the polynomial $x^4 - 4x^3 + 2x^2 + 4x + 4$, but he got a letter from Euler in 1742 in which he was told that his polynomial happened to be equal to

$$(x^2 - (2+\alpha)x + 1 + \sqrt{7} + \alpha)(x^2 - (2-\alpha)x + 1 + \sqrt{7} - \alpha),$$

where α is the square root of $4 + 2\sqrt{7}$. Also, Euler mentioned that

$$x^4 + a^4 = (x^2 + a\sqrt{2}\cdot x + a^2)(x^2 - a\sqrt{2}\cdot x + a^2).$$

A first attempt at proving the theorem was made by d'Alembert in 1746, but his proof was incomplete. Among other problems, it assumed implicitly a theorem (now known as Puiseux's theorem) which would not be proved until more than a century later, and furthermore the proof assumed the fundamental theorem of algebra. Other attempts were made by Euler (1749), de Foncenex (1759), Lagrange (1772), and Laplace (1795). These last four attempts assumed implicitly Girard's assertion; to be more precise, the existence of solutions was assumed and all that remained to be proved was that their form was $a + bi$ for some real numbers a and b. In modern terms, Euler, de Foncenex, Lagrange, and Laplace were assuming the existence of a splitting field of the polynomial $p(z)$.

At the end of the 18th century, two new proofs were published which did not assume the existence of roots, but neither of which was complete. One of them, due to James Wood and mainly algebraic, was published in 1798 and it was totally ignored. Wood's proof had an algebraic gap. The other one was published by Gauss in 1799 and it was mainly geometric, but it had a topological gap, filled by Alexander Ostrowski in 1920, as discussed in Smale 1981 (Smale writes, "...I wish to point out what an immense gap Gauss' proof contained. It is a subtle point even today that a real algebraic plane curve cannot enter a disk without leaving. In fact even though Gauss redid this proof 50 years later, the gap remained. It was not until 1920 that Gauss' proof was completed. In the reference Gauss, A. Ostrowski has a paper which does this and gives an excellent discussion of the problem as well..."). A rigorous proof was first published by Argand in 1806 (and revisited in 1813); it was here that, for the first time, the fundamental theorem of algebra was stated for polynomials with complex coefficients, rather than just real coefficients. Gauss produced two other proofs in 1816 and another version of his original proof in 1849.

The first textbook containing a proof of the theorem was Cauchy's *Cours d'analyse de l'École Royale Polytechnique* (1821). It contained Argand's proof, although Argand is not credited for it.

None of the proofs mentioned so far is constructive. It was Weierstrass who raised for the first time, in the middle of the 19th century, the problem of finding a constructive proof of the fundamental theorem of algebra. He presented his solution, that amounts in modern terms to a combination of the Durand–Kerner method with the homotopy continuation principle, in 1891. Another proof of this kind was obtained by Hellmuth Kneser in 1940 and simplified by his son Martin Kneser in 1981.

Without using countable choice, it is not possible to constructively prove the fundamental theorem of algebra for complex numbers based on the Dedekind real numbers (which are not constructively equivalent to the Cauchy real numbers without countable choice). However, Fred Richman proved a reformulated version of the theorem that does work.

Proofs

All proofs below involve some analysis, or at least the topological concept of continuity of real or complex functions. Some also use differentiable or even analytic functions. This fact has led to the remark that the Fundamental Theorem of Algebra is neither fundamental, nor a theorem of algebra.

Some proofs of the theorem only prove that any non-constant polynomial with real coefficients has some complex root. This is enough to establish the theorem in the general case because, given a non-constant polynomial $p(z)$ with complex coefficients, the polynomial

$$q(z) = p(z)\overline{p(\overline{z})}$$

has only real coefficients and, if z is a zero of $q(z)$, then either z or its conjugate is a root of $p(z)$.

A large number of non-algebraic proofs of the theorem use the fact (sometimes called "growth lemma") that an n-th degree polynomial function $p(z)$ whose dominant coefficient is 1 behaves like z^n when $|z|$ is large enough. A more precise statement is: there is some positive real number R such that:

$$\tfrac{1}{2}|z^n| < |p(z)| < \tfrac{3}{2}|z^n|$$

when $|z| > R$.

Complex-analytic Proofs

Find a closed disk D of radius r centered at the origin such that $|p(z)| > |p(0)|$ whenever $|z| \geq r$. The minimum of $|p(z)|$ on D, which must exist since D is compact, is therefore achieved at some point z_0 in the interior of D, but not at any point of its boundary. The Maximum modulus principle (applied to $1/p(z)$) implies then that $p(z_0) = 0$. In other words, z_0 is a zero of $p(z)$.

A variation of this proof does not require the use of the maximum modulus principle (in fact, the same argument with minor changes also gives a proof of the maximum modulus principle for holomorphic functions). If we assume by contradiction that $a := p(z_0) \neq 0$, then, expanding $p(z)$ in powers of $z - z_0$ we can write

$$p(z) = a + c_k(z - z_0)^k + c_{k+1}(z - z_0)^{k+1} + \ldots + c_n(z - z_0)^n.$$

Here, the c_j are simply the coefficients of the polynomial $z \to p(z + z_0)$, and we let k be the index of the first coefficient following the constant term that is non-zero. But now we see that for z sufficiently close to z_0 this has behavior asymptotically similar to the simpler polynomial $q(z) = a + c_k(z - z_0)^k$,

in the sense that (as is easy to check) the function $\left| \dfrac{p(z) - q(z)}{(z - z_0)^{k+1}} \right|$ is bounded by some positive constant M in some neighborhood of z_0. Therefore if we define $\theta_0 = (\arg(a) + \pi - \arg(c_k))/k$ and let $z = z_0 + re^{i\theta_0}$,, then for any sufficiently small positive number r (so that the bound M mentioned above holds), using the triangle inequality we see that

$$|p(z)| < |q(z)| + r^{k+1}\left|\frac{p(z)-q(z)}{r^{k+1}}\right|$$

$$\leq \left|a + (-1)c_k r^k e^{i(\arg(a)-\arg(c_k))}\right| + Mr^{k+1}$$

$$= |a| - |c_k|r^k + Mr^{k+1}.$$

When r is sufficiently close to 0 this upper bound for $|p(z)|$ is strictly smaller than $|a|$, in contradiction to the definition of z_0. (Geometrically, we have found an explicit direction θ_0 such that if one approaches z_0 from that direction one can obtain values $p(z)$ smaller in absolute value than $|p(z_0)|$.)

Another analytic proof can be obtained along this line of thought observing that, since $|p(z)| > |p(0)|$ outside D, the minimum of $|p(z)|$ on the whole complex plane is achieved at z_0. If $|p(z_0)| > 0$, then $1/p$ is a bounded holomorphic function in the entire complex plane since, for each complex number z, $|1/p(z)| \leq |1/p(z_0)|$. Applying Liouville's theorem, which states that a bounded entire function must be constant, this would imply that $1/p$ is constant and therefore that p is constant. This gives a contradiction, and hence $p(z_0) = 0$.

Yet another analytic proof uses the argument principle. Let R be a positive real number large enough so that every root of $p(z)$ has absolute value smaller than R; such a number must exist because every non-constant polynomial function of degree n has at most n zeros. For each $r > R$, consider the number

$$\frac{1}{2\pi i}\int_{c(r)}\frac{p'(z)}{p(z)}dz,$$

where $c(r)$ is the circle centered at 0 with radius r oriented counterclockwise; then the argument principle says that this number is the number N of zeros of $p(z)$ in the open ball centered at 0 with radius r, which, since $r > R$, is the total number of zeros of $p(z)$. On the other hand, the integral of n/z along $c(r)$ divided by $2\pi i$ is equal to n. But the difference between the two numbers is

$$\frac{1}{2\pi i}\int_{c(r)}\left(\frac{p'(z)}{p(z)}-\frac{n}{z}\right)dz = \frac{1}{2\pi i}\int_{c(r)}\frac{zp'(z)-np(z)}{zp(z)}dz.$$

The numerator of the rational expression being integrated has degree at most n - 1 and the degree of the denominator is $n + 1$. Therefore, the number above tends to 0 as $r \to +\infty$. But the number is also equal to $N - n$ and so $N = n$.

Still another complex-analytic proof can be given by combining linear algebra with the Cauchy theorem. To establish that every complex polynomial of degree $n > 0$ has a zero, it suffices to show that every complex square matrix of size $n > 0$ has a (complex) eigenvalue. The proof of the latter statement is by contradiction.

Let A be a complex square matrix of size $n > 0$ and let I_n be the unit matrix of the same size. Assume A has no eigenvalues. Consider the resolvent function

$$R(z) = (zI_n - A)^{-1},$$

which is a meromorphic function on the complex plane with values in the vector space of matrices. The eigenvalues of A are precisely the poles of $R(z)$. Since, by assumption, A has no eigenvalues, the function $R(z)$ is an entire function and Cauchy theorem implies that

$$\int_{c(r)} R(z)dz = 0.$$

On the other hand, $R(z)$ expanded as a geometric series gives:

$$R(z) = z^{-1}(I_n - z^{-1}A)^{-1} = z^{-1}\sum_{k=0}^{\infty}\frac{1}{z^k}A^k.$$

This formula is valid outside the closed disc of radius $||A||$ (the operator norm of A). Let $r > ||A||$. Then

$$\int_{c(r)} R(z)dz = \sum_{k=0}^{\infty}\int_{c(r)}\frac{dz}{z^{k+1}}A^k = 2\pi i I_n$$

(in which only the summand $k = 0$ has a nonzero integral). This is a contradiction, and so A has an eigenvalue.

Finally, Rouché's theorem gives perhaps the shortest proof of the theorem.

Topological Proofs

Let $z_0 \in C$ be such that the minimum of $|p(z)|$ on the whole complex plane is achieved at z_0; it was seen at the proof which uses Liouville's theorem that such a number must exist. We can write $p(z)$ as a polynomial in $z - z_0$: there is some natural number k and there are some complex numbers c_k, c_{k+1}, ..., c_n such that $c_k \neq 0$ and that

$$p(z) = p(z_0) + c_k(z - z_0)^k + c_{k+1}(z - z_0)^{k+1} + \cdots + c_n(z - z_0)^n.$$

In the case that $p(z_0)$ is nonzero, it follows that if a is a kth root of $-p(z_0)/c_k$ and if t is positive and sufficiently small, then $|p(z_0 + ta)| < |p(z_0)|$, which is impossible, since $|p(z_0)|$ is the minimum of $|p|$ on D.

For another topological proof by contradiction, suppose that $p(z)$ has no zeros. Choose a large positive number R such that, for $|z| = R$, the leading term z^n of $p(z)$ dominates all other terms combined; in other words, such that $|z|^n > |a_{n-1}z^{n-1} + \cdots + a_0|$. As z traverses the circle given by the equation $|z| = R$ once counter-clockwise, $p(z)$, like z^n, winds n times counter-clockwise around 0. At the other extreme, with $|z| = 0$, the "curve" $p(z)$ is simply the single (nonzero) point $p(0)$, whose winding number is clearly 0. If the loop followed by z is continuously deformed between these extremes, the path of $p(z)$ also deforms continuously. We can explicitly write such a deformation as $H(Re^{i\theta}, t) = p((1 - t)Re^{i\theta})$, where $0 \leq t \leq 1$. If one views the variable t as time, then at time zero the curve is $p(z)$ and at time one the curve is $p(0)$. Clearly at every point t, $p(z)$ cannot be zero by the original assumption, therefore during the deformation, the curve never crosses zero. Therefore the winding number of the curve around zero should never change. However, given that the winding number started as n and ended as 0, this is absurd. Therefore, $p(z)$ has at least one zero.

Algebraic Proofs

These proofs use two facts about real numbers that require only a small amount of analysis (more precisely, the intermediate value theorem):

- every polynomial with odd degree and real coefficients has some real root;

- every non-negative real number has a square root.

The second fact, together with the quadratic formula, implies the theorem for real quadratic polynomials. In other words, algebraic proofs of the fundamental theorem actually show that if R is any real-closed field, then its extension $C = R(\sqrt{-1})$ is algebraically closed.

As mentioned above, it suffices to check the statement "every non-constant polynomial $p(z)$ with real coefficients has a complex root". This statement can be proved by induction on the greatest non-negative integer k such that 2^k divides the degree n of $p(z)$. Let a be the coefficient of z^n in $p(z)$ and let F be a splitting field of $p(z)$ over C; in other words, the field F contains C and there are elements $z_1, z_2, ..., z_n$ in F such that

$$p(z) = a(z - z_1)(z - z_2) \cdots (z - z_n).$$

If $k = 0$, then n is odd, and therefore $p(z)$ has a real root. Now, suppose that $n = 2^k m$ (with m odd and $k > 0$) and that the theorem is already proved when the degree of the polynomial has the form $2^{k-1}m'$ with m' odd. For a real number t, define:

$$q_t(z) = \prod_{1 \le i < j \le n} \left(z - z_i - z_j - t z_i z_j \right).$$

Then the coefficients of $q_t(z)$ are symmetric polynomials in the z_i's with real coefficients. Therefore, they can be expressed as polynomials with real coefficients in the elementary symmetric polynomials, that is, in $-a_1, a_2, ..., (-1)^n a_n$. So $q_t(z)$ has in fact *real* coefficients. Furthermore, the degree of $q_t(z)$ is $n(n-1)/2 = 2^{k-1}m(n-1)$, and $m(n-1)$ is an odd number. So, using the induction hypothesis, q_t has at least one complex root; in other words, $z_i + z_j + t z_i z_j$ is complex for two distinct elements i and j from $\{1, ..., n\}$. Since there are more real numbers than pairs (i, j), one can find distinct real numbers t and s such that $z_i + z_j + t z_i z_j$ and $z_i + z_j + s z_i z_j$ are complex (for the same i and j). So, both $z_i + z_j$ and $z_i z_j$ are complex numbers. It is easy to check that every complex number has a complex square root, thus every complex polynomial of degree 2 has a complex root by the quadratic formula. It follows that z_i and z_j are complex numbers, since they are roots of the quadratic polynomial $z^2 - (z_i + z_j)z + z_i z_j$.

Joseph Shipman showed in 2007 that the assumption that odd degree polynomials have roots is stronger than necessary; any field in which polynomials of prime degree have roots is algebraically closed (so "odd" can be replaced by "odd prime" and furthermore this holds for fields of all characteristics). For axiomatization of algebraically closed fields, this is the best possible, as there are counterexamples if a single prime is excluded. However, these counterexamples rely on -1 having a square root. If we take a field where -1 has no square root, and every polynomial of degree $n \in I$ has a root, where I is any fixed infinite set of odd numbers, then every polynomial $f(x)$ of odd degree has a root (since $(x^2 + 1)^k f(x)$ has a root, where k is chosen so that $\deg(f) + 2k \in I$). Mohsen Aliabadi generalized Shipman's result for any field in 2013, proving that the sufficient condition for an arbitrary field (of any characteristic) to be algebraically closed is having a root for any polynomial of prime degree.

Another algebraic proof of the fundamental theorem can be given using Galois theory. It suffices to show that C has no proper finite field extension. Let K/C be a finite extension. Since the normal closure of K over R still has a finite degree over C (or R), we may assume without loss of generality that K is a normal extension of R (hence it is a Galois extension, as every algebraic extension of a field of characteristic 0 is separable). Let G be the Galois group of this extension, and let H be a Sylow 2-subgroup of G, so that the order of H is a power of 2, and the index of H in G is odd. By the fundamental theorem of Galois theory, there exists a subextension L of K/R such that $\mathrm{Gal}(K/L) = H$. As $[L{:}R] = [G{:}H]$ is odd, and there are no nonlinear irreducible real polynomials of odd degree, we must have $L = R$, thus $[K{:}R]$ and $[K{:}C]$ are powers of 2. Assuming by way of contradiction that $[K{:}C] > 1$, we conclude that the 2-group $\mathrm{Gal}(K/C)$ contains a subgroup of index 2, so there exists a subextension M of C of degree 2. However, C has no extension of degree 2, because every quadratic complex polynomial has a complex root, as mentioned above. This shows that $[K{:}C] = 1$, and therefore $K = C$, which completes the proof.

Geometric Proofs

There exists still another way to approach the fundamental theorem of algebra, due to J. M. Almira and A. Romero: by Riemannian Geometric arguments. The main idea here is to prove that the existence of a non-constant polynomial $p(z)$ without zeros implies the existence of a flat Riemannian metric over the sphere S^2. This leads to a contradiction, since the sphere is not flat.

Recall that a Riemannian surface (M, g) is said to be flat if its Gaussian curvature, which we denote by K_g, is identically null. Now, Gauss–Bonnet theorem, when applied to the sphere S^2, claims that

$$\int_{S^2} K_g = 4\pi,$$

which proves that the sphere is not flat.

Let us now assume that $n > 0$ and $p(z) = a_0 + a_1 z + \cdots + a_n z^n \neq 0$ for each complex number z. Let us define $p^*(z) = z^n p(1/z) = a_0 z^n + a_1 z^{n-1} + \cdots + a_n$. Obviously, $p^*(z) \neq 0$ for all z in C. Consider the polynomial $f(z) = p(z)p^*(z)$. Then $f(z) \neq 0$ for each z in C. Furthermore,

$$f\left(\tfrac{1}{w}\right) = p\left(\tfrac{1}{w}\right)p^*\left(\tfrac{1}{w}\right) = w^{-2n} p^*(w)p(w) = w^{-2n} f(w).$$

We can use this functional equation to prove that g, given by

$$g = \frac{1}{|f(w)|^{\frac{2}{n}}} |dw|^2$$

for w in C, and

$$g = \frac{1}{|f(1/w)|^{\frac{2}{n}}} |d(1/w)|^2$$

for $w \in S^2 \setminus \{0\}$, is a well defined Riemannian metric over the sphere S^2 (which we identify with the extended complex plane $C \cup \{\infty\}$).

Now, a simple computation shows that

$$\forall w \in \mathbf{C}: \frac{1}{|f(w)|^{\frac{1}{n}}} K_g = \frac{1}{n} \Delta \log |f(w)| = \frac{1}{n} \Delta \mathrm{Re}(\log f(w)) = 0,$$

since the real part of an analytic function is harmonic. This proves that $K_g = 0$.

Corollaries

Since the fundamental theorem of algebra can be seen as the statement that the field of complex numbers is algebraically closed, it follows that any theorem concerning algebraically closed fields applies to the field of complex numbers. Here are a few more consequences of the theorem, which are either about the field of real numbers or about the relationship between the field of real numbers and the field of complex numbers:

- The field of complex numbers is the algebraic closure of the field of real numbers.

- Every polynomial in one variable z with complex coefficients is the product of a complex constant and polynomials of the form $z + a$ with a complex.

- Every polynomial in one variable x with real coefficients can be uniquely written as the product of a constant, polynomials of the form $x + a$ with a real, and polynomials of the form $x^2 + ax + b$ with a and b real and $a^2 - 4b < 0$ (which is the same thing as saying that the polynomial $x^2 + ax + b$ has no real roots). This implies that the number of non-real complex roots (up to multiplicity) is always even.

- Every rational function in one variable x, with real coefficients, can be written as the sum of a polynomial function with rational functions of the form $a/(x - b)^n$ (where n is a natural number, and a and b are real numbers), and rational functions of the form $(ax + b)/(x^2 + cx + d)^n$ (where n is a natural number, and a, b, c, and d are real numbers such that $c^2 - 4d < 0$). A corollary of this is that every rational function in one variable and real coefficients has an elementary primitive.

- Every algebraic extension of the real field is isomorphic either to the real field or to the complex field.

Abel−Ruffini Theorem

In algebra, the Abel−Ruffini theorem (also known as Abel's impossibility theorem) states that there is no algebraic solution—that is, solution in radicals—to the general polynomial equations of degree five or higher with arbitrary coefficients. The theorem is named after Paolo Ruffini, who made an incomplete proof in 1799, and Niels Henrik Abel, who provided a proof in 1824.

Interpretation

The theorem does *not* assert that some higher-degree polynomial equations have *no* solution. In fact, the opposite is true: *every* non-constant polynomial equation in one unknown, with real or complex coefficients, has at least one complex number as a solution (and thus, by polynomial division, as many complex roots as its degree, counting repeated roots); this is the fundamental theorem of algebra. These solutions can be computed to any desired degree of accuracy using numerical methods such as the Newton–Raphson method or the Laguerre method, and in this way they are no different from solutions to polynomial equations of the second, third, or fourth degrees. The theorem only shows that there is no *general* solution in radicals that applies to *all* equations of a given degree greater than 4.

The solution of any second-degree polynomial equation can be expressed in terms of its coefficients, using only addition, subtraction, multiplication, division, and square roots, in the familiar quadratic formula: the roots of the equation $ax^2 + bx + c = 0$ (with $a \neq 0$) are

$$\frac{-b \pm \sqrt{b^2 - 4ac}}{2a}.$$

Analogous formulas for third-degree equations and fourth-degree equations (using square roots and cube roots) have been known since the 16th century. What the Abel–Ruffini theorem says is that there is no similar formula for general equations of fifth degree or higher. In principle, it could be that the equations of the fifth degree could be split in several types and, for each one of these types, there could be some algebraic solution valid within that type. Or, as Ian Stewart wrote, "for all that Abel's methods could prove, every particular quintic equation might be soluble, with a special formula for each equation." However, this is not so, but this impossibility lies outside the scope of the Abel–Ruffini theorem and is part of the Galois theory.

Proof

The following proof is based on Galois theory and it is valid for any field of characteristic 0. Historically, Ruffini and Abel's proofs precede Galois theory. For a modern presentation of Abel's proof see the article of Rosen or the books of Tignol or Pesic.

One of the fundamental theorems of Galois theory states that a polynomial $P(x) \in F[x]$ is solvable by radicals over F if and only if its splitting field K over F has a solvable Galois group, so the proof of the Abel–Ruffini theorem comes down to computing the Galois group of the general polynomial of the fifth degree.

Consider five indeterminates $y_1, y_2, y_3, y_4,$ and y_5, let $E = Q(y_1, y_2, y_3, y_4, y_5)$, and let

$$P(x) = (x - y_1)(x - y_2)(x - y_3)(x - y_4)(x - y_5) \in E[x].$$

Expanding $P(x)$ out yields the elementary symmetric functions of the y_i:

$$s_1 = y_1 + y_2 + y_3 + y_4 + y_5,$$

$$s_2 = y_1 y_2 + y_1 y_3 + y_1 y_4 + y_1 y_5 + y_2 y_3 + y_2 y_4 + y_2 y_5 + y_3 y_4 + y_3 y_5 + y_4 y_5,$$

$$s_3 = y_1y_2y_3 + y_1y_2y_4 + y_1y_2y_5 + y_1y_3y_4 + y_1y_3y_5 + y_1y_4y_5 + y_2y_3y_4 + y_2y_3y_5 + y_2y_4y_5 + y_3y_4y_5,$$

$$s_4 = y_1y_2y_3y_4 + y_1y_2y_3y_5 + y_1y_2y_4y_5 + y_1y_3y_4y_5 + y_2y_3y_4y_5,$$

$$s_5 = y_1y_2y_3y_4y_5.$$

The coefficient of x^n in $P(x)$ is thus $(-1)^{5-n}s_{5-n}$. Let $F = Q(s_1, s_2, s_3, s_4, s_5)$ be the field obtained by ad-joining the symmetric functions to the rationals. Then $P(x) \in F[x]$. Because the y_i's are indetermi-nates, every permutation σ in the symmetric group on 5 letters S_5 induces a distinct automorphism σ' on E that leaves Q fixed and permutes the elements y_i. Since an arbitrary rearrangement of the roots of the product form still produces the same polynomial, e.g.

$$(x - y_3)(x - y_1)(x - y_2)(x - y_5)(x - y_4)$$

is the same polynomial as

$$(x - y_1)(x - y_2)(x - y_3)(x - y_4)(x - y_5),$$

the automorphisms σ' also leave F fixed, so they are elements of the Galois group Gal(E/F). Therefore, we have shown that $S_5 \subseteq$ Gal(E/F); however there could possibly be automorphisms there that are not in S_5. But, since the Galois group of the splitting field of a quintic polynomial has at most 5! elements, and since E is a splitting field of $P(x)$, it follows that Gal(E/F) is isomorphic to S_5. Generalizing this ar-gument shows that the Galois group of every general polynomial of degree n is isomorphic to S_n.

And what of S_5? The only composition series of S_5 is $S_5 \geq A_5 \geq \{e\}$ (where A_5 is the alternating group on five letters, also known as the icosahedral group). However, the quotient group $A_5/\{e\}$ (isomorphic to A_5 itself) is not abelian, and so S_5 is not solvable, so it must be that the general polynomial of the fifth degree has no solution in radicals. Since the first nontrivial normal subgroup of the symmetric group on n letters is always the alternating group on n letters, and since the alternating groups on n letters for $n \geq 5$ are always simple and non-abelian, and hence not solvable, it also says that the general polyno-mials of all degrees higher than the fifth also have no solution in radicals. Q.E.D.

The above construction of the Galois group for a fifth degree polynomial only applies to the *general polynomial*; specific polynomials of the fifth degree may have different Galois groups with quite differ-ent properties, e.g. $x^5 - 1$ has a splitting field generated by a primitive 5th root of unity, and hence its Galois group is abelian and the equation itself solvable by radicals; moreover, the argument does not provide any rational-valued quintic that has S_5 or A_5 as its Galois group. However, since the result is on the general polynomial, it does say that a general "quintic formula" for the roots of a quintic using only a finite combination of the arithmetic operations and radicals in terms of the coefficients is impossible.v

The proof is not valid if applied to polynomials whose degree is less than 5. Indeed:v

- the group A_4 is *not* simple, because the subgroup $\{e, (12)(34), (13)(24), (14)(23)\}$ is a nor-mal subgroup;

- the groups A_2 and A_3 *are* simple, but since they are abelian too (A_2 is the trivial group and A_3 is the cyclic group of order 3), that is not a problem.

The proof remains valid if, instead of working with five indeterminates, one works with five con-crete algebraically independent complex numbers, because, by the same argument, Gal(E/F) = S_5.

History

Around 1770, Joseph Louis Lagrange began the groundwork that unified the many different tricks that had been used up to that point to solve equations, relating them to the theory of groups of permutations, in the form of Lagrange resolvents. This innovative work by Lagrange was a precursor to Galois theory, and its failure to develop solutions for equations of fifth and higher degrees hinted that such solutions might be impossible, but it did not provide conclusive proof. The first person who conjectured that the problem of solving quintics by radicals might be impossible to solve was Carl Friedrich Gauss, who wrote in 1798 in section 359 of his book *Disquisitiones Arithmeticae* (which would be published only in 1801) that "there is little doubt that this problem does not so much defy modern methods of analysis as that it proposes the impossible". The next year, in his thesis, he wrote "After the labors of many geometers left little hope of ever arriving at the resolution of the general equation algebraically, it appears more and more likely that this resolution is impossible and contradictory." And he added "Perhaps it will not be so difficult to prove, with all rigor, the impossibility for the fifth degree. I shall set forth my investigations of this at greater length in another place." Actually, Gauss published nothing else on this subject.

TEORIA GENERALE

DELLE

EQUAZIONI,

IN CUI SI DIMOSTRA IMPOSSIBILE

LA SOLUZIONE ALGEBRAICA DELLE

EQUAZIONI GENERALI DI GRADO

SUPERIORE AL QUARTO

DI

PAOLO RUFFINI.

PARTE PRIMA.

BOLOGNA MDCCXCVIIII.

NELLA STAMPERIA DI S. TOMMASO D' AQUINO.

Teoria generale delle equazioni, 1799

The theorem was first nearly proved by Paolo Ruffini in 1799. He sent his proof to several mathematicians to get it acknowledged, amongst them Lagrange (who did not reply) and Augustin-Louis Cauchy, who sent him a letter saying: "Your memoir on the general solution of equations is a work which I have always believed should be kept in mind by mathematicians and which, in my opinion, proves conclusively the algebraic unsolvability of general equations of higher than fourth degree." However, in general, Ruffini's proof was not considered convincing. Abel wrote: "The first and, if I am not mistaken, the only one who, before me, has sought to prove the impossibility of the algebraic solution of general equations is the mathematician Ruffini. But his memoir is so complicated that it is very difficult to determine the validity of his argument. It seems to me that his argument is not completely satisfying."

The proof also, as it was discovered later, was incomplete. Ruffini assumed that all radicals that he was dealing with could be expressed from the roots of the polynomial using field operations alone; in modern terms, he assumed that the radicals belonged to the splitting field of the polynomial. To see why this is really an extra assumption, consider, for instance, the polynomial $P(x) = x^3 - 15x - 20$. According to Cardano's formula, one of its roots (all of them, actually) can be expressed as the sum of a cube root of $10 + 5i$ with a cube root of $10 - 5i$. On the other hand, since $P(-3) < 0$, $P(-2) > 0$, $P(-1) < 0$, and $P(5) > 0$, the roots r_1, r_2, and r_3 of $P(x)$ are all real and therefore the field $Q(r_1, r_2, r_3)$ is a subfield of R. But then the numbers $10 \pm 5i$ cannot belong to $Q(r_1, r_2, r_3)$. While Cauchy either did not notice Ruffini's assumption or felt that it was a minor one, most historians believe that the proof was not complete until Abel proved the theorem on natural irrationalities, which asserts that the assumption holds in the case of general polynomials. The Abel-Ruffini theorem is thus generally credited to Abel, who published a proof in just six pages in 1824. However, this short number of pages was obtained at the cost of writing in a very terse style. This was due to the fact that he had the proof printed at his own expenses and he needed to save paper and money. A more elaborated version of the proof would be published in 1826.

Proving that the general quintic (and higher) equations were unsolvable by radicals did not completely settle the matter, because the Abel–Ruffini theorem does not provide necessary and sufficient conditions for saying precisely which quintic (and higher) equations are unsolvable by radicals. Abel was working on a complete characterization when he died in 1829.

According to Nathan Jacobson, "The proofs of Ruffini and of Abel [...] were soon superseded by the crowning achievement of this line of research: Galois' discoveries in the theory of equations." In 1830, Galois (at the age of 18) submitted to the Paris Academy of Sciences a memoir on his theory of solvability by radicals, which was ultimately rejected in 1831 as being too sketchy and for giving a condition in terms of the roots of the equation instead of its coefficients. Galois was aware of the contributions of Ruffini and Abel, since he wrote "It is a common truth, today, that the general equation of degree greater than 4 cannot be solved by radicals... this truth has become common (by hearsay) despite the fact that geometers have ignored the proofs of Abel and Ruffini..." Galois then died in 1832 and his paper *Mémoire sur les conditions de resolubilité des équations par radicaux* remained unpublished until 1846, when it was published by Joseph Liouville accompanied by some of his own explanations. Prior to this publication, Liouville announced Galois' result to the Academy in a speech he gave on 4 July 1843. A simplification of Abel's proof was published by Pierre Wantzel in 1845. When he published it, he was already aware of the contributions by Galois and he mentions that, whereas Abel's proof is valid only for general polynomials, Galois' approach can be used to provide a concrete polynomial of degree 5 whose roots cannot be expressed in radicals from its coefficients.

In 1963, Vladimir Arnold discovered a topological proof of the Abel–Ruffini theorem, which served as a starting point for topological Galois theory.

Binomial Theorem

```
              1
            1   1
          1   2   1
        1   3   3   1
      1   4   6   4   1
    1   5   10  10   5   1
```

The binomial coefficients appear as the entries of Pascal's triangle where each entry is the sum of the two above it.

In elementary algebra, the binomial theorem (or binomial expansion) describes the algebraic expansion of powers of a binomial. According to the theorem, it is possible to expand the power $(x + y)^n$ into a sum involving terms of the form $ax^b y^c$, where the exponents b and c are nonnegative integers with $b + c = n$, and the coefficient a of each term is a specific positive integer depending on n and b. For example,

$$(x + y)^4 = x^4 + 4x^3 y + 6x^2 y^2 + 4xy^3 + y^4.$$

The coefficient a in the term of $ax^b y^c$ is known as the binomial coefficient $\binom{n}{b}$ or $\binom{n}{c}$ (the two have the same value). These coefficients for varying n and b can be arranged to form Pascal's triangle. These numbers also arise in combinatorics, where $\binom{n}{b}$ gives the number of different combinations of b elements that can be chosen from an n-element set.

History

Special cases of the binomial theorem were known from ancient times. The 4th century B.C. Greek mathematician Euclid mentioned the special case of the binomial theorem for exponent 2. There is evidence that the binomial theorem for cubes was known by the 6th century in India.

Binomial coefficients, as combinatorial quantities expressing the number of ways of selecting k objects out of n without replacement, were of interest to the ancient Hindus. The earliest known reference to this combinatorial problem is the *Chanda śāstra* by the Hindu lyricist Pingala (c. 200 B.C.), which contains a method for its solution. The commentator Halayudha from the 10th century A.D. explains this method using what is now known as Pascal's triangle. By the 6th century A.D., the Hindu mathematicians probably knew how to express this as a quotient $\dfrac{n!}{(n-k)!k!}$, and a clear statement of this rule can be found in the 12th century text *Lilavati* by Bhaskara.

The binomial theorem as such can be found in the work of 11th-century Persian mathematician Al-Karaji, who described the triangular pattern of the binomial coefficients. He also provided a mathematical proof of both the binomial theorem and Pascal's triangle, using a primitive form of mathematical induction. The Persian poet and mathematician Omar Khayyam was probably familiar with the formula to higher orders, although many of his mathematical works are lost. The binomial expansions of small degrees were known in the 13th century mathematical works of Yang Hui and also Chu Shih-Chieh. Yang Hui attributes the method to a much earlier 11th century text of Jia Xian, although those writings are now also lost.

In 1544, Michael Stifel introduced the term "binomial coefficient" and showed how to use them to express $(1 + a)^n$ in terms of $(1 + a)^{n-1}$, via "Pascal's triangle". Blaise Pascal studied the eponymous triangle comprehensively in the treatise *Traité du triangle arithmétique* (1653). However, the pattern of numbers was already known to the European mathematicians of the late Renaissance, including Stifel, Niccolò Fontana Tartaglia, and Simon Stevin.

Isaac Newton is generally credited with the generalized binomial theorem, valid for any rational exponent.

Theorem Statement

According to the theorem, it is possible to expand any power of $x + y$ into a sum of the form

$$(x+y)^n = \binom{n}{0}x^n y^0 + \binom{n}{1}x^{n-1}y^1 + \binom{n}{2}x^{n-2}y^2 + \cdots + \binom{n}{n-1}x^1 y^{n-1} + \binom{n}{n}x^0 y^n,$$

where each $\binom{n}{k}$ is a specific positive integer known as a binomial coefficient. (When an exponent is zero, the corresponding power expression is taken to be 1 and this multiplicative factor is often omitted from the term. Hence one often sees the right side written as $\binom{n}{0}x^n + \ldots.$).) This formula is also referred to as the binomial formula or the binomial identity. Using summation notation, it can be written as

$$(x+y)^n = \sum_{k=0}^{n}\binom{n}{k}x^{n-k}y^k = \sum_{k=0}^{n}\binom{n}{k}x^k y^{n-k}.$$

The final expression follows from the previous one by the symmetry of x and y in the first expression, and by comparison it follows that the sequence of binomial coefficients in the formula is symmetrical. A simple variant of the binomial formula is obtained by substituting 1 for y, so that it involves only a single variable. In this form, the formula reads

$$(1+x)^n = \binom{n}{0}x^0 + \binom{n}{1}x^1 + \binom{n}{2}x^2 + \cdots + \binom{n}{n-1}x^{n-1} + \binom{n}{n}x^n,$$

or equivalently

$$(1+x)^n = \sum_{k=0}^{n}\binom{n}{k}x^k.$$

Examples

$$
\begin{array}{ccccccccccccccc}
 & & & & & & & 1 & & & & & & & \\
 & & & & & & 1 & & 1 & & & & & & \\
 & & & & & 1 & & 2 & & 1 & & & & & \\
 & & & & 1 & & 3 & & 3 & & 1 & & & & \\
 & & & 1 & & 4 & & 6 & & 4 & & 1 & & & \\
 & & 1 & & 5 & & 10 & & 10 & & 5 & & 1 & & \\
 & 1 & & 6 & & 15 & & 20 & & 15 & & 6 & & 1 & \\
1 & & 7 & & 21 & & 35 & & 35 & & 21 & & 7 & & 1 \\
\end{array}
$$

Pascal's triangle

The most basic example of the binomial theorem is the formula for the square of $x + y$:

$$(x+y)^2 = x^2 + 2xy + y^2.$$

The binomial coefficients 1, 2, 1 appearing in this expansion correspond to the second row of Pascal's triangle. (Note that the top "1" of the triangle is considered to be row 0, by convention.) The coefficients of higher powers of $x + y$ correspond to lower rows of the triangle:

$$(x+y)^3 = x^3 + 3x^2 y + 3xy^2 + y^3,$$
$$(x+y)^4 = x^4 + 4x^3 y + 6x^2 y^2 + 4xy^3 + y^4,$$
$$(x+y)^5 = x^5 + 5x^4 y + 10x^3 y^2 + 10x^2 y^3 + 5xy^4 + y^5,$$
$$(x+y)^6 = x^6 + 6x^5 y + 15x^4 y^2 + 20x^3 y^3 + 15x^2 y^4 + 6xy^5 + y^6,$$
$$(x+y)^7 = x^7 + 7x^6 y + 21x^5 y^2 + 35x^4 y^3 + 35x^3 y^4 + 21x^2 y^5 + 7xy^6 + y^7.$$

Several patterns can be observed from these examples. In general, for the expansion $(x + y)^n$:

1. the powers of x start at n and decrease by 1 in each term until they reach 0 (with $x^0 = 1$ often unwritten);

2. the powers of y start at 0 and increase by 1 until they reach n;

3. the nth row of Pascal's Triangle will be the coefficients of the expanded binomial when the terms are arranged in this way;

4. the number of terms in the expansion before like terms are combined is the sum of the coefficients and is equal to 2^n; and

5. there will be $n + 1$ terms in the expression after combining like terms in the expansion.

The binomial theorem can be applied to the powers of any binomial. For example,

$$(x+2)^3 = x^3 + 3x^2(2) + 3x(2)^2 + 2^3$$
$$= x^3 + 6x^2 + 12x + 8.$$

For a binomial involving subtraction, the theorem can be applied by using the form $(x - y)^n = (x + (-y))^n$. This has the effect of changing the sign of every other term in the expansion:

$$(x-y)^3 = (x+(-y))^3 = x^3 + 3x^2(-y) + 3x(-y)^2 + (-y)^3 = x^3 - 3x^2 y + 3xy^2 - y^3.$$

Geometric Explanation

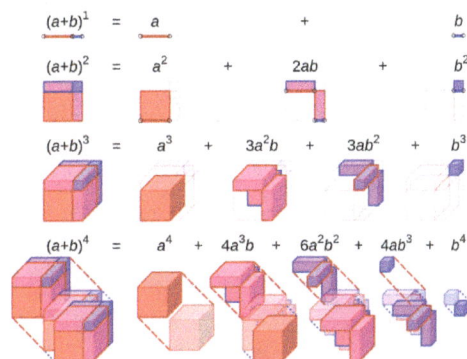

Visualisation of binomial expansion up to the 4th power

For positive values of a and b, the binomial theorem with $n = 2$ is the geometrically evident fact that a square of side $a + b$ can be cut into a square of side a, a square of side b, and two rectangles with sides a and b. With $n = 3$, the theorem states that a cube of side $a + b$ can be cut into a cube of side a, a cube of side b, three $a \times a \times b$ rectangular boxes, and three $a \times b \times b$ rectangular boxes.

In calculus, this picture also gives a geometric proof of the derivative $(x^n)' = nx^{n-1}$: if one sets $a = x$ and $b = \Delta x$, interpreting b as an infinitesimal change in a, then this picture shows the infinitesimal change in the volume of an n-dimensional hypercube, $(x + \Delta x)^n$, where the coefficient of the linear term (in Δx) is nx^{n-1}, the area of the n faces, each of dimension $(n-1)$:

$$(x + \Delta x)^n = x^n + nx^{n-1}\Delta x + \binom{n}{2}x^{n-2}(\Delta x)^2 + \cdots.$$

Substituting this into the definition of the derivative via a difference quotient and taking limits means that the higher order terms – $(\Delta x)^2$ and higher – become negligible, and yields the formula $(x^n)' = nx^{n-1}$, interpreted as

> "the infinitesimal rate of change in volume of an n-cube as side length varies is the area of n of its $(n-1)$-dimensional faces".

If one integrates this picture, which corresponds to applying the fundamental theorem of calculus, one obtains Cavalieri's quadrature formula, the integral $\int x^{n-1}\,dx = \frac{1}{n}x^n$ – see proof of Cavalieri's quadrature formula for details.

Binomial Coefficients

The coefficients that appear in the binomial expansion are called binomial coefficients. These are usually written $\binom{n}{k}$, and pronounced "n choose k".

Formulae

The coefficient of $x^{n-k}y^k$ is given by the formula

$$\binom{n}{k} = \frac{n!}{k!(n-k)!}$$

which is defined in terms of the factorial function $n!$. Equivalently, this formula can be written

$$\binom{n}{k} = \frac{n(n-1)\cdots(n-k+1)}{k(k-1)\cdots 1} = \prod_{\ell=1}^{k} \frac{n-\ell+1}{\ell} = \prod_{\ell=0}^{k-1} \frac{n-\ell}{k-\ell}$$

with k factors in both the numerator and denominator of the fraction. Note that, although this formula involves a fraction, the binomial coefficient $\binom{n}{k}$ is actually an integer.

Combinatorial Interpretation

The binomial coefficient $\binom{n}{k}$ can be interpreted as the number of ways to choose k elements from an n-element set. This is related to binomials for the following reason: if we write $(x + y)^n$ as a product

$$(x + y)(x + y)(x + y)\cdots(x + y),$$

then, according to the distributive law, there will be one term in the expansion for each choice of either x or y from each of the binomials of the product. For example, there will only be one term x^n, corresponding to choosing x from each binomial. However, there will be several terms of the form $x^{n-2}y^2$, one for each way of choosing exactly two binomials to contribute a y. Therefore, after combining like terms, the coefficient of $x^{n-2}y^2$ will be equal to the number of ways to choose exactly 2 elements from an n-element set.

Proofs

Combinatorial Proof

Example

The coefficient of xy^2 in

$$
\begin{aligned}
(x+y)^3 &= (x+y)(x+y)(x+y) \\
&= xxx + xxy + xyx + \underline{xyy} + yxx + \underline{yxy} + \underline{yyx} + yyy \\
&= x^3 + 3x^2y + \underline{3xy^2} + y^3.
\end{aligned}
$$

equals $\binom{3}{2} = 3$ because there are three x,y strings of length 3 with exactly two y's, namely,

$$xyy,\ yxy,\ yyx,$$

corresponding to the three 2-element subsets of $\{1, 2, 3\}$, namely,

$$\{2,3\},\ \{1,3\},\ \{1,2\},$$

where each subset specifies the positions of the y in a corresponding string.

General Case

Expanding $(x + y)^n$ yields the sum of the 2^n products of the form $e_1e_2 \dots e_n$ where each e_i is x or y. Rearranging factors shows that each product equals $x^{n-k}y^k$ for some k between 0 and n. For a given k, the following are proved equal in succession:

- the number of copies of $x^{n-k}y^k$ in the expansion
- the number of n-character x,y strings having y in exactly k positions
- the number of k-element subsets of $\{1, 2, \dots, n\}$
- $\binom{n}{k}$ (this is either by definition, or by a short combinatorial argument if one is defining

$\binom{n}{k}$ as $\dfrac{n!}{k!(n-k)!}$).

This proves the binomial theorem.

Inductive Proof

Induction yields another proof of the binomial theorem. When $n = 0$, both sides equal 1, since $x^0 = 1$ and $\binom{0}{0} = 1$. Now suppose that the equality holds for a given n; we will prove it for $n + 1$. For $j, k \geq 0$, let $[f(x, y)]_{j,k}$ denote the coefficient of $x^j y^k$ in the polynomial $f(x, y)$. By the inductive hypothesis, $(x + y)^n$ is a polynomial in x and y such that $[(x + y)^n]_{j,k}$ is $\binom{n}{k}$ if $j + k = n$, and 0 otherwise. The identity

$$(x + y)^{n+1} = x(x + y)^n + y(x + y)^n$$

shows that $(x + y)^{n+1}$ also is a polynomial in x and y, and

$$[(x + y)^{n+1}]_{j,k} = [(x + y)^n]_{j-1,k} + [(x + y)^n]_{j,k-1},$$

since if $j + k = n + 1$, then $(j - 1) + k = n$ and $j + (k - 1) = n$. Now, the right hand side is

$$\binom{n}{k} + \binom{n}{k-1} = \binom{n+1}{k},$$

by Pascal's identity. On the other hand, if $j + k \neq n + 1$, then $(j - 1) + k \neq n$ and $j + (k - 1) \neq n$, so we get $0 + 0 = 0$. Thus

$$(x + y)^{n+1} = \sum_{k=0}^{n+1} \binom{n+1}{k} x^{n+1-k} y^k,$$

which is the inductive hypothesis with $n + 1$ substituted for n and so completes the inductive step.

Generalizations

Newton's Generalized Binomial Theorem

Around 1665, Isaac Newton generalized the binomial theorem to allow real exponents other than nonnegative integers. (The same generalization also applies to complex exponents.) In this generalization, the finite sum is replaced by an infinite series. In order to do this, one needs to give meaning to binomial coefficients with an arbitrary upper index, which cannot be done using the usual formula with factorials. However, for an arbitrary number r, one can define

$$\binom{r}{k} = \frac{r(r-1)\cdots(r-k+1)}{k!} = \frac{(r)_k}{k!},$$

where $(\cdot)_k$ is the Pochhammer symbol, here standing for a falling factorial. This agrees with the usual definitions when r is a nonnegative integer. Then, if x and y are real numbers with $|x| > |y|$,[Note 1] and r is any complex number, one has

$$(x + y)^r = \sum_{k=0}^{\infty} \binom{r}{k} x^{r-k} y^k$$

$$= x^r + rx^{r-1}y + \frac{r(r-1)}{2!} x^{r-2} y^2 + \frac{r(r-1)(r-2)}{3!} x^{r-3} y^3 + \cdots.$$

When r is a nonnegative integer, the binomial coefficients for $k > r$ are zero, so this equation reduces to the usual binomial theorem, and there are at most $r + 1$ nonzero terms. For other values of r, the series typically has infinitely many nonzero terms.

For example, with $r = 1/2$ gives the following series for the square root:

$$\sqrt{1+x} = 1 + \frac{1}{2}x - \frac{1}{8}x^2 + \frac{1}{16}x^3 - \frac{5}{128}x^4 + \frac{7}{256}x^5 - \cdots$$

Taking $r = -1$, the generalized binomial series gives the geometric series formula, valid for $|x| < 1$:

$$(1+x)^{-1} = \frac{1}{1+x} = 1 - x + x^2 - x^3 + x^4 - x^5 + \cdots$$

More generally, with $r = -s$:

$$\frac{1}{(1-x)^s} = \sum_{k=0}^{\infty} \binom{s+k-1}{k} x^k \equiv \sum_{k=0}^{\infty} \binom{s+k-1}{s-1} x^k.$$

So, for instance, when $s = 1/2$,

$$\frac{1}{\sqrt{1+x}} = 1 - \frac{1}{2}x + \frac{3}{8}x^2 - \frac{5}{16}x^3 + \frac{35}{128}x^4 - \frac{63}{256}x^5 + \cdots$$

Further Generalizations

The generalized binomial theorem can be extended to the case where x and y are complex numbers. For this version, one should again assume $|x| > |y|$ and define the powers of $x + y$ and x using a holomorphic branch of log defined on an open disk of radius $|x|$ centered at x.

The generalized binomial theorem is valid also for elements x and y of a Banach algebra as long as $xy = yx$, x is invertible, and $||y/x|| < 1$.

Multinomial Theorem

The binomial theorem can be generalized to include powers of sums with more than two terms. The general version is

$$(x_1 + x_2 + \cdots + x_m)^n = \sum_{k_1+k_2+\cdots+k_m=n} \binom{n}{k_1, k_2, \ldots, k_m} x_1^{k_1} x_2^{k_2} \cdots x_m^{k_m}.$$

where the summation is taken over all sequences of nonnegative integer indices k_1 through k_m such that the sum of all k_i is n. (For each term in the expansion, the exponents must add up to n). The coefficients $\binom{n}{k_1,\ldots,k_m}$ are known as multinomial coefficients, and can be computed by the formula

$$\binom{n}{k_1, k_2, \ldots, k_m} = \frac{n!}{k_1! \cdot k_2! \cdots k_m!}.$$

Combinatorially, the multinomial coefficient $\binom{n}{k_1,\cdots,k_m}$ counts the number of different ways to partition an n-element set into disjoint subsets of sizes k_1, ..., k_m.

Multi-binomial Theorem

It is often useful when working in more dimensions, to deal with products of binomial expressions. By the binomial theorem this is equal to

$$(x_1 + y_1)^{n_1} \cdots (x_d + y_d)^{n_d} = \sum_{k_1=0}^{n_1} \cdots \sum_{k_d=0}^{n_d} \binom{n_1}{k_1} x_1^{k_1} y_1^{n_1-k_1} \cdots \binom{n_d}{k_d} x_d^{k_d} y_d^{n_d-k_d}.$$

This may be written more concisely, by multi-index notation, as

$$(x+y)^\alpha = \sum_{v \le \alpha} \binom{\alpha}{v} x^v y^{\alpha-v}.$$

Applications

Multiple-angle Identities

For the complex numbers the binomial theorem can be combined with De Moivre's formula to yield multiple-angle formulas for the sine and cosine. According to De Moivre's formula,

$$\cos(nx) + i\sin(nx) = (\cos x + i\sin x)^n.$$

Using the binomial theorem, the expression on the right can be expanded, and then the real and imaginary parts can be taken to yield formulas for $\cos(nx)$ and $\sin(nx)$. For example, since

$$(\cos x + i\sin x)^2 = \cos^2 x + 2i\cos x \sin x - \sin^2 x,$$

De Moivre's formula tells us that

$$\cos(2x) = \cos^2 x - \sin^2 x \quad \text{and} \quad \sin(2x) = 2\cos x \sin x,$$

which are the usual double-angle identities. Similarly, since

$$(\cos x + i\sin x)^3 = \cos^3 x + 3i\cos^2 x \sin x - 3\cos x \sin^2 x - i\sin^3 x,$$

De Moivre's formula yields

$$\cos(3x) = \cos^3 x - 3\cos x \sin^2 x \quad \text{and} \quad \sin(3x) = 3\cos^2 x \sin x - \sin^3 x.$$

In general,

$$\cos(nx) = \sum_{k \text{ even}} (-1)^{k/2} \binom{n}{k} \cos^{n-k} x \sin^k x$$

and

$$\sin(nx) = \sum_{k \text{ odd}} (-1)^{(k-1)/2} \binom{n}{k} \cos^{n-k} x \sin^k x.$$

Series for e

The number e is often defined by the formula

$$e = \lim_{n \to \infty} \left(1 + \frac{1}{n}\right)^n.$$

Applying the binomial theorem to this expression yields the usual infinite series for e. In particular:

$$\left(1 + \frac{1}{n}\right)^n = 1 + \binom{n}{1}\frac{1}{n} + \binom{n}{2}\frac{1}{n^2} + \binom{n}{3}\frac{1}{n^3} + \cdots + \binom{n}{n}\frac{1}{n^n}.$$

The kth term of this sum is

$$\binom{n}{k}\frac{1}{n^k} = \frac{1}{k!} \cdot \frac{n(n-1)(n-2)\cdots(n-k+1)}{n^k}$$

As $n \to \infty$, the rational expression on the right approaches one, and therefore

$$\lim_{n \to \infty} \binom{n}{k}\frac{1}{n^k} = \frac{1}{k!}.$$

This indicates that e can be written as a series:

$$e = \sum_{k=0}^{\infty} \frac{1}{k!} = \frac{1}{0!} + \frac{1}{1!} + \frac{1}{2!} + \frac{1}{3!} + \cdots.$$

Indeed, since each term of the binomial expansion is an increasing function of n, it follows from the monotone convergence theorem for series that the sum of this infinite series is equal to e.

Derivative of the Power Function

In finding the derivative of the power function $f(x) = x^n$ for integer n using the definition of derivative, one can expand the binomial $(x + h)^n$.

Nth Derivative of a Product

To indicate the formula for the derivative of order n of the product of two functions, the formula of the binomial theorem is used symbolically.

Probability

The binomial theorem is closely related to the probability mass function of the Negative binomial distribution. The probability of a (countable) collection of independent Bernoulli trials $\{X_t\}_{t \in S}$ with probability of success $p \in [0,1]$ all not happening is

$$P(\cap_{t \in S} X_t^C) = (1-p)^{|S|} = \sum_{n=0}^{|S|} \binom{|S|}{n}(-p)^n$$

A useful upper bound for this quantity is e^{-pn}.

The Binomial Theorem in Abstract Algebra

Formula (1) is valid more generally for any elements x and y of a semiring satisfying $xy = yx$. The theorem is true even more generally: alternativity suffices in place of associativity.

The binomial theorem can be stated by saying that the polynomial sequence { 1, x, x^2, x^3, ... } is of binomial type.

In Popular Culture

- The binomial theorem is mentioned in the Major-General's Song in the comic opera The Pirates of Penzance.

- Professor Moriarty is described by Sherlock Holmes as having written a treatise on the binomial theorem.

- The Portuguese poet Fernando Pessoa, using the heteronym Álvaro de Campos, wrote that "Newton's Binomial is as beautiful as the Venus de Milo. The truth is that few people notice it."

Chevalley–Warning Theorem

In algebra, the Chevalley–Warning theorem implies that certain polynomial equations in sufficiently many variables over a finite field have solutions. It was proved by Ewald Warning (1935) and a slightly weaker form of the theorem, known as Chevalley's theorem, was proved by Chevalley (1935). Chevalley's theorem implied Artin's and Dickson's conjecture that finite fields are quasi-algebraically closed fields (Artin 1982, page x).

Statement of the Theorems

Let $\mathbb{F}$ be a finite field and $\{f_j\}_{j=1}^r \subseteq \mathbb{F}[X_1,...,X_n]$ be a set of polynomials such that the number of variables satisfies

$$n > \sum_{j=1}^{r} d_j$$

where d_j is the total degree of f_j. The theorems are statements about the solutions of the following system of polynomial equations

$$f_j(x_1,\ldots,x_n) = 0 \quad \text{for } j = 1,\ldots,r.$$

- *Chevalley–Warning theorem* states that the number of common solutions is divisible by the characteristic p of $\mathbb{F}$. Or in other words, the cardinality of the vanishing set of $\{f_j\}_{j=1}^{r}$ is 0 modulo p.

- *Chevalley's theorem* states that if the system has the trivial solution $(0,\ldots,0) \in \mathbb{F}^n$, i.e. if the polynomials have no constant terms, then the system also has a non-trivial solution $(a_1,\ldots,a_n) \in \mathbb{F}^n \setminus \{(0,\ldots,0)\}$.

Chevalley's theorem is an immediate consequence of the Chevalley–Warning theorem since p is at least 2.

Both theorems are best possible in the sense that, given any n, the list $f_j = x_j, j = 1,\ldots,n$ has total degree n and only the trivial solution. Alternatively, using just one polynomial, we can take f_1 to be the degree n polynomial given by the norm of $x_1 a_1 + \ldots + x_n a_n$ where the elements a form a basis of the finite field of order p^n.

Warning proved another theorem, known as Warning's second theorem, which states that if the system of polynomial equations has the trivial solution, then it has at least q^{n-d} solutions where q is the size of the finite field and $d := d_1 + \ldots + d_r$. Chevalley's theorem also follows directly from this.

Proof of Warning's Theorem

Remark: If $i < q-1$ then

$$\sum_{x \in \mathbb{F}} x^i = 0$$

so the sum over $\mathbb{F}^n$ of any polynomial in $x_1,\ldots,x_n$ of degree less than $n(q-1)$ also vanishes.

The total number of common solutions modulo p of $f_1,\ldots,f_r = 0$ is equal to

$$\sum_{x \in \mathbb{F}^n} (1 - f_1^{q-1}(x)) \cdot \ldots \cdot (1 - f_r^{q-1}(x))$$

because each term is 1 for a solution and 0 otherwise. If the sum of the degrees of the polynomials f_i is less than n then this vanishes by the remark above.

Artin's Conjecture

It is a consequence of Chevalley's theorem that finite fields are quasi-algebraically closed. This had been conjectured by Emil Artin in 1935. The motivation behind Artin's conjecture was his observation that quasi-algebraically closed fields have trivial Brauer group, together with the fact that finite fields have trivial Brauer group by Wedderburn's theorem.

The Ax–Katz Theorem

The Ax–Katz theorem, named after James Ax and Nicholas Katz, determines more accurately a power q^b of the cardinality q of $\mathbb{F}$ dividing the number of solutions; here, if d is the largest of the d_j, then the exponent b can be taken as the ceiling function of

$$\frac{n - \sum_j d_j}{d}.$$

The Ax–Katz result has an interpretation in étale cohomology as a divisibility result for the (reciprocals of) the zeroes and poles of the local zeta-function. Namely, the same power of divides each of these algebraic integers.

Boolean Prime Ideal Theorem

In mathematics, a prime ideal theorem guarantees the existence of certain types of subsets in a given algebra. A common example is the Boolean prime ideal theorem, which states that ideals in a Boolean algebra can be extended to prime ideals. A variation of this statement for filters on sets is known as the ultrafilter lemma. Other theorems are obtained by considering different mathematical structures with appropriate notions of ideals, for example, rings and prime ideals (of ring theory), or distributive lattices and *maximal* ideals (of order theory). This article focuses on prime ideal theorems from order theory.

Although the various prime ideal theorems may appear simple and intuitive, they cannot be derived in general from the axioms of Zermelo–Fraenkel set theory without the axiom of choice (abbreviated ZF). Instead, some of the statements turn out to be equivalent to the axiom of choice (AC), while others—the Boolean prime ideal theorem, for instance—represent a property that is strictly weaker than AC. It is due to this intermediate status between ZF and ZF + AC (ZFC) that the Boolean prime ideal theorem is often taken as an axiom of set theory. The abbreviations BPI or PIT (for Boolean algebras) are sometimes used to refer to this additional axiom.

Prime Ideal Theorems

Recall that an order ideal is a (non-empty) directed lower set. If the considered poset has binary suprema (a.k.a. joins), as do the posets within this article, then this is equivalently characterized as a non-empty lower set I that is closed for binary suprema (i.e. x, y in I imply xy in I). An ideal I is prime if its set-theoretic complement in the poset is a filter. Ideals are proper if they are not equal to the whole poset.

Historically, the first statement relating to later prime ideal theorems was in fact referring to filters—subsets that are ideals with respect to the dual order. The ultrafilter lemma states that every filter on a set is contained within some maximal (proper) filter—an *ultrafilter*. Recall that filters on sets are proper filters of the Boolean algebra of its powerset. In this special case, maximal filters (i.e. filters that are not strict subsets of any proper filter) and prime filters (i.e. filters that with each union of subsets X and Y contain also X or Y) coincide. The dual of this statement thus assures that every ideal of a powerset is contained in a prime ideal.

The above statement led to various generalized prime ideal theorems, each of which exists in a weak and in a strong form. *Weak prime ideal theorems* state that every *non-trivial* algebra of a certain class has at least one prime ideal. In contrast, *strong prime ideal theorems* require that every ideal that is disjoint from a given filter can be extended to a prime ideal that is still disjoint from that filter. In the case of algebras that are not posets, one uses different substructures instead of filters. Many forms of these theorems are actually known to be equivalent, so that the assertion that "PIT" holds is usually taken as the assertion that the corresponding statement for Boolean algebras (BPI) is valid.

Another variation of similar theorems is obtained by replacing each occurrence of *prime ideal* by *maximal ideal*. The corresponding maximal ideal theorems (MIT) are often—though not always—stronger than their PIT equivalents.

Boolean Prime Ideal Theorem

The Boolean prime ideal theorem is the strong prime ideal theorem for Boolean algebras. Thus the formal statement is:

> Let B be a Boolean algebra, let I be an ideal and let F be a filter of B, such that I and F are disjoint. Then I is contained in some prime ideal of B that is disjoint from F.

The weak prime ideal theorem for Boolean algebras simply states:

> Every Boolean algebra contains a prime ideal.

We refer to these statements as the weak and strong *BPI*. The two are equivalent, as the strong BPI clearly implies the weak BPI, and the reverse implication can be achieved by using the weak BPI to find prime ideals in the appropriate quotient algebra.

The BPI can be expressed in various ways. For this purpose, recall the following theorem:

For any ideal I of a Boolean algebra B, the following are equivalent:

- I is a prime ideal.

- I is a maximal ideal, i.e. for any proper ideal J, if I is contained in J then $I = J$.

- For every element a of B, I contains exactly one of $\{a, \neg a\}$.

This theorem is a well-known fact for Boolean algebras. Its dual establishes the equivalence of prime filters and ultrafilters. Note that the last property is in fact self-dual—only the prior assumption that I is an ideal gives the full characterization. All of the implications within this theorem can be proven in ZF.

Thus the following (strong) maximal ideal theorem (MIT) for Boolean algebras is equivalent to BPI:

> Let B be a Boolean algebra, let I be an ideal and let F be a filter of B, such that I and F are disjoint. Then I is contained in some maximal ideal of B that is disjoint from F.

Note that one requires "global" maximality, not just maximality with respect to being disjoint from F. Yet, this variation yields another equivalent characterization of BPI:

> Let B be a Boolean algebra, let I be an ideal and let F be a filter of B, such that I and F are disjoint. Then I is contained in some ideal of B that is maximal among all ideals disjoint from F.

The fact that this statement is equivalent to BPI is easily established by noting the following theorem: For any distributive lattice L, if an ideal I is maximal among all ideals of L that are disjoint to a given filter F, then I is a prime ideal. The proof for this statement (which can again be carried out in ZF set theory) is included in the article on ideals. Since any Boolean algebra is a distributive lattice, this shows the desired implication.

All of the above statements are now easily seen to be equivalent. Going even further, one can exploit the fact the dual orders of Boolean algebras are exactly the Boolean algebras themselves. Hence, when taking the equivalent duals of all former statements, one ends up with a number of theorems that equally apply to Boolean algebras, but where every occurrence of *ideal* is replaced by *filter*. It is worth noting that for the special case where the Boolean algebra under consideration is a powerset with the subset ordering, the "maximal filter theorem" is called the ultrafilter lemma.

Summing up, for Boolean algebras, the weak and strong MIT, the weak and strong PIT, and these statements with filters in place of ideals are all equivalent. It is known that all of these statements are consequences of the Axiom of Choice, *AC*, (the easy proof makes use of Zorn's lemma), but cannot be proven in ZF (Zermelo-Fraenkel set theory without *AC*), if ZF is consistent. Yet, the BPI is strictly weaker than the axiom of choice, though the proof of this statement, due to J. D. Halpern and Azriel Lévy is rather non-trivial.

Further Prime Ideal Theorems

The prototypical properties that were discussed for Boolean algebras in the above section can easily be modified to include more general lattices, such as distributive lattices or Heyting algebras. However, in these cases maximal ideals are different from prime ideals, and the relation between PITs and MITs is not obvious.

Indeed, it turns out that the MITs for distributive lattices and even for Heyting algebras are equivalent to the axiom of choice. On the other hand, it is known that the strong PIT for distributive lattices is equivalent to BPI (i.e. to the MIT and PIT for Boolean algebras). Hence this statement is strictly weaker than the axiom of choice. Furthermore, observe that Heyting algebras are not self dual, and thus using filters in place of ideals yields different theorems in this setting. Maybe surprisingly, the MIT for the duals of Heyting algebras is not stronger than BPI, which is in sharp contrast to the abovementioned MIT for Heyting algebras.

Finally, prime ideal theorems do also exist for other (not order-theoretical) abstract algebras. For example, the MIT for rings implies the axiom of choice. This situation requires to replace the order-theoretic term "filter" by other concepts—for rings a "multiplicatively closed subset" is appropriate.

The Ultrafilter Lemma

A filter on a set X is a nonempty collection of nonempty subsets of X that is closed under finite intersection and under superset. An ultrafilter is a maximal filter. The ultrafilter lemma states that every filter on a set X is a subset of some ultrafilter on X. This lemma is most often used in the study of topology. An ultrafilter that does not contain finite sets is called non-principal. The ultrafilter lemma, and in particular the existence of non-principal ultrafilters (consider the filter of all sets with finite complements), follows easily from Zorn's lemma.

The ultrafilter lemma is equivalent to the Boolean prime ideal theorem, with the equivalence provable in ZF set theory without the axiom of choice. The idea behind the proof is that the subsets of any set form a Boolean algebra partially ordered by inclusion, and any Boolean algebra is representable as an algebra of sets by Stone's representation theorem.

Applications

Intuitively, the Boolean prime ideal theorem states that there are "enough" prime ideals in a Boolean algebra in the sense that we can extend *every* ideal to a maximal one. This is of practical importance for proving Stone's representation theorem for Boolean algebras, a special case of Stone duality, in which one equips the set of all prime ideals with a certain topology and can indeed regain the original Boolean algebra (up to isomorphism) from this data. Furthermore, it turns out that in applications one can freely choose either to work with prime ideals or with prime filters, because every ideal uniquely determines a filter: the set of all Boolean complements of its elements. Both approaches are found in the literature.

Many other theorems of general topology that are often said to rely on the axiom of choice are in fact equivalent to BPI. For example, the theorem that a product of compact Hausdorff spaces is compact is equivalent to it. If we leave out "Hausdorff" we get a theorem equivalent to the full axiom of choice.

A not too well known application of the Boolean prime ideal theorem is the existence of a non-measurable set (the example usually given is the Vitali set, which requires the axiom of choice). From this and the fact that the BPI is strictly weaker than the axiom of choice, it follows that the existence of non-measurable sets is strictly weaker than the axiom of choice.

In linear algebra, the Boolean prime ideal theorem can be used to prove that any two bases of a given vector space have the same cardinality.

Rational Root Theorem

In algebra, the rational root theorem (or rational root test, rational zero theorem, rational zero test or p/q theorem) states a constraint on rational solutions of a polynomial equation

$$a_n x^n + a_{n-1} x^{n-1} + \cdots + a_0 = 0$$

with integer coefficients. These solutions are the possible roots (equivalently, zeroes) of the polynomial on the left side of the equation.

If a_0 and a_n are nonzero, then each rational solution x, when written as a fraction $x = p/q$ in lowest terms (i.e., the greatest common divisor of p and q is 1), satisfies

- p is an integer factor of the constant term a_0, and

- q is an integer factor of the leading coefficient a_n.

The rational root theorem is a special case (for a single linear factor) of Gauss's lemma on the factorization of polynomials. The integral root theorem is a special case of the rational root theorem if the leading coefficient $a_n = 1$.

Application

The theorem is used in order to determine whether a polynomial has any rational roots, and if so to find them. Since the theorem gives constraints on the numerator and denominator of the fully reduced rational roots as being divisors of certain numbers, all possible combinations of divisors can be checked and either the rational roots will be found or it will be determined that there are none. If one or more are found, they can be factored out of the polynomial, resulting in a polynomial of lower degree whose roots are also roots of the original polynomial.

Cubic Equation

The general cubic equation

$$ax^3 + bx^2 + cx + d = 0$$

with integer coefficients has three solutions in the complex plane. If it is found by the rational root test that there are no rational solutions, then the only way to express the solutions algebraically is to use cube roots. But if the test finds three rational solutions, then the cube roots are avoided. And if precisely one rational solution r is found to exist, then $(x - r)$ can be factored out of the cubic polynomial using polynomial long division, leaving a quadratic polynomial whose two roots are the remaining two roots of the cubic; and these can be found using the quadratic formula, again avoiding the use of cube roots.

Proofs

First Proof

Let $P(x) = a_n x^n + a_{n-1} x^{n-1} + \ldots + a_1 x + a_0$ for some $a_0, \ldots, a_n \in \mathbb{Z}$, and suppose $P(p/q) = 0$ for some coprime $p, q \in \mathbb{Z}$:

$$P\left(\tfrac{p}{q}\right) = a_n \left(\tfrac{p}{q}\right)^n + a_{n-1} \left(\tfrac{p}{q}\right)^{n-1} + \cdots + a_1 \left(\tfrac{p}{q}\right) + a_0 = 0.$$

If we multiply both sides by q^n, shift the constant term to the right hand side, and factor out p on the left hand side, we get

$$p(a_n p^{n-1} + a_{n-1} q p^{n-2} + \cdots + a_1 q^{n-1}) = -a_0 q^n.$$

We see that p times the integer quantity in parentheses equals $-a_0 q^n$, so p divides $a_0 q^n$. But p is coprime to q and therefore to q^n, so by (the generalized form of) Euclid's lemma it must divide the remaining factor a_0 of the product.

If we instead shift the leading term to the right hand side and factor out q on the left hand side, we get

$$q(a_{n-1}p^{n-1} + a_{n-2}qp^{n-2} + \cdots + a_0q^{n-1}) = -a_np^n.$$

And for similar reasons, we can conclude that q divides a_n.

Proof using Gauss's Lemma

Should there be a nontrivial factor dividing all the coefficients of the polynomial, then one can divide by the greatest common divisor of the coefficients so as to obtain a primitive polynomial in the sense of Gauss's lemma; this does not alter the set of rational roots and only strengthens the divisibility conditions. That lemma says that if the polynomial factors in Q[X], then it also factors in Z[X] as a product of primitive polynomials. Now any rational root p/q corresponds to a factor of degree 1 in Q[X] of the polynomial, and its primitive representative is then $qx - p$, assuming that p and q are coprime. But any multiple in Z[X] of $qx - p$ has leading term divisible by q and constant term divisible by p, which proves the statement. This argument shows that more generally, any irreducible factor of P can be supposed to have integer coefficients, and leading and constant coefficients dividing the corresponding coefficients of P.

Examples

First

In the polynomial

$$2x^3 + x - 1,$$

any rational root fully reduced would have to have a numerator that divides evenly into 1 and a denominator that divides evenly into 2. Hence the only possible rational roots are ±1/2 and ±1; since neither of these equates the polynomial to zero, it has no rational roots.

Second

In the polynomial

$$x^3 - 7x + 6$$

the only possible rational roots would have a numerator that divides 6 and a denominator that divides 1, limiting the possibilities to ±1, ±2, ±3, and ±6. of these, 1, 2, and −3 equate the polynomial to zero, and hence are its rational roots. (In fact these are its only roots since a cubic has only three roots; in general a polynomial could have some rational and some irrational roots.)

Third

Every rational root of the polynomial

$$3x^3 - 5x^2 + 5x - 2$$

must be among the numbers symbolically indicated by

$$\pm \frac{1,2}{1,3},$$

which gives the list of 8 possible answers:

$$\pm\{1,2,\frac{1}{3},\frac{2}{3}\}.$$

These root candidates can be tested using Horner's method (for instance). In this particular case there is exactly one rational root. If a root candidate does not cause the polynomial to equal zero, it can be used to shorten the list of remaining candidates. For example, $x = 1$ does not work, as the polynomial then equals 1. This means that substituting $x = 1 + t$ yields a polynomial in t with constant term 1, while the coefficient of t^3 remains the same as the coefficient of x^3. Applying the rational root theorem thus yields the following possible roots for t:

$$t = \pm \frac{1}{1,3}.$$

Therefore,

$$x = 1 + t = 2, 0, \frac{4}{3}, \frac{2}{3}.$$

Root candidates that do not occur on both lists are ruled out. The list of rational root candidates has thus shrunk to just $x = 2$ and $x = 2/3$.

If k rational roots are found ($k \geq 1$), Horner's method will also yield a polynomial of degree $n - k$ whose roots, together with the rational roots, are exactly the roots of the original polynomial. It may also be the case that none of the candidates is a solution; in this case the equation setting the polynomial equal to 0 has no rational solution. If the equation lacks a constant term a_0, then 0 is one of the rational solutions of the equation.

Factor Theorem

In algebra, the factor theorem is a theorem linking factors and zeros of a polynomial. It is a special case of the polynomial remainder theorem.

The factor theorem states that a polynomial $f(x)$ has a factor $(x - k)$ if and only if $f(x) = 0$ (i.e. k is a root).

Factorization of Polynomials

Two problems where the factor theorem is commonly applied are those of factoring a polynomial and finding the roots of a polynomial equation; it is a direct consequence of the theorem that these problems are essentially equivalent.

The factor theorem is also used to remove known zeros from a polynomial while leaving all unknown zeros intact, thus producing a lower degree polynomial whose zeros may be easier to find. Abstractly, the method is as follows:

1. "Guess" a zero a of the polynomial f. (In general, this can be *very hard*, but maths textbook problems that involve solving a polynomial equation are often designed so that some roots are easy to discover.)

2. Use the factor theorem to conclude that $(x-a)$ is a factor of $f(x)$.

3. Compute the polynomial $g(x) = f(x) / (x-a)$, for example using polynomial long division or synthetic division.

4. Conclude that any root $x \neq a$ of $f(x) = 0$ is a root of $g(x) = 0$. Since the polynomial degree of g is one less than that of f, it is "simpler" to find the remaining zeros by studying g.

Example

Find the factors at

$$x^3 + 7x^2 + 8x + 2.$$

To do this you would use trial and error (or the rational root theorem) to find the first x value that causes the expression to equal zero. To find out if $(x-1)$ is a factor, substitute $x = 1$ into the polynomial above:

$$x^3 + 7x^2 + 8x + 2 = (1)^3 + 7(1)^2 + 8(1) + 2$$

$$= 1 + 7 + 8 + 2$$

$$= 18.$$

As this is equal to 18 and not 0 this means $(x-1)$ is not a factor of $x^3 + 7x^2 + 8x + 2$. So, we next try $(x+1)$ (substituting $x = -1$ into the polynomial):

$$(-1)^3 + 7(-1)^2 + 8(-1) + 2.$$

This is equal to 0. Therefore $x - (-1)$, which is to say $x+1$, is a factor, and -1 is a root of $x^3 + 7x^2 + 8x + 2$.

The next two roots can be found by algebraically dividing $x^3 + 7x^2 + 8x + 2$ by $(x+1)$ to get a quadratic, which can be solved directly, by the factor theorem or by the quadratic formula.

$$\frac{x^3 + 7x^2 + 8x + 2}{x+1} = x^2 + 6x + 2$$

and therefore $(x+1)$ and $x^2 + 6x + 2$ are the factors of $x^3 + 7x^2 + 8x + 2$.

In algebra, the Krull–Akizuki theorem states the following: let A be a one-dimensional reduced noetherian ring, K its total ring of fractions. If B is a subring of a finite extension L of K containing A and is not a field, then B is a one-dimensional noetherian ring. Furthermore, for every nonzero ideal I of B, $B >$ is finite over A.

Note that the theorem does not say that B is finite over A. The theorem does not extend to higher dimension. One important consequence of the theorem is that the integral closure of a Dedekind domain A in a finite extension of the field of fractions of A is again a Dedekind domain. This consequence does generalize to a higher dimension: the Mori–Nagata theorem states that the integral closure of a noetherian domain is a Krull domain.

Proof

Here, we give a proof when $L = K$. Let $\mathfrak{p}_i$ be minimal prime ideals of A; there are finitely many of them. Let K_i be the field of fractions of $A / \mathfrak{p}_i$ and I_i the kernel of the natural map $B \to K \to K_i$. Then we have:

$$A / \mathfrak{p}_i \subset B / I_i \subset K_i.$$

Now, if the theorem holds when A is a domain, then this implies that B is a one-dimensional noetherian domain since each B / I_i is and since $B = \prod B / I_i$. Hence, we reduced the proof to the case A is a domain. Let $0 \neq I \subset B$ be an ideal and let a be a nonzero element in the nonzero ideal $I \cap A$. Set $I_n = a^n B \cap A + aA$. Since $A >$ is a zero-dim noetherian ring; thus, artinian, there is an l such that $I_n = I_l$ for all $n \geq l$. We claim

$$a^l B \subset a^{l+1} B + A.$$

Since it suffices to establish the inclusion locally, we may assume A is a local ring with the maximal ideal $\mathfrak{m}$. Let x be a nonzero element in B. Then, since A is noetherian, there is an n such that $\mathfrak{m}^{n+1} \subset x^{-1} A$ and so $a^{n+1} x \in a^{n+1} B \cap A \subset I_{n+2}$. Thus,

$$a^n x \in a^{n+1} B \cap A + A.$$

Now, assume n is a minimum integer such that $n \geq l$ and the last inclusion holds. If , then we easily see that $a^n x \in I_{n+1}$. But then the above inclusion holds for $n-1$, contradiction. Hence, we have $n = l$ and this establishes the claim. It now follows:

$$B / aB \simeq a^l B / a^{l+1} B \subset (a^{l+1} B + A) / a^{l+1} B \simeq A / a^{l+1} B \cap A.$$

Hence, $B >$ has finite length as A-module. In particular, the image of I there is finitely generated and so I is finitely generated. Finally, the above shows that $B >$ has zero dimension and so B has dimension one.

References

- Tignol, Jean-Pierre (2016), "Ruffini and Abel on general equations", Galois' Theory of algebraic equations (2nd ed.), World Scientific, ISBN 978-981-4704-69-4.

- Pesic, Peter (2004), Abel's proof. An essay on the sources and meaning of mathematical unsolvability, MIT Press, ISBN 0-262-66182-9.

- Tignol, Jean-Pierre (2016), "Galois", Galois' Theory of algebraic equations (2nd ed.), World Scientific, ISBN 978-981-4704-69-4.

- Alekseev, V. B. (2004), Abel's theorem in problems and solutions. Based on the lectures of Professor V. I. Arnold, Kluwer Academic Publishers, ISBN 1-4020-2186-0.

- Khovanskii, Askold (2014), Topological Galois Theory: Solvability and Unsolvability of Equations in Finite Terms, Springer Monographs in Mathematics, Springer-Verlag, doi:10.1007/978-3-642-38871-2, ISBN 978-3-642-38870-5

- Bourbaki, N. (18 November 1998). Elements of the History of Mathematics Paperback. J. Meldrum (Translator). ISBN 978-3-540-64767-6.

- Cover, Thomas M.; Thomas, Joy A. (2001-01-01). Data Compression. John Wiley & Sons, Inc. p. 320. doi:10.1002/0471200611.ch5. ISBN 9780471200611.

Functions Related to Algebra

The chapter serves as a source to understand the basic functions related to algebra. Some of the basic functions are algebraic function, cubic function, quantic function and quartic function. Algebraic functions are functions that can be expressed as the root of a polynomial equation. The following text will provide an integrated understanding of algebra.

Algebraic Function

In mathematics, an algebraic function is a function that can be defined as the root of a polynomial equation. Quite often algebraic functions can be expressed using a finite number of terms, involving only the algebraic operations addition, subtraction, multiplication, division, and raising to a fractional power.

Common examples of such functions are:

$$f(x) = 1/x$$

$$f(x) = \sqrt{x}$$

$$f(x) = \frac{\sqrt{1+x^3}}{x^{3/7} - \sqrt{7}x^{1/3}}$$

Some algebraic functions, however, cannot be expressed by such finite expressions (this is Abel–Ruffini theorem). This is the case, for example, of the Bring radical, which is the function implicitly defined by

$$f(x)^5 + f(x) + x = 0.$$

In more precise terms, an algebraic function of degree n in one variable x is a function $y = f(x)$ that satisfies a polynomial equation

$$a_n(x)y^n + a_{n-1}(x)y^{n-1} + \cdots + a_0(x) = 0$$

where the coefficients $a_i(x)$ are polynomial functions of x, with coefficients belonging to a set S. Quite often, $S = \mathbb{Q}$, and one then talks about "function algebraic over $\mathbb{Q}$", and the evaluation at a given rational value of such an algebraic function gives an algebraic number.

A function which is not algebraic is called a transcendental function, as it is for example the case of

$\exp(x), \tan(x), \ln(x), \Gamma(x)$. A composition of transcendental functions can give an algebraic function: $f(x) = \cos(\arcsin(x)) = \sqrt{1-x^2}$.

As an equation of degree n has n roots, a polynomial equation does not implicitly define a single function, but n functions, sometimes also called branches. Consider for example the equation of the unit circle: $y^2 + x^2 = 1$. This determines y, except only up to an overall sign; accordingly, it has two branches: $y = \pm\sqrt{1-x^2}$.

An algebraic function in **m** variables is similarly defined as a function y which solves a polynomial equation in $m + 1$ variables:

$$p(y, x_1, x_2, \ldots, x_m) = 0.$$

It is normally assumed that p should be an irreducible polynomial. The existence of an algebraic function is then guaranteed by the implicit function theorem.

Formally, an algebraic function in m variables over the field K is an element of the algebraic closure of the field of rational functions $K(x_1, \ldots, x_m)$.

Algebraic Functions in One Variable

Introduction and Overview

The informal definition of an algebraic function provides a number of clues about the properties of algebraic functions. To gain an intuitive understanding, it may be helpful to regard algebraic functions as functions which can be formed by the usual algebraic operations: addition, multiplication, division, and taking an nth root. Of course, this is something of an oversimplification; because of casus irreducibilis (and more generally the fundamental theorem of Galois theory), algebraic functions need not be expressible by radicals.

First, note that any polynomial function $y - p(x)$ is an algebraic function, since it is simply the solution y to the equation

$$y - p(x) = 0.$$

More generally, any rational function $y = \dfrac{p(x)}{q(x)}$ is algebraic, being the solution to

$$q(x)y - p(x) = 0.$$

Moreover, the nth root of any polynomial $y = \sqrt[n]{p(x)}$ is an algebraic function, solving the equation

$$y^n - p(x) = 0.$$

Surprisingly, the inverse function of an algebraic function is an algebraic function. For supposing that y is a solution to

$$a_n(x)y^n + \cdots + a_0(x) = 0,$$

for each value of x, then x is also a solution of this equation for each value of y. Indeed, interchanging the roles of x and y and gathering terms,

$$b_m(y)x^m + b_{m-1}(y)x^{m-1} + \cdots + b_0(y) = 0.$$

Writing x as a function of y gives the inverse function, also an algebraic function.

However, not every function has an inverse. For example, $y = x^2$ fails the horizontal line test: it fails to be one-to-one. The inverse is the algebraic "function" $x = \pm\sqrt{y}$. Another way to understand this, is that the set of branches of the polynomial equation defining our algebraic function is the graph of an algebraic curve.

The Role of Complex Numbers

From an algebraic perspective, complex numbers enter quite naturally into the study of algebraic functions. First of all, by the fundamental theorem of algebra, the complex numbers are an algebraically closed field. Hence any polynomial relation $p(y, x) = 0$ is guaranteed to have at least one solution (and in general a number of solutions not exceeding the degree of p in x) for y at each point x, provided we allow y to assume complex as well as real values. Thus, problems to do with the domain of an algebraic function can safely be minimized.

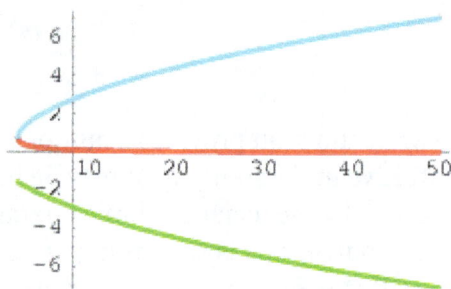

A graph of three branches of the algebraic function y, where $y^3 - xy + 1 = 0$, over the domain $3/2^{2/3} < x < 50$.

Furthermore, even if one is ultimately interested in real algebraic functions, there may be no means to express the function in terms of addition, multiplication, division and taking *nth* roots without resorting to complex numbers. For example, consider the algebraic function determined by the equation

$$y^3 - xy + 1 = 0.$$

Using the cubic formula, we get

$$y = -\frac{2x}{\sqrt[3]{-108 + 12\sqrt{81 - 12x^3}}} + \frac{\sqrt[3]{-108 + 12\sqrt{81 - 12x^3}}}{6}.$$

For $x > \dfrac{3}{\sqrt[3]{4}}$, the square root is real and the cubic root is thus well defined, providing the unique real root. On the other hand, for $x > \dfrac{3}{\sqrt[3]{4}}$, the square root is not real, and one has to choose, for the square root, either non real-square root. Thus the cubic root has to be chosen among three non-real numbers. If the same choices are done in the two terms of the formula, the three choices for the cubic root provide the three branches shown, in the accompanying image.

It may be proven that there is no way to express this function in terms *nth* roots using real numbers only, even though the resulting function is real-valued on the domain of the graph shown.

On a more significant theoretical level, using complex numbers allows one to use the powerful techniques of complex analysis to discuss algebraic functions. In particular, the argument principle can be used to show that any algebraic function is in fact an analytic function, at least in the multiple-valued sense.

Formally, let $p(x, y)$ be a complex polynomial in the complex variables x and y. Suppose that $x_0 \in C$ is such that the polynomial $p(x_0, y)$ of y has n distinct zeros. We shall show that the algebraic function is analytic in a neighborhood of x_0. Choose a system of n non-overlapping discs Δ_i containing each of these zeros. Then by the argument principle

$$\frac{1}{2\pi i} \oint_{\partial \Delta_i} \frac{p_y(x_0, y)}{p(x_0, y)} dy = 1.$$

By continuity, this also holds for all x in a neighborhood of x_0. In particular, $p(x,y)$ has only one root in Δ_i, given by the residue theorem:

$$f_i(x) = \frac{1}{2\pi i} \oint_{\partial \Delta_i} y \frac{p_y(x, y)}{p(x, y)} dy$$

which is an analytic function.

Monodromy

Note that the foregoing proof of analyticity derived an expression for a system of n different function elements $f_i(x)$, provided that x is not a critical point of $p(x, y)$. A *critical point* is a point where the number of distinct zeros is smaller than the degree of p, and this occurs only where the highest degree term of p vanishes, and where the discriminant vanishes. Hence there are only finitely many such points $c_1, ..., c_m$.

A close analysis of the properties of the function elements f_i near the critical points can be used to show that the monodromy cover is ramified over the critical points (and possibly the point at infinity). Thus the entire function associated to the f_i has at worst algebraic poles and ordinary algebraic branchings over the critical points.

Note that, away from the critical points, we have

$$p(x, y) = a_n(x)(y - f_1(x))(y - f_2(x)) \cdots (y - f_n(x))$$

since the f_i are by definition the distinct zeros of p. The monodromy group acts by permuting the factors, and thus forms the monodromy representation of the Galois group of p. (The monodromy action on the universal covering space is related but different notion in the theory of Riemann surfaces.)

History

The ideas surrounding algebraic functions go back at least as far as René Descartes. The first discussion of algebraic functions appears to have been in Edward Waring's 1794 *An Essay on the Principles of Human Knowledge* in which he writes:

let a quantity denoting the ordinate, be an algebraic function of the abscissa x, by the common methods of division and extraction of roots, reduce it into an infinite series ascending or descending according to the dimensions of x, and then find the integral of each of the resulting terms.

Cubic Function

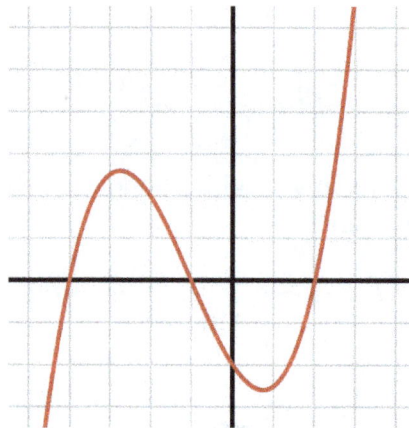

Graph of a cubic function with 3 real roots (where the curve crosses the horizontal axis—where $y = 0$). The case shown has two critical points. Here the function is $f(x) = (x^3 + 3x^2 - 6x - 8)/4$.

In algebra, a cubic function is a function of the form

$$f(x) = ax^3 + bx^2 + cx + d,$$

where a is nonzero.

Setting $f(x) = 0$ produces a cubic equation of the form:

$$ax^3 + bx^2 + cx + d = 0.$$

The solutions of this equation are called roots of the polynomial $f(x)$. If all of the coefficients a, b, c, and d of the cubic equation are real numbers then there will be at least one real root (this is true for all odd degree polynomials). All of the roots of the cubic equation can be found algebraically. (This is also true of a quadratic or quartic (fourth degree) equation, but no higher-degree equation, by the Abel–Ruffini theorem). The roots can also be found trigonometrically. Alternatively, numerical approximations of the roots can be found using root-finding algorithms like Newton's method.

The coefficients do not need to be complex numbers. Much of what is covered below is valid for coefficients of any field with characteristic 0 or greater than 3. The solutions of the cubic equation do not necessarily belong to the same field as the coefficients. For example, some cubic equations with real coefficients have roots that are complex numbers.

History

Cubic equations were known to the ancient Babylonians, Greeks, Chinese, Indians, and Egyptians. Babylonian (20th to 16th centuries BC) cuneiform tablets have been found with tables for calculating cubes and cube roots. The Babylonians could have used the tables to solve cubic equations, but no evidence exists to confirm that they did. The problem of doubling the cube involves the simplest and oldest studied cubic equation, and one for which the ancient Egyptians did not believe a solution existed. In the 5th century BC, Hippocrates reduced this problem to that of finding two mean proportionals between one line and another of twice its length, but could not solve this with a compass and straightedge construction, a task which is now known to be impossible. Methods for solving cubic equations appear in *The Nine Chapters on the Mathematical Art*, a Chinese mathematical text compiled around the 2nd century BC and commented on by Liu Hui in the 3rd century. In the 3rd century, the Greek mathematician Diophantus found integer or rational solutions for some bivariate cubic equations (Diophantine equations). Hippocrates, Menaechmus and Archimedes are believed to have come close to solving the problem of doubling the cube using intersecting conic sections, though historians such as Reviel Netz dispute whether the Greeks were thinking about cubic equations or just problems that can lead to cubic equations. Some others like T. L. Heath, who translated all Archimedes' works, disagree, putting forward evidence that Archimedes really solved cubic equations using intersections of two conics, but also discussed the conditions where the roots are 0, 1 or 2.

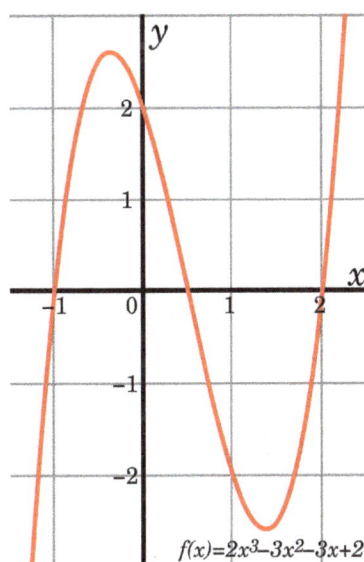

Two-dimensional graph of a cubic, the polynomial $f(x) = 2x^3 - 3x^2 - 3x + 2$.

In the 7th century, the Tang dynasty astronomer mathematician Wang Xiaotong in his mathematical treatise titled Jigu Suanjing systematically established and solved numerically 25 cubic equations of the form $x^3 + px^2 + qx = N$, 23 of them with $p, q \neq 0$, and two of them with $q = 0$.

In the 11th century, the Persian poet-mathematician, Omar Khayyám (1048–1131), made significant progress in the theory of cubic equations. In an early paper, he discovered that a cubic equation can have more than one solution and stated that it cannot be solved using compass and straightedge constructions. He also found a geometric solution. In his later work, the *Treatise on Demonstration of Problems of Algebra*, he wrote a complete classification of cubic equations with general geometric solutions found by means of intersecting conic sections.

In the 12th century, the Indian mathematician Bhaskara II attempted the solution of cubic equations without general success. However, he gave one example of a cubic equation: $x^3 + 12x = 6x^2 + 35$. In the 12th century, another Persian mathematician, Sharaf al-Dīn al-Tūsī (1135–1213), wrote the *Al-Muʿādalāt* (*Treatise on Equations*), which dealt with eight types of cubic equations with positive solutions and five types of cubic equations which may not have positive solutions. He used what would later be known as the "Ruffini-Horner method" to numerically approximate the root of a cubic equation. He also developed the concepts of a derivative function and the maxima and minima of curves in order to solve cubic equations which may not have positive solutions. He understood the importance of the discriminant of the cubic equation to find algebraic solutions to certain types of cubic equations.

Leonardo de Pisa, also known as Fibonacci (1170–1250), was able to closely approximate the positive solution to the cubic equation $x^3 + 2x^2 + 10x = 20$, using the Babylonian numerals. He gave the result as 1,22,7,42,33,4,40 (equivalent to $1 + 22/60 + 7/60^2 + 42/60^3 + 33/60^4 + 4/60^5 + 40/60^6$) , which differs from the correct value by only about three trillionths.

In the early 16th century, the Italian mathematician Scipione del Ferro (1465–1526) found a method for solving a class of cubic equations, namely those of the form $x^3 + mx = n$. In fact, all cubic equations can be reduced to this form if we allow m and n to be negative, but negative numbers were not known to him at that time. Del Ferro kept his achievement secret until just before his death, when he told his student Antonio Fiore about it.

Niccolò Fontana Tartaglia

In 1530, Niccolò Tartaglia (1500–1557) received two problems in cubic equations from Zuanne da Coi and announced that he could solve them. He was soon challenged by Fiore, which led to a famous contest between the two. Each contestant had to put up a certain amount of money and to propose a number of problems for his rival to solve. Whoever solved more problems within 30 days would get all the money. Tartaglia received questions in the form $x^3 + mx = n$, for which he had worked out a general method. Fiore received questions in the form $x^3 + mx^2 = n$, which proved to be too difficult for him to solve, and Tartaglia won the contest.

Later, Tartaglia was persuaded by Gerolamo Cardano (1501–1576) to reveal his secret for solving cubic equations. In 1539, Tartaglia did so only on the condition that Cardano would never reveal it and that if he did write a book about cubics, he would give Tartaglia time to publish. Some years later, Cardano learned about Ferro's prior work and published Ferro's method in his book *Ars Magna* in 1545, meaning Cardano gave Tartaglia six years to publish his results (with credit given to Tartaglia for an independent solution). Cardano's promise with Tartaglia stated that he not publish Tartaglia's work, and Cardano felt he was publishing del Ferro's, so as to get around the promise. Nevertheless, this led to a challenge to Cardano by Tartaglia, which Cardano denied. The challenge was eventually accepted by Cardano's student Lodovico Ferrari (1522–1565). Ferrari did better than Tartaglia in the competition, and Tartaglia lost both his prestige and income.

Cardano noticed that Tartaglia's method sometimes required him to extract the square root of a negative number. He even included a calculation with these complex numbers in *Ars Magna*, but he did not really understand it. Rafael Bombelli studied this issue in detail and is therefore often considered as the discoverer of complex numbers.

François Viète (1540–1603) independently derived the trigonometric solution for the cubic with three real roots, and René Descartes (1596–1650) extended the work of Viète.

Critical Points and Inflection Point of a Cubic Function

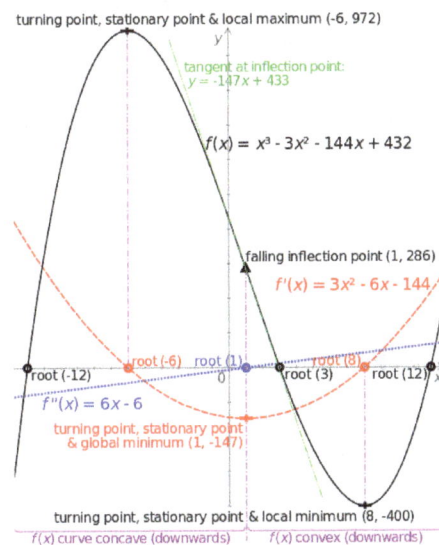

The roots, turning points, stationary points, inflection point and concavity of a cubic polynomial $x^3 - 3x^2 - 144x + 432$ (black line) and its first and second derivatives (red and blue).

The critical points of a function are those values of x where the slope of the function is zero. The critical points of a cubic function f defined by $f(x) = ax^3 + bx^2 + cx + d$, occur at values of x such that the first derivative of the cubic is zero:

$$3ax^2 + 2bx + c = 0.$$

The solutions of that equation are the critical points of the cubic equation and are given, using the quadratic formula, by

$$x_{\text{critical}} = \frac{-b \pm \sqrt{b^2 - 3ac}}{3a}.$$

The expression inside of the square root,

$$\Delta_0 = b^2 - 3ac,$$

determines what type of critical points the function has. If $\Delta_0 > 0$, then the cubic function has a local maximum and a local minimum. If $\Delta_0 = 0$, then the cubic's inflection point is the only critical point. If $\Delta_0 < 0$, then there are no critical points. In cases where $\Delta_0 \leq 0$, the cubic function is strictly monotonic. The diagram to the right is an example of the case where $\Delta_0 > 0$. The other two cases do not have the local maximum or the local minimum but still have an inflection point.

The value of Δ_0 also plays an important role in determining the nature of the roots of the cubic equation and in the calculation of those roots.

The inflection point of a function is where that function changes concavity. The inflection point of our cubic function occurs at:

$$x_{\text{inflection}} = -\frac{b}{3a},$$

a value that is also important in solving the cubic equation. The cubic function has point symmetry about its inflection point.

All of the above assumes that the coefficients are real as well as the variable x.

General Solution to the Cubic Equation with Real Coefficients

This section is about how to solve the cubic equation using various methods. The general cubic equation has the form:

$$ax^3 + bx^2 + cx + d = 0$$

with $a \neq 0$.

Algebraic Solution

The algebraic solution of the cubic equation can be derived in a number of different ways.

The Discriminant

The number and types of roots is determined by the discriminant of the cubic equation,

$$\Delta = 18abcd - 4b^3d + b^2c^2 - 4ac^3 - 27a^2d^2.$$

It turns out that:

- If $\Delta > 0$, then the equation has three distinct real roots.

- If $\Delta = 0$, then the equation has a multiple root and all its roots are real.

- If $\Delta < 0$, then the equation has one real root and two non-real complex conjugate roots.

General formula

The general solution of the cubic equation involves first calculating:

$$\Delta_0 = b^2 - 3ac,$$

$$\Delta_1 = 2b^3 - 9abc + 27a^2d, \text{ and}$$

$$C = \sqrt[3]{\frac{\Delta_1 \pm \sqrt{\Delta_1^2 - 4\Delta_0^3}}{2}}.$$

(If the discriminant Δ has already been calculated then the equality $\Delta_1^2 - 4\Delta_0^3 = -27a^2\Delta$, can be used to simplify the calculation of C.) There are three possible cube roots implied by the expression, of which at least two are non-real complex numbers; any of these may be chosen when defining C. (In addition either sign in front of the square root may be chosen unless $\Delta_0 = 0$ in which case the sign must be chosen so that the two terms inside the cube root do not cancel.)

The general formula for one of the roots, in terms of the coefficients, is as follows:

$$x = -\frac{1}{3a}\left(b + C + \frac{\Delta_0}{C}\right).$$

Note that, while this equality is valid for all non-zero C, it is not the most convenient form for multiple roots ($\Delta = 0$), which is covered in the next section. (The case when $C = 0$ only occurs when both Δ and Δ_0 are equal to 0 and is also covered in the next section.)

The other two roots of the cubic equation can be determined using the same equality, using the other two choices for the cube root in the equation for C: denoting the first choice simply as C, the others can be written as $\left(-\frac{1}{2} + \frac{1}{2}\sqrt{3}i\right) C$ and $\left(-\frac{1}{2} - \frac{1}{2}\sqrt{3}i\right) C$.

The above equality can be expressed compactly including all 3 roots as follows:

$$x_k = -\frac{1}{3a}\left(b + \zeta^k C + \frac{\Delta_0}{\zeta^k C}\right), \qquad k \in \{0,1,2\},$$

where $\zeta = -\frac{1}{2} + \frac{1}{2}\sqrt{3}i$ (which is a cube root of unity). In the case of three real roots, this solution expresses them in terms of non-real complex terms (since any choice of C is non-real) whose imaginary components offset each other but cannot be eliminated from the formula.

This formula for the three roots applies even when the coefficients in the cubic are non-real, although the analysis of the sign of Δ does not hold since Δ is then not real in general.

Multiple Roots, $\Delta = 0$

If both Δ and Δ_0 are equal to 0, then the equation has a single root (which is a triple root):

$$-\frac{b}{3a}.$$

If $\Delta = 0$ and $\Delta_0 \neq 0$, then there are both a double root,

$$\frac{9ad - bc}{2\Delta_0},$$

and a simple root,

$$\frac{4abc - 9a^2d - b^3}{a\Delta_0}.$$

Trigonometric and Hyperbolic Solutions

Reduction to a Depressed Cubic

Dividing $ax^3 + bx^2 + cx + d = 0$ by a and substituting $t - b/3a$ for x we get the equation

$$t^3 + pt + q = 0 \tag{2}$$

where

$$p = \frac{3ac - b^2}{3a^2},$$

$$q = \frac{2b^3 - 9abc + 27a^2d}{27a^3}.$$

The left hand side of equation (2) is a monic trinomial called a depressed cubic, because the quadratic term has coefficient 0.

Any formula for the roots of a depressed cubic may be transformed into a formula for the roots of equation (1) by substituting the above values for p and q and using the relation $x = t - b/3a$.

Therefore, only equation (2) is considered in the following.

Trigonometric Solution for Three Real Roots

When a cubic equation has three real roots, the formulas expressing these roots in terms of radicals involve complex numbers. It has been proved that when none of the three real roots is rational—

the *casus irreducibilis*— one cannot express the roots in terms of real radicals. Nevertheless, purely real expressions of the solutions may be obtained using hypergeometric functions, or more elementarily in terms of trigonometric functions, specifically in terms of the cosine and arccosine functions.

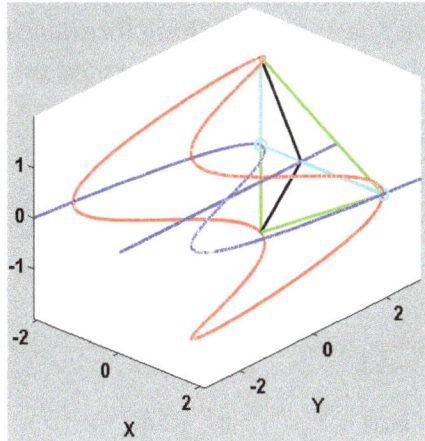

This animation shows the roots of x^3-3*x=cos(3*t) for different values of t; the roots are [2*cos(t), 2*cos(t+2*pi/3), 2*cos(t-2*pi/3)]

The formulas which follow, due to François Viète, are true in general (except when $p = 0$), and are purely real when the equation has three real roots, but involve complex cosines and arccosines when there is only one real root.

Starting from equation (2), $t^3 + pt + q = 0$, let us set $t = u \cos \theta$. The idea is to choose u to make equation (2) coincide with the identity

$$4\cos^3 \theta - 3\cos \theta - \cos(3\theta) = 0.$$

In fact, choosing $u = 2\sqrt{-\dfrac{p}{3}}$ and dividing equation (2) by $\dfrac{u^3}{4}$ we get

$$4\cos^3 \theta - 3\cos \theta - \frac{3q}{2p}\sqrt{\frac{-3}{p}} = 0.$$

Combining with the above identity, we get

$$\cos(3\theta) = \frac{3q}{2p}\sqrt{\frac{-3}{p}}$$

and thus the roots are

$$t_k = 2\sqrt{-\frac{p}{3}}\cos\left(\frac{1}{3}\arccos\left(\frac{3q}{2p}\sqrt{\frac{-3}{p}}\right) - \frac{2\pi k}{3}\right) \quad \text{for} \quad k = 0, 1, 2.$$

This formula involves only real terms if $p < 0$ and the argument of the arccosine is between -1 and 1. The last condition is equivalent to $4p^3 + 27q^2 \le 0$, which itself implies $p < 0$. Thus the above formula for the roots involves only real terms if and only if the three roots are real.

Denoting by $C(p, q)$ the above value of t_0, and using the inequalities $0 \leq \arccos(u) \leq \pi$ for a real number u such that $-1 \leq u \leq 1$, the three roots may also be expressed as

$$t_0 = C(p,q), \qquad t_2 = -C(p,-q), \qquad t_1 = -t_0 - t_2.$$

If the three roots are real, we have $t_0 \geq t_1 \geq t_2$. All these formulas may be straightforwardly transformed into formulas for the roots of the general cubic equation (1), using the back substitution described above.

Hyperbolic Solution for One Real Root

When there is only one real root (and $p \neq 0$), it may be similarly represented using hyperbolic functions, as

$$t_0 = -2\frac{|q|}{q}\sqrt{-\frac{p}{3}}\cosh\left(\frac{1}{3}\text{arcosh}\left(\frac{-3|q|}{2p}\sqrt{\frac{-3}{p}}\right)\right) \quad \text{if} \quad 4p^3 + 27q^2 > 0 \text{ and } p < 0,$$

$$t_0 = -2\sqrt{\frac{p}{3}}\sinh\left(\frac{1}{3}\text{arsinh}\left(\frac{3q}{2p}\sqrt{\frac{3}{p}}\right)\right) \quad \text{if} \quad p > 0.$$

If $p \neq 0$ and the inequalities on the right are not satisfied (the case of three real roots), the formulas remain valid but involve complex quantities.

When $p = \pm 3$, the above values of t_0 are sometimes called the Chebyshev cube root. More precisely, the values involving cosines and hyperbolic cosines define, when $p = -3$, the same analytic function denoted $C_{1/3}(q)$, which is the proper Chebyshev cube root. The value involving hyperbolic sines is similarly denoted $S_{1/3}(q)$, when $p = -3$.

Factorization

If the cubic equation $ax^3 + bx^2 + cx + d = 0$ with integer coefficients has a rational root, it can be found using the rational root test: If the root $r = {}^m/_n$ is fully reduced, then m is a factor of d and n is a factor of a, so all possible combinations of values for m and n (both positive and negative for one of them) can be checked for whether they satisfy the cubic equation.

The rational root test may also be used for a cubic equation with rational coefficients: by multiplication by the lowest common denominator of the coefficients, one gets an equation with integer coefficients which has exactly the same roots.

The rational root test is particularly useful when there are three real roots because the algebraic solution unhelpfully expresses the real roots in terms of complex entities; if the test yields a rational root, it can be factored out and the remaining roots can be found by solving a quadratic. The rational root test is also helpful in the presence of one real and two complex roots because again, if it yields a rational root, it allows all of the roots to be written without the use of cube roots: If r is any root of the cubic, then we may factor out $x - r$ using polynomial long division to obtain

$$ax^3 + bx^2 + cx + d = (x - r)\left(ax^2 + (b + ar)x + c + br + ar^2\right).$$

Hence if we know one root, perhaps from the rational root test, we can find the other two by using the quadratic formula to find the roots of the quadratic $ax^2 + (b + ar)x + c + br + ar^2$, giving

$$\frac{-b - ra \pm \sqrt{b^2 - 4ac - 2abr - 3a^2r^2}}{2a}$$

for the other two roots.

Geometric Solutions

Omar Khayyám's Solution

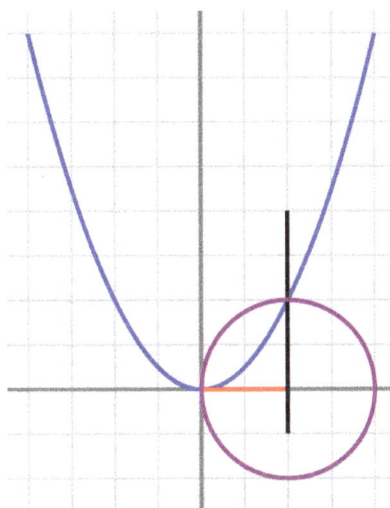

Omar Khayyám's geometric solution of a cubic equation, for the case $m = 2$, $n = 16$, giving the root 2. The fact that the vertical line intersects the x-axis at the center of the circle is specific to this particular example

As shown in this graph, to solve the third-degree equation $x^3 + m^2x = n$ where $n > 0$, Omar Khayyám constructed the parabola $y = x^2/m$, the circle which has as a diameter the line segment $[0, n/m^2]$ on the positive x-axis, and a vertical line through the point above the x-axis where the circle and parabola intersect. The solution is given by the length of the horizontal line segment from the origin to the intersection of the vertical line and the x-axis.

A simple modern proof of the method is the following: multiplying the equation by x and regrouping the terms gives

$$\frac{x^4}{m^2} = x\left(\frac{n}{m^2} - x\right).$$

The left-hand side is the value of y^2 on the parabola. The equation of the circle being $y^2 + x(x - n/m^2) = 0$, the right hand side is the value of y^2 on the circle.

Solution with Angle Trisector

A cubic equation with real coefficients can be solved geometrically using compass, straightedge, and an angle trisector if and only if it has three real roots.

Nature of the Roots in the Case of Real Coefficients

Algebraic Nature of the Roots

Every cubic equation (1), $ax^3 + bx^2 + cx + d = 0$, with real coefficients and $a \neq 0$, has three solutions (some of which may equal each other if they are real, and two of which may be complex non-real numbers) and at least one real solution r_1, this last assertion being a consequence of the intermediate value theorem. If $x - r_1$ is factored out of the cubic polynomial, what remains is a quadratic polynomial whose roots r_2 and r_3 are roots of the cubic; by the quadratic formula, these roots are either both real (giving a total of three real roots for the cubic) or are complex conjugates, in which case the cubic has one real and two non-real roots.

It was explained above how to use the sign of the discriminant in order to distinguish between these cases. In fact,

$$\Delta = \left(a^2(r_1 - r_2)(r_1 - r_3)(r_2 - r_3)\right)^2, \tag{3}$$

because a straightforward computation shows that

$$\left(a^2(r_1 - r_2)(r_1 - r_3)(r_2 - r_3)\right)^2 = a^4\big(18(r_1 + r_2 + r_3)(r_1r_2 + r_1r_3 + r_2r_3)r_1r_2r_3 +$$
$$-4(r_1 + r_2 + r_3)^3 r_1r_2r_3 + (r_1 + r_2 + r_3)^2(r_1r_2 + r_1r_3 + r_2r_3)^2 +$$
$$-4(r_1r_2 + r_1r_3 + r_2r_3)^3 - 27(r_1r_2r_3)^2\big)$$

and, by Vieta's formulas, the right hand side of this equality is equal to

$$a^4\left(18\frac{b}{a}\cdot\frac{c}{a}\cdot\frac{d}{a} - 4\left(\frac{b}{a}\right)^3\frac{d}{a} + \left(\frac{b}{a}\right)^2\left(\frac{c}{a}\right)^2 - 4\left(\frac{c}{a}\right)^3 - 27\left(\frac{d}{a}\right)^2\right) = \Delta.$$

The equality (3) shows that $\Delta = 0$ if and only if the equation has a multiple root. This cannot possibly be the case when r_2 and r_3 are non-real complex numbers, because the fact that r_1 is real assures that r_1 is different from r_2 and from r_3 and, on the other hand, the fact that r_2 and r_3 are non-real and that each of them is the conjugate of the other one assures that $r_2 \neq r_3$.

If r_2 and r_3 are non-real, then

$$(r_1 - r_2)(r_1 - r_3)(r_2 - r_3) = (r_1 - r_2)\left(r_1 - \overline{r_2}\right)\left(r_2 - \overline{r_2}\right)$$
$$= (r_1 - r_2)\overline{(r_1 - r_2)}2\,\mathrm{Im}(r_2)i$$
$$= 2\,|r_1 - r_2|^2\,\mathrm{Im}(r_2)i$$

Since this is the product of a non-zero real number by i, its square is a real number less than 0 and therefore $\Delta < 0$. Finally, if the numbers r_1, r_2, and r_3 are three distinct real numbers, then the product $(r_1 - r_2)(r_1 - r_3)(r_2 - r_3)$ is a non-zero real number, and so $\Delta > 0$.

Geometric Interpretation of the Roots

Three Real Roots

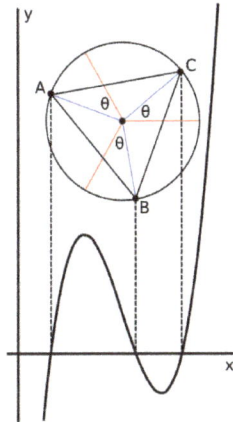

For the cubic (1) with three real roots, the roots are the projection on the x-axis of the vertices A, B, and C of an equilateral triangle. The center of the triangle has the same abscissa as the inflection point.

Viète's trigonometric expression of the roots in the three-real-roots case lends itself to a geometric interpretation in terms of a circle. When the cubic is written in depressed form (2), $t^3 + pt + q = 0$, as shown above, the solution can be expressed as

$$t_k = 2\sqrt{-\frac{p}{3}}\cos\left(\frac{1}{3}\arccos\left(\frac{3q}{2p}\sqrt{\frac{-3}{p}}\right) - k\frac{2\pi}{3}\right) \quad \text{for} \quad k = 0, 1, 2.$$

Here $\arccos\left(\dfrac{3q}{2p}\sqrt{\dfrac{-3}{p}}\right)$ is an angle in the unit circle; taking $1/3$ of that angle corresponds to taking a cube root of a complex number; adding $-k2\pi/3$ for $k = 1, 2$ finds the other cube roots; and multiplying the cosines of these resulting angles by $2\sqrt{-\dfrac{p}{3}}$ corrects for scale.

For the non-depressed case (1), the depressed case as indicated previously is obtained by defining t such that $x = t - b/3a$ so $t = x + b/3a$. Graphically this corresponds to simply shifting the graph horizontally when changing between the variables t and x, without changing the angle relationships. This shift moves the point of inflection and the centre of the circle onto the y-axis. Consequently, the roots of the equation in t sum to zero.

One Real and Two Complex Roots

In the Cartesian Plane

If a cubic is plotted in the Cartesian plane, the real root can be seen graphically as the horizontal intercept of the curve. But further, if the complex conjugate roots are written as $g \pm hi$ then g is the abscissa (the positive or negative horizontal distance from the origin) of the tangency point of a line that is tangent to the cubic curve and intersects the horizontal axis at the same place as does the cubic curve; and h is the square root of the tangent of the angle between this line and the horizontal axis.

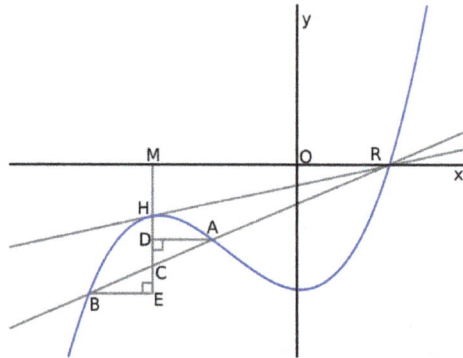

The slope of line RA is twice that of RH. Denoting the complex roots of the cubic as $g \pm hi$, $g = OM$ (negative here) and $h = \sqrt{\tan ORH} = \sqrt{\text{slope of line } RH} = BE = DA$.

In the Complex Plane

With one real and two complex roots, the three roots can be represented as points in the complex plane, as can the two roots of the cubic's derivative. There is an interesting geometrical relationship among all these roots.

The points in the complex plane representing the three roots serve as the vertices of an isosceles triangle. (The triangle is isosceles because one root is on the horizontal (real) axis and the other two roots, being complex conjugates, appear symmetrically above and below the real axis.) Marden's Theorem says that the points representing the roots of the derivative of the cubic are the foci of the Steiner inellipse of the triangle—the unique ellipse that is tangent to the triangle at the midpoints of its sides. If the angle at the vertex on the real axis is less than $\pi/3$ then the major axis of the ellipse lies on the real axis, as do its foci and hence the roots of the derivative. If that angle is greater than $\pi/3$, the major axis is vertical and its foci, the roots of the derivative, are complex conjugates. And if that angle is $\pi/3$, the triangle is equilateral, the Steiner inellipse is simply the triangle's incircle, its foci coincide with each other at the incenter, which lies on the real axis, and hence the derivative has duplicate real roots.

Derivation of the Roots

Cardano's Method

The solutions can be found with the following method due to Scipione del Ferro and Tartaglia, published by Gerolamo Cardano in 1545 in his book *Ars Magna*. This method applies to the depressed cubic (2), $t^3 + pt + q = 0$. We introduce two variables u and v linked by the condition $u + v = t$ and substitute this in the depressed cubic (2), giving

$$u^3 + v^3 + (3uv + p)(u+v) + q = 0.$$

At this point Cardano imposed a second condition for the variables u and v: $3uv + p = 0$. As the first parenthesis vanishes in previous equality, we get $u^3 + v^3 = -q$ and $u^3 v^3 = -p^3/27$. The combination of these two equations leads to a quadratic equation (since they are the sum and the product of u^3 and v^3). Thus u^3 and v^3 are the two roots of the quadratic equation $z^2 + qz - p^3/27 = 0$. Cardano assumed that $q^2/4 + p^3/27 \geq 0$. He suggested that his readers consult another of his books, *De Re-*

gula Aliza, which was published only in 1570, for the case in which $q^2/4 + p^3/27 < 0$. Solving this quadratic equation and using the fact that u and v may be exchanged, we find

$$u^3 = -\frac{q}{2} + \sqrt{\frac{q^2}{4} + \frac{p^3}{27}} \text{ and } v^3 = -\frac{q}{2} - \sqrt{\frac{q^2}{4} + \frac{p^3}{27}}.$$

Therefore, $u + v$ is equal to:

$$t = \sqrt[3]{-\frac{q}{2} + \sqrt{\frac{q^2}{4} + \frac{p^3}{27}}} + \sqrt[3]{-\frac{q}{2} - \sqrt{\frac{q^2}{4} + \frac{p^3}{27}}}. \qquad \textit{(Cardano's formula)}$$

In his book *L'Algebra*, published in 1572, Rafael Bombelli explained that what was done above still works, with a small difference, when $q^{2/4 + p3/27 < 0}$, as long as one knows how to use complex numbers. The small difference is due to the fact that a non-zero complex number has 3 cube roots and not just one. Therefore, although the equality $uv = -p/3$ implies that $u^3v^3 = -p^{3/27}$, this is *not* an equivalence. So, we do not simply take *any* cube root of

$$-\frac{q}{2} + \sqrt{\frac{q^2}{4} + \frac{p^3}{27}} \qquad\qquad (4)$$

and add it to *any* cube root of

$$-\frac{q}{2} - \sqrt{\frac{q^2}{4} + \frac{p^3}{27}} \qquad\qquad (5)$$

(unless one of them is 0); besides, that would provide 9 solutions to equation (2). Instead, since we want to have $3uv = -p$, Cardano's formula means the sum of *a* cube root u of (4) with $-p/3u$ (or, if (4) is equal to 0, the sum of *a* cube root v of (5) with $-p/3v$). This only fails if *both* numbers (4) and (5) are equal to 0, in which case $p = q = 0$ and Cardano's formula simply means $t = \sqrt[3]{0} + \sqrt[3]{0}$ (= 0), which is compatible with the fact that, since $p = q = 0$, (2) simplifies to $t^3 = 0$.

Actually, it is not necessary to compute the three cube roots of (4). To see why, let $\xi = -1/2 + 1/2\sqrt{3}i$. Then ξ and $\bar\xi$ (= $-1/2 - 1/2\sqrt{3}i = \xi^2$) are the non-real cube roots of 1. If u is a cube root of (4), and v is a cube root of (5) such that $3uv = -p$, then the roots of (2) are $u + v$, $\xi u + \bar\xi v$, and $\bar\xi u + \xi v$, since, in each case, we have the sum of a cube root of (4) with a cube root (5) and moreover the product of these two roots is, in each case, equal to $-p/3$.

Note that $\xi^3 = 1$ and that $\xi^4 = \xi$. Thus Cardano's formula, written unambiguously to give the three roots, is

$$t_k = \zeta^k \sqrt[3]{-\frac{q}{2} + \sqrt{\frac{q^2}{4} + \frac{p^3}{27}}} + \zeta^{2k} \sqrt[3]{-\frac{q}{2} - \sqrt{\frac{q^2}{4} + \frac{p^3}{27}}}, \quad k = 0,1,2,$$

where the cube roots expressed as radicals are defined to be any pair of cube roots whose product is $-p/3$. If p and q are real and $q^2/4 + p^3/27 < 0$, this is the same thing as requiring that the cube roots be complex conjugates, while if p and q are real and $q^2/4 + p^3/27 \geq 0$, the real cube roots can be chosen.

Cardano's formula, interpreted in this way, is equivalent to the general solution given earlier when the coefficient of the quadratic term is 0.

We will examine certain particular cases. Before that, it is convenient to note that, if u is a cube root of (4), if v is a cube root of (5), and if both numbers p and $u \times v$ are real, then it is automatically true that $u \times v = -p/3$. This is so because

$$(u \times v)^3 = u^3 \times v^3 = q^2/4 - (q^2/4 + p^3/27) = -p^3/27 = (-p/3)^3.$$

Let us now see the particular cases.

- If p and q are real numbers and $q^2/4 + p^3/27 > 0$, let u be the real cube root of (4) and let v be the real cube root of (5). Then $u \times v = -p/3$, because $u \times v$ is real. So, the roots of the equation (2) are $u + v$ (which is a real number), $\xi u + \bar{\xi} v$, and $\bar{\xi} u + \xi v$. Each of the second and the third roots is the conjugate of the other one. This can be used to prove that they are non-real. Indeed, two real numbers are the conjugates of each other if and only if they are the same real number. But

$$\xi u + \bar{\xi} v = \bar{\xi} u + \xi v \Leftrightarrow (\xi - \bar{\xi})(u - v) = 0,$$

and this last assertion is false, since $\xi \neq \bar{\xi}$ and $u \neq v$ (because u and v are real numbers whose cubes are distinct).

- If p and q are real numbers and $q^2/4 + p^3/27 < 0$, then the numbers (4) and (5) are complex numbers each of which is the conjugate of the other one. Let u be a cube root of (4) and let v be the conjugate of u. Then v^3 is the conjugate of u^3 and this proves that v^3 is equal to (5). So, again, since $u \times v$ is real we have $u \times v = -p/3$, and therefore the roots of the equation (2) are $u + \bar{u}$, $\xi u + \bar{\xi} \bar{u}$, and $\bar{\xi} u + \xi \bar{u}$. In this case all roots are real, since each one of them is the sum of a complex number with its conjugate.

- If $p = 0$, then the roots of the equation (2) are the cube roots of $-q$. This is compatible with Cardano's formula, because one of (4) or (5) is 0 and the other is $-q$.

- If $q = 0$, then the roots of (2) are 0 and the square roots of $-p$. Again, this is compatible with Cardano's formula, because if u is a square root of $p/3$, then u^3 is a square root of $p^3/27$, and this square root is equal to (4) or to (5), since we are assuming that $q = 0$. If v is the other square root of $p/3$ then, by the same reason, v^3 is equal to (4) or to (5) and furthermore if u is equal to (4) then v is equal to (5) and vice versa. But then $v = -u$ and so $u + v = u - u = 0$. On the other hand, $\xi u + \bar{\xi} v = (\xi - \bar{\xi}) \times u$ and so $(\xi u + \bar{\xi} v)^2 = (\xi - \bar{\xi})^2 \times u^2 = (-3) \times p/3 = -p$, which means that $\xi u + \bar{\xi} v$ is a square root of $-p$. Finally, $\bar{\xi} u + \xi v = (\bar{\xi} - \xi) \times u = -(\xi u + \bar{\xi} v)$, and so it must be the other square root of $-p$.

- If $q^2/4 + p^3/27 = 0$ (but p and q are not 0), then (2) has a simple root, which is $3q/p$, and a double root, which is $-3q/2p$. Again, this is compatible with Cardano's formula. To see why, note that asserting that $q^2/4 + p^3/27 = 0$ is equivalent to asserting that $27q^2/4p^3 = -1$. If $u = v = 3q/2p$, then $u^3 = v^3 = 27q^3/8p^2 = -q/2$ and $3uv = 27q^2/4p^2 = -p$. Thus, Cardano's formula says that the roots of (2) are $u + v = 2u = 3q/p$, $\xi u + \bar{\xi} v = (\xi + \bar{\xi}) \times u = -3q/2p$, and $\bar{\xi} u + \xi v = (\bar{\xi} + \xi) \times u = -3q/2p$.

In this last case (that is, when $q^2/4 + p^3/27 = 0$ but p and q are not 0), although the computations made above do suggest that $3q/p$ is a simple root of (2) whereas $-3q/2p$ is a double root (having been obtained in two different ways), they don't really prove it. However, this can be easily confirmed. Just note that

$$\left(x-\frac{3q}{p}\right)\left(x+\frac{3q}{2p}\right)^2 = x^3 - \frac{27q^2}{4p^2}x - \frac{27q^3}{4p^3}$$

$$= x^3 - \frac{27q^2}{4p^3}px - \frac{27q^2}{4p^3}q$$

$$= x^3 + px + q,$$

since $27q^2/4p^3 = -1$.

The numbers provided by Cardano's formula are solutions of the equation (2), but there might be other solutions besides these. However, this does not occur. Let u and v be numbers such that $u^3 + v^3 = -q$ and $3uv = -p$. In order to see that $u + v$, $\xi u + \bar{\xi} v$, and $\bar{\xi} u + \xi v$ are the only roots of the polynomial $t^3 + pt + q$, it is enough to notice that

$$\left(t-(u+v)\right)\left(t-\left(\zeta u+\bar{\zeta} v\right)\right)\left(t-\left(\bar{\zeta} u+\zeta v\right)\right) = t^3 - 3uvt - u^3 - v^3$$

$$= t^3 + pt + q.$$

Therefore, $t^3 + pt + q = 0$ if and only if $t = u + v$, $t = \xi u + \bar{\xi} v$ or $t = \bar{\xi} u + \xi v$.

Vieta's Substitution

Starting from the depressed cubic (2), $t^3 + pt + q = 0$, we make the substitution $t = w - p/3w$, known as Vieta's substitution. This results in the equation

$$w^3 + q - \frac{p^3}{27w^3} = 0.$$

Multiplying by w^3, it becomes a sextic equation in w, which is in fact a quadratic equation in w^3:

$$w^6 + qw^3 - \frac{p^3}{27} = 0.$$

(6)

The quadratic formula allows equation (6) to be solved for w^3. If w_1, w_2 and w_3 are the three cube roots of one of the solutions in w^3, then the roots of the original depressed cubic are $w_1 - p/3w_1$, $w_2 - p/3w_2$, and $w_3 - p/3w_3$. Another way of expressing the roots is to take $\xi = -1/2 + 1/2\sqrt{3}i$; then the roots of the original depressed cubic are $w_1 - p/3w_1$, $\xi w_1 - p/3\xi w_1$, and, $\xi^2 w_1 - p/3\xi^2 w_1$. This method only fails when both roots of the equation (6) are equal to 0, but this only happens when $p = q = 0$, in which case the only solution of equation (2) is 0.

Actually, the substitution originally used by Vieta (in a text published posthumously in 1615) was $t = p/3w - w$, but it leads to similar computations. More precisely, Vieta introduced a new variable w and he imposed the condition $w(t + w) = p/3$.

As far as formulae are concerned, Vieta's approach leads to the same result as Cardano's method. However, it is theoretically simpler, for two reasons:

- Each root of the equation (2) is expressed from the start by an expression that involves a single cube root. Therefore, there is no ambiguity as in Cardano's formula.

- It is nearly trivial that there are no other roots besides the ones obtained by this method. This follows from the fact that any complex number can be written as $w - p/3w$ for some other complex number w.

Lagrange's Method

In his paper *Réflexions sur la résolution algébrique des équations* ("Thoughts on the algebraic solving of equations"), Joseph Louis Lagrange introduced a new method to solve equations of low degree.

This method works well for cubic and quartic equations, but Lagrange did not succeed in applying it to a quintic equation, because it requires solving a resolvent polynomial of degree at least six. This is explained by the Abel–Ruffini theorem, which proves that such polynomials cannot be solved by radicals. Nevertheless, the modern methods for solving solvable quintic equations are mainly based on Lagrange's method.

In the case of cubic equations, Lagrange's method gives the same solution as Cardano's. By drawing attention to a geometrical problem that involves two cubes of different size Cardano explains in his book *Ars Magna* how he arrived at the idea of considering the unknown of the cubic equation as a sum of two other quantities. Lagrange's method may also be applied directly to the general cubic equation (1), $ax^3 + bx^2 + cx + d = 0$, without using the reduction to the depressed cubic equation (2), $t^3 + pt + q = 0$. Nevertheless, the computation is much easier with this reduced equation.

Suppose that x_0, x_1 and x_2 are the roots of equation (1) or (2), and define $\xi = -1/2 + 1/2\sqrt{3}i$ (a complex cube root of 1, i.e. a primitive third root of unity) which satisfies the relation $\xi^2 + \xi + 1 = 0$. We now set

$$s_0 = x_0 + x_1 + x_2,$$

$$s_1 = x_0 + \zeta x_1 + \zeta^2 x_2,$$

$$s_2 = x_0 + \zeta^2 x_1 + \zeta x_2.$$

This is the discrete Fourier transform of the roots: observe that while the coefficients of the polynomial are symmetric in the roots, in this formula an *order* has been chosen on the roots, so these are not symmetric in the roots. The roots may then be recovered from the three s_i by inverting the above linear transformation via the inverse discrete Fourier transform, giving

$$x_0 = \tfrac{1}{3}(s_0 + s_1 + s_2),$$

$$x_1 = \tfrac{1}{3}(s_0 + \zeta^2 s_1 + \zeta s_2),$$

$$x_2 = \tfrac{1}{3}(s_0 + \zeta s_1 + \zeta^2 s_2).$$

The polynomial s_0 is equal, by Vieta's formulas, to $-b/a$ in case of equation (1) and to 0 in case of equation (2), so we only need to seek values for the other two.

The polynomials s_1 and s_2 are not symmetric functions of the roots: s_0 is invariant, while the two non-trivial cyclic permutations of the roots send s_1 to ξs_1 and s_2 to $\xi^2 s_2$, or s_1 to $\xi^2 s_1$ and s_2 to ξs_2 (depending on which permutation), while transposing x_1 and x_2 switches s_1 and s_2; other transpositions switch these roots and multiply them by a power of ξ.

Thus s_1^3, s_2^3 and $s_1 s_2$ are left invariant by the cyclic permutations of the roots, which multiply them by $\xi^3 = 1$. Also $s_1 s_2$ and $s_1^3 + s_2^3$ are left invariant by the transposition of x_1 and x_2 which exchanges s_1 and s_2. As the permutation group S_3 of the roots is generated by these permutations, it follows that $s_1^3 + s_2^3$ and $s_1 s_2$ are symmetric functions of the roots and may thus be written as polynomials in the elementary symmetric polynomials and thus as rational functions of the coefficients of the equation. Let $s_1^3 + s_2^3 = A$ and $s_1 s_2 = B$ in these expressions, which will be explicitly computed below.

We have that s_1^3 and s_2^3 are the two roots of the quadratic equation $z^2 - Az + B^3 = 0$. Thus the resolution of the equation may be finished exactly as described for Cardano's method, with s_1 and s_2 in place of u and v.

Computation of A and B

Setting $E_1 = x_0 + x_1 + x_2$, $E_2 = x_0 x_1 + x_1 x_2 + x_2 x_0$ and $E_3 = x_0 x_1 x_2$, the elementary symmetric polynomials, we have, using that $\xi^3 = 1$:

$$s_1^3 = x_0^3 + x_1^3 + x_2^3 + 3\zeta(x_0^2 x_1 + x_1^2 x_2 + x_2^2 x_0) + 3\zeta^2(x_0 x_1^2 + x_1 x_2^2 + x_2 x_0^2) + 6x_0 x_1 x_2.$$

The expression for s_2^3 is the same with ξ and ξ^2 exchanged. Thus, using $\xi^2 + \xi = -1$ we get

$$A = s_1^3 + s_2^3 = 2(x_0^3 + x_1^3 + x_2^3) - 3(x_0^2 x_1 + x_1^2 x_2 + x_2^2 x_0 + x_0 x_1^2 + x_1 x_2^2 + x_2 x_0^2) + 12x_0 x_1 x_2,$$

and a straightforward computation gives

$$A = s_1^3 + s_2^3 = 2E_1^3 - 9E_1 E_2 + 27E_3.$$

Similarly we have

$$B = s_1 s_2 = x_0^2 + x_1^2 + x_2^2 + (\zeta + \zeta^2)(x_0 x_1 + x_1 x_2 + x_2 x_0) = E_1^2 - 3E_2.$$

When solving equation (1) we have $E_1 = -b/a$, $E_2 = c/a$ and $E_3 = -d/a$. With equation (2), we have $E_1 = 0$, $E_2 = p$ and $E_3 = -q$ and thus $A = -27q$ and $B = -3p$.

Note that with equation (2), we have $x_0 = 1/3(s_1 + s_2)$ and $s_1 s_2 = -3p$, while in Cardano's method we have set $x_0 = u + v$ and $uv = -1/3p$. Thus we have, up to the exchange of u and v, $s_1 = 3u$ and $s_2 = 3v$. In other words, in this case, Cardano's method and Lagrange's method compute exactly the same things, up to a factor of three in the auxiliary variables, the main difference being that Lagrange's method explains why these auxiliary variables appear in the problem.

General Solution to the Cubic Equation with Arbitrary Coefficients

If we are dealing with a cubic equation whose coefficients belong to some field k (whose characteristic is either 0 or greater than 3), then what was done above algebraically still works, with one exception: the results concerning the sign of the discriminant, since they make no sense for general fields, although the fact that the equation has a multiple root if and only if $\Delta = 0$ is still true (and for the same reason). In this more general case, we work with an extension K of k in which every non-zero element has two square roots and three cube roots. For instance, if $k = Q$, we can take $K = \bar{Q}$, the field of algebraic numbers.

In particular, all that was done above algebraically still works if $k = K = C$. Therefore, every cubic equation with complex coefficients has some complex root, which is a particular case of the fundamental theorem of algebra.

In this general context, the formulae for roots in the case in which $\Delta = 0$ show that these roots also belong to the field k.

In a field k whose characteristic is either 2 or 3, this approach does not work because then the formulae for the roots became meaningless, since they involve division by 2 and 3.

Galois Groups of Irreducible Cubics

The Galois group of an irreducible separable polynomial of degree n is a transitive subgroup of S_n. In particular, the Galois group of an irreducible separable cubic is a transitive subgroup of S_3 and there are only two such subgroups: S_3 and A_3. There is a simple way of determining the Galois group of a concrete irreducible cubic $f(x)$ over a field k: it is A_3 if the discriminant of the cubic is the square of an element of k and S_3 otherwise. Indeed, if Δ is not the square of an element of k, then $k[\sqrt{\Delta}]$ is an extension of degree 2 of k. On the other hand, if r_1, r_2, and r_3 are the roots of $f(x)$, then, since the equality (3) holds, that is, since $\Delta = (a^2(r_1 - r_2)(r_1 - r_3)(r_2 - r_3))^2$, $k[\sqrt{\Delta}] \subset k[r_1, r_2, r_3]$, and so, by the multiplicativity formula for degrees, the degree of $k[r_1, r_2, r_3]$ over k (that is, the order of the Galois group of $f(x)$) must be a multiple of the degree of $k[\sqrt{\Delta}]$, which is 2. Therefore, it must be an even number, and so the Galois group can only be S_3.

On the other hand, if Δ *is* the square of an element of k, then, again by the equality (3), we have $(r_1 - r_2)(r_1 - r_3)(r_2 - r_3) \in k$. Therefore, if σ belongs to the Galois group of $f(x)$, then σ maps $(r_1 - r_2)(r_1 - r_3)(r_2 - r_3)$ into itself. But then σ cannot act on the set $\{r_1, r_2, r_3\}$ as the transposition that exchanges r_1 and r_2 and leaves r_3 fixed, because then σ would map $(r_1 - r_2)(r_1 - r_3)(r_2 - r_3)$ into $-(r_1 - r_2)(r_1 - r_3)(r_2 - r_3)$. So, in this case, the Galois group of $f(x)$ is not S_3 and therefore it must be A_3.

It is clear from this criterion that, if we are working over the field Q, the Galois group of most irreducible cubic polynomials is S_3. An example of an irreducible cubic polynomial with rational coefficients whose Galois group is A_3 is $p(x) = x^3 - 3x - 1$, whose discriminant is $81 = 9^2$. The polynomial $p(x)$ is uscd in thc standard proof of the impossibility of trisecting arbitrary angles using straightedge and compass only.

Collinearities

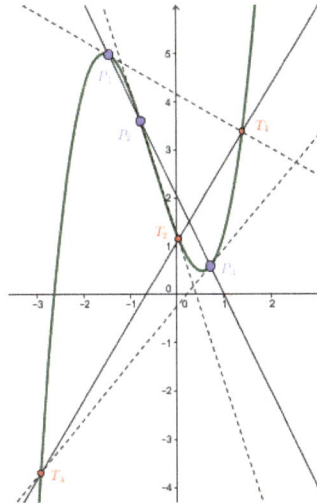

The points P_1, P_2, and P_3 (in blue) are collinear and belong to the graph of $x^3 + 3/2x^2 - 5/2x + 5/4$. The points T_1, T_2, and T_3 (in red) are the intersections of the (dotted) tangent lines to the graph at these points with the graph itself. They are collinear too.

The tangent lines to a cubic at three collinear points intercept the cubic again at collinear points. This can be seen as follows. If the cubic is defined by $f(x) = ax^3 + bx^2 + cx + d$ and if α is a real number, then the tangent to the graph of f at the point $(\alpha, f(\alpha))$ is the line

$$\{(x, f(\alpha) + (x - \alpha)f'(\alpha)) : x \in \mathbf{R}\}.$$

So, the intersection point between this line and the graph of f can be obtained solving the equation $f(x) = f(\alpha) + (x - \alpha)f'(\alpha)$. This is a cubic equation, but it is clear that α is a root, and in fact a double root, since the line is tangent to the graph. The remaining root is $-b/a - 2\alpha$. So, the other intersection point between the tangent line and the graph of f is the point

$$\left(-\frac{b}{a} - 2\alpha, f\left(-\frac{b}{a} - 2\alpha\right)\right) = \left(-\frac{b}{a} - 2\alpha, -8a\alpha^3 - 8b\alpha^2 - 2\left(\frac{b^2}{a} + c\right)\alpha - \frac{bc}{a} + d\right)$$

$$= \left(-\frac{b}{a} - 2\alpha, -\frac{bc}{a} + 9d + \left(6c - 2\frac{b^2}{a}\right)\alpha - 8f(\alpha)\right).$$

Therefore, if P is a point of the graph of f, the other intersection point between the tangent line at P and the graph is the point $A(P)$, where A is the map defined by

$$A: \quad \mathbf{R}^2 \quad \longrightarrow \quad \mathbf{R}^2$$

$$(x, y) \quad \mapsto \quad \left(-\frac{b}{a} - 2x, -\frac{bc}{a} + 9d + \left(6c - 2\frac{b^2}{a}\right)x - 8y\right).$$

Since A is an affine map, if P_1, P_2, and P_3 are collinear, then so are the points $A(P_1)$, $A(P_2)$, and $A(P_3)$.

Symmetry

The graph of a cubic function has 180° rotational or point symmetry about its inflection point. The inflection point of a general cubic polynomial,

$$f(x) = ax^3 + bx^2 + cx + d$$

occurs at a point $(x_0, f(x_0))$ such that $f''(x_0) = 0$. Since $f''(x) = 6ax + 2b$, the inflection point is $(-b/3a, 2b^3/27a^2 - bc/3a + d)$. Translating the function so that the inflection point is at the origin, one obtains the function f_T defined by:

$$f_T(x) = f(x + x_0) - y_0$$

$$= a\left(x - \frac{b}{3a}\right)^3 + b\left(x - \frac{b}{3a}\right)^2 + c\left(x - \frac{b}{3a}\right) + d - \left(\frac{2b^3}{27a^2} - \frac{bc}{3a} + d\right)$$

$$= ax^3 + \left(c - \frac{b^2}{3a}\right)x.$$

As all terms are odd powers of x, $f_T(-x) = -f_T(x)$ proving that all cubic functions are rotationally symmetrical about their inflection points.

Applications

Cubic equations arise in various other contexts.

Marden's theorem states that the foci of the Steiner inellipse of any triangle can be found by using the cubic function whose roots are the coordinates in the complex plane of the triangle's three vertices. The roots of the first derivative of this cubic are the complex coordinates of those foci.

The area of a regular heptagon can be expressed in terms of the roots of a cubic. Further, the ratios of the long diagonal to the side, the side to the short diagonal, and the negative of the short diagonal to the long diagonal all satisfy a particular cubic equation. In addition, the ratio of the inradius to the circumradius of a heptagonal triangle is one of the solutions of a cubic equation.

Given the cosine (or other trigonometric function) of an arbitrary angle, the cosine of one-third of that angle is one of the roots of a cubic.

The solution of the general quartic equation relies on the solution of its resolvent cubic.

The eigenvalues of a 3×3 matrix are the roots of a cubic polynomial which is the characteristic polynomial of the matrix.

The characteristic equation of a third-order linear difference equation or differential equation is a cubic equation.

In analytical chemistry, the Charlot equation, which can be used to find the pH of buffer solutions, can be solved using a cubic equation.

In chemical engineering and thermodynamics, cubic equations of state are used to model the PVT (pressure, volume, temperature) behavior of substances.

Kinematic equations involving changing rates of acceleration are cubic.

Quintic Function

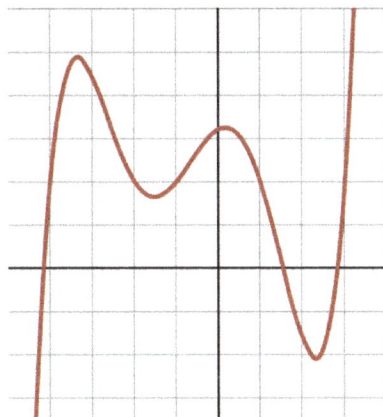

Graph of a polynomial of degree 5, with 3 real zeros (roots) and 4 critical points.

In algebra, a quintic function is a function of the form

$$g(x) = ax^5 + bx^4 + cx^3 + dx^2 + ex + f,$$

where a, b, c, d, e and f are members of a field, typically the rational numbers, the real numbers or the complex numbers, and a is nonzero. In other words, a quintic function is defined by a polynomial of degree five.

If a is zero but one of the coefficients b, c, d, or e is non-zero, the function is classified as either a quartic function, cubic function, quadratic function or linear function.

Because they have an odd degree, normal quintic functions appear similar to normal cubic functions when graphed, except they may possess an additional local maximum and local minimum each. The derivative of a quintic function is a quartic function.

Setting $g(x) = 0$ and assuming $a \neq 0$ produces a quintic equation of the form:

$$ax^5 + bx^4 + cx^3 + dx^2 + ex + f = 0.$$

Solving quintic equations in terms of radicals was a major problem in algebra, from the 16th century, when cubic and quartic equations were solved, until the first half of the 19th century, when the impossibility of such a general solution was proved (Abel–Ruffini theorem).

Finding Roots of a Quintic Equation

Finding the roots of a given polynomial has been a prominent mathematical problem.

Solving linear, quadratic, cubic and quartic equations by factorization into radicals can always be done, no matter whether the roots are rational or irrational, real or complex; there are formulae that yield the required solutions. However, there is no algebraic expression for general quintic equations over the rationals in terms of radicals; this statement is known as the Abel–Ruffini theorem, first published in 1799. This result also holds for equations of higher degrees. An example of a quintic whose roots cannot be expressed in terms of radicals is $x^5 - x + 1 = 0$. This quintic is in Bring–Jerrard normal form.

Some quintics may be solved in terms of radicals. However, the solution is generally too complex to be used in practice. Instead, numerical approximations are calculated using root-finding algorithm for polynomials.

Solvable Quintics

Some quintic equations can be solved in terms of radicals. These include the quintic equations defined by a polynomial that is reducible, such as $x^5 - x^4 - x + 1 = (x^2 + 1)(x + 1)(x - 1)^2$. As solving reducible quintic equations reduces immediately to solving polynomials of lower degree, only irreducible quintic equations are considered in this section, and the term "quintic" will refer only to irreducible quintics. A solvable quintic is thus an irreducible quintic polynomial whose roots may be expressed in terms of radicals.

For characterizing solvable quintics, and more generally solvable polynomials of higher degree, Évariste Galois developed techniques which gave rise to group theory and Galois theory. Applying these techniques, Arthur Cayley found a general criterion for determining whether any given quintic is solvable. This criterion is the following.

Given the equation

$$ax^5 + bx^4 + cx^3 + dx^2 + ex + f = 0,$$

the Tschirnhaus transformation $x = y - b/5a$, which depresses the quintic (that means removes the term of degree four), gives the equation

$$y^5 + py^3 + qy^2 + ry + s = 0,,$$

where

$$p = \frac{5ac - 2b^2}{5a^2}$$

$$q = \frac{25a^2d - 15abc + 4b^3}{25a^3}$$

$$r = \frac{125a^3e - 50a^2bd + 15ab^2c - 3b^4}{125a^4}$$

$$s = \frac{3125a^4f - 625a^3be + 125a^2b^2d - 25ab^3c + 4b^5}{3125a^5}$$

Both quintics are solvable by radicals if and only if either they are factorisable in equations of lower degrees with rational coefficients or the polynomial $P^2 - 1024z\Delta$, named *Cayley's resolvent*, has a rational root in z, where

$$P = z^3 - z^2(20r + 3p^2) - z(8p^2r - 16pq^2 - 240r^2 + 400sq - 3p^4)$$

$$-p^6 + 28p^4r - 16p^3q^2 - 176p^2r^2 - 80p^2sq + 224prq^2 - 64q^4$$

$$+4000ps^2 + 320r^3 - 1600rsq$$

and

$$\Delta = -128p^2r^4 + 3125s^4 - 72p^4qrs + 560p^2qr^2s + 16p^4r^3 + 256r^5 + 108p^5s^2$$

$$-1600qr^3s + 144pq^2r^3 - 900p^3rs^2 + 2000pr^2s^2 - 3750pqs^3 + 825p^2q^2s^2$$

$$+2250q^2rs^2 + 108q^5s - 27q^4r^2 - 630pq^3rs + 16p^3q^3s - 4p^3q^2r^2.$$

Cayley's result allows us to test if a quintic is solvable. If it is the case, finding its roots is a more difficult problem, which consists of expressing the roots in terms of radicals involving the coefficients of the quintic and the rational root of Cayley's resolvent.

In 1888, George Paxton Young described how to solve a solvable quintic equation, without providing an explicit formula; Daniel Lazard wrote out a three-page formula (Lazard (2004)   .

Quintics in Bring–Jerrard Form

There are several parametric representations of solvable quintics of the form $x^5 + ax + b = 0$, called the Bring–Jerrard form.

During the second half of 19th century, John Stuart Glashan, George Paxton Young, and Carl Runge gave such a parameterization: an irreducible quintic with rational coefficients in Bring–Jerrard form is solvable if and only if either $a = 0$ or it may be written

$$x^5 + \frac{5\mu^4(4v + 3)}{v^2 + 1}x + \frac{4\mu^5(2v + 1)(4v + 3)}{v^2 + 1} = 0$$

where μ and v are rational.

In 1994, Blair Spearman and Kenneth S. Williams gave an alternative,

$$x^5 + \frac{5e^4(4c + 3)}{c^2 + 1}x + \frac{-4e^5(2c - 11)}{c^2 + 1} = 0.$$

The relationship between the 1885 and 1994 parameterizations can be seen by defining the expression

$$b = \frac{4}{5}\left(a + 20 \pm 2\sqrt{(20-a)(5+a)}\right)$$

where $a = 5(4v + 3)/v^2 + 1$. Using the negative case of the square root yields, after scaling variables, the first parametrization while the positive case gives the second.

The substitution $c = -m/l^5$, $e = 1/l$ in the Spearman-Williams parameterization allows one to not exclude the special case $a = 0$, giving the following result:

If a and b are rational numbers, the equation $x^5 + ax + b = 0$ is solvable by radicals if either its left-hand side is a product of polynomials of degree less than 5 with rational coefficients or there exist two rational numbers l and m such that

$$a = \frac{5l(3l^5 - 4m)}{m^2 + l^{10}} \qquad b = \frac{4(11l^5 + 2m)}{m^2 + l^{10}}.$$

Roots of a Solvable Quintic

A polynomial equation is solvable by radicals if its Galois group is a solvable group. In the case of irreducible quintics, the Galois group is a subgroup of the symmetric group S_5 of all permutations of a five element set, which is solvable if and only if it is a subgroup of the group F_5, of order 20, generated by the cyclic permutations (1 2 3 4 5) and (1 2 4 3).

If the quintic is solvable, one of the solutions may be represented by an algebraic expression involving a fifth root and at most two square roots, generally nested. The other solutions may then obtained either by changing of fifth root or by multiplying all the occurrences of the fifth root by the same power of a primitive 5th root of unity

$$\frac{\sqrt{-10 - 2\sqrt{5}} + \sqrt{5} - 1}{4}.$$

(The other primitive 5th roots of unity may be deduced by changing the signs of the square roots.)

It follows that one may need four different square roots for writing all the roots of a solvable quintic. Even for the first root that involves at most two square roots, the expression of the solutions in terms of radicals is usually huge. However, when no square root is needed, the form of the first solution may be rather simple, as for the equation $x^5 - 5x^4 + 30x^3 - 50x^2 + 55x - 21 = 0$, for which the only real solution is

$$x = 1 + \sqrt[5]{2} - \left(\sqrt[5]{2}\right)^2 + \left(\sqrt[5]{2}\right)^3 - \left(\sqrt[5]{2}\right)^4.$$

An example of a more complex (although small enough to be written here) solution is the unique real root of $x^5 - 5x + 12 = 0$. Let $a = \sqrt{2}\varphi^{-1}$, $b = \sqrt{2}\varphi$, and $c = \sqrt[4]{5}$, where $\varphi = 1 + \sqrt{5}/2$ is the golden ratio. Then the only real solution $x = -1.84208...$ is given by

$$-cx = \sqrt[5]{(a+c)^2(b-c)} + \sqrt[5]{(-a+c)(b-c)^2} + \sqrt[5]{(a+c)(b+c)^2} - \sqrt[5]{(-a+c)^2(b+c)},$$

or, equivalently, by

$$x = \sqrt[5]{y_1} + \sqrt[5]{y_2} + \sqrt[5]{y_3} + \sqrt[5]{y_4},$$

where the y_i are the four roots of the quartic equation

$$y^4 + 4y^3 + \frac{4}{5}y^2 - \frac{8}{5^3}y - \frac{1}{5^5} = 0.$$

More generally, if an equation $P(x) = 0$ of prime degree p with rational coefficients is solvable in radicals, then one can define an auxiliary equation $Q(y) = 0$ of degree $p - 1$, also with rational coefficients, such that each root of P is the sum of p-th roots of the roots of Q. These p-th roots have been introduced by Joseph-Louis Lagrange, and their product by p are commonly called Lagrange resolvents. The computation of Q and its roots can be used to solve $P(x) = 0$. However these p-th roots may not be computed independently (this would provide p^{p-1} roots instead of p). Thus a correct solution needs to express all these p-roots in term of one of them. Galois theory shows that this is always theoretically possible, even if the resulting formula may be too large to be of any use.

It is possible that some of the roots of Q are rational (as in the first example of this section) or some are zero. In these cases, the formula for the roots is much simpler, as for the solvable de Moivre quintic

$$x^5 + 5ax^3 + 5a^2x + b = 0,$$

where the auxiliary equation has two zero roots and reduces, by factoring them out, to the quadratic equation

$$y^2 + by - a^5 = 0,$$

such that the five roots of the de Moivre quintic are given by

$$x_k = \omega^k \sqrt[5]{y_i} - \frac{a}{\omega^k \sqrt[5]{y_i}},$$

where y_i is any root of the auxiliary quadratic equation and ω is any of the four primitive 5th roots of unity. This can be easily generalized to construct a solvable septic and other odd degrees, not necessarily prime.

Other Solvable Quintics

There are infinitely many solvable quintics in Bring-Jerrard form which have been parameterized in a preceding section.

Up to the scaling of the variable, there are exactly five solvable quintics of the shape $x^5 + ax^2 + b$, which are (where s is a scaling factor):

$$x^5 - 2s^3x^2 - \frac{s^5}{5}$$

$$x^5 - 100s^3x^2 - 1000s^5$$

$$x^5 - 5s^3x^2 - 3s^5$$

$$x^5 - 5s^3x^2 + 15s^5$$

$$x^5 - 25s^3x^2 - 300s^5$$

Paxton Young (1888) gave a number of examples, some of them being reducible, having a rational root:

$x^5 - 10x^3 - 20x^2 - 1505x - 7412$

$x^5 + \dfrac{625}{4}x + 3750$

$x^5 - \dfrac{22}{5}x^3 - \dfrac{11}{25}x^2 + \dfrac{462}{125}x + \dfrac{979}{3125}$

$x^5 + 20x^3 + 20x^2 + 30x + 10$ Solution: $\sqrt[5]{2} - \sqrt[5]{2}^2 + \sqrt[5]{2}^3 - \sqrt[5]{2}^4$

$x^5 + 320x^2 - 1000x + 4288$ Reducible: −8 is a root

$x^5 + 40x^2 - 69x + 108$ Reducible: −4 is a root

$x^5 - 20x^3 + 250x - 400$

$x^5 - 5x^3 + \dfrac{85}{8}x - 13/2$

$x^5 + \dfrac{20}{17}x + \dfrac{21}{17}$

$x^5 - \dfrac{4}{13}x + \dfrac{29}{65}$

$x^5 + \dfrac{10}{13}x + \dfrac{3}{13}$

$x^5 + 110(5x^3 + 60x^2 + 800x + 8320)$

$x^5 - 20x^3 - 80x^2 - 150x - 656$

$x^5 - 40x^3 + 160x^2 + 1000x - 5888$

$x^5 - 50x^3 - 600x^2 - 2000x - 11200$

$x^5 + 110(5x^3 + 20x^2 - 360x + 800)$

$x^5 - 20x^3 + 320x^2 + 540x + 6368$ Reducible : -8 is a root

$$x^5 - 20x^3 - 160x^2 - 420x - 8928 \qquad\qquad \text{Reducible : -8 is a root}$$

$$x^5 - 20x^3 + 170x + 208$$

An infinite sequence of solvable quintics may be constructed, whose roots are sums of n-th roots of unity, with $n = 10k + 1$ being a prime number:

$x^5 + x^4 - 4x^3 - 3x^2 + 3x + 1$	Roots: $2\cos(\dfrac{2k\pi}{11})$
$x^5 + x^4 - 12x^3 - 21x^2 + x + 5$	Root: $\displaystyle\sum_{k=0}^{5} e^{\frac{2i\pi 6^k}{31}}$
$y^5 + y^4 - 16y^3 + 5y^2 + 21y - 9$	Root: $\displaystyle\sum_{k=0}^{7} e^{\frac{2i\pi 3^k}{41}}$
$y^5 + y^4 - 24y^3 - 17y^2 + 41y - 13$	Root: $\displaystyle\sum_{k=0}^{11} e^{\frac{2i\pi (21)^k}{61}}$
$y^5 + y^4 - 28y^3 + 37y^2 + 25y + 1$	Root: $\displaystyle\sum_{k=0}^{13} e^{\frac{2i\pi (23)^k}{71}}$

There are also two parameterized families of solvable quintics: The Kondo–Brumer quintic,

$$x^5 + (a-3)x^4 + (-a+b+3)x^3 + (a^2 - a - 1 - 2b)x^2 + bx + a = 0$$

and the family depending on the parameters a, l, m

$$x^5 - 5p(2x^3 + ax^2 + bx) - pc = 0$$

where

$$p = \frac{l^2(4m^2 + a^2) - m^2}{4}, \qquad b = l(4m^2 + a^2) - 5p - 2m^2, \qquad c = \frac{b(a + 4m) - p(a - 4m) - a^2 m}{2}$$

Casus Irreducibilis

Analogously to cubic equations, there are solvable quintics which have five real roots all of whose solutions in radicals involve roots of complex numbers. This is *casus irreducibilis* for the quintic, which is discussed in Dummit.

Beyond Radicals

About 1835, Jerrard demonstrated that quintics can be solved by using ultraradicals (also known as Bring radicals), the unique real root of $t^5 + t - a = 0$ for real numbers a. In 1858

Charles Hermite showed that the Bring radical could be characterized in terms of the Jacobi theta functions and their associated elliptic modular functions, using an approach similar to the more familiar approach of solving cubic equations by means of trigonometric functions. At around the same time, Leopold Kronecker, using group theory, developed a simpler way of deriving Hermite's result, as had Francesco Brioschi. Later, Felix Klein came up with a method that relates the symmetries of the icosahedron, Galois theory, and the elliptic modular functions that are featured in Hermite's solution, giving an explanation for why they should appear at all, and developed his own solution in terms of generalized hypergeometric functions. Similar phenomena occur in degree 7 (septic equations) and 11, as studied by Klein and discussed in icosahedral symmetry: related geometries.

Solving Through Bring Radical

A Tschirnhaus transformation, which may be computed by solving a quartic equation, reduces the general quintic equation of the form

$$x^5 + a_4x^4 + a_3x^3 + a_2x^2 + a_1x + a_0 = 0$$

to the Bring–Jerrard normal form $x^5 - x + t = 0$.

The roots of this equation cannot be expressed by radicals. However, in 1858, Charles Hermite published the first known solution of this equation in terms of elliptic functions. At around the same time Francesco Brioschi and Leopold Kronecker came upon equivalent solutions.

Application to Celestial Mechanics

Solving for the locations of the Lagrangian points of an astronomical orbit in which the masses of both objects are non-negligible involves solving a quintic.

More precisely, the locations of L_2 and L_1 are the solutions to the following equations, where the gravitational forces of two masses on a third (e.g. Sun and Earth on satellites such as Gaia at L_2 and SOHO at L_1) provide the satellite's centripetal force necessary to be in a synchronous orbit with Earth around the Sun:

$$\frac{GmM_S}{(R \pm r)^2} \pm \frac{GmM_E}{r^2} = m\omega^2(R \pm r)$$

The $\pm$ sign corresponds to L_2 and L_1, respectively; G is the gravitational constant, ω the angular velocity, r the distance of the satellite to Earth, R the distance Sun to Earth (i.e. the semi-major axis of Earth's orbit), and m, M_S, and M_E are the respective masses of satellite, Earth, and Sun.

Using Kepler's Third Law $\omega^2 = \frac{4\pi^2}{P^2} = \frac{G(M_S + M_E)}{R^3}$ and rearranging all terms yields the quintic

$$ar^5 + br^4 + cr^3 + dr^2 + er + f = 0$$

with $a = \pm(M_S + M_E)$, $b = +(M_S + M_E)3R$, $c = \pm(M_S + M_E)3R^2$, $d = +(M_E \mp M_E)R^3$ (thus d = 0 for L_2), $e = -M_E 2R^4$, $f = \mp M_E R^5$.

Solving these two quintics yields $r = 1.501 \times 10^9\, m$ for L_2 and $r = 1.491 \times 10^9\, m$ for L_1. The Sun–Earth Lagrangian points L_2 and L_1 are usually given as 1.5 million km from Earth.

Quartic Function

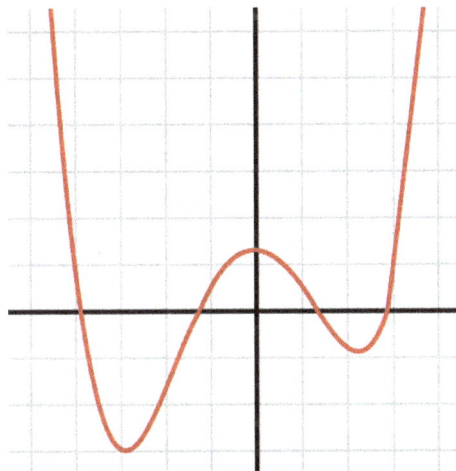

Graph of a polynomial of degree 4, with 3 critical points and four real roots (crossings of the x axis) (and thus no complex roots). If one or the other of the local minima were above the x axis, or if the local maximum were below it, or if there were no local maximum and one minimum below the x axis, there would only be two real roots (and two complex roots). If all three local extrema were above the x axis, or if there were no local maximum and one minimum above the x axis, there would be no real root (and four complex roots).

In algebra, a quartic function is a function of the form

$$f(x) = ax^4 + bx^3 + cx^2 + dx + e,$$

where a is nonzero, which is defined by a polynomial of degree four, called quartic polynomial.

Sometimes the term biquadratic is used instead of *quartic*, but, usually, biquadratic function refers to a quadratic function of a square (or, equivalently, to the function defined by a quartic polynomial without terms of odd degree), having the form

$$f(x) = ax^4 + cx^2 + e.$$

A quartic equation, or equation of the fourth degree, is an equation that equates a quartic polynomial to zero, of the form

$$ax^4 + bx^3 + cx^2 + dx + e = 0,$$

where $a \neq 0$.

The derivative of a quartic function is a cubic function.

Since a quartic function is defined by a polynomial of even degree, it has the same infinite limit when the argument goes to positive or negative infinity. If a is positive, then the function increases to positive infinity at both ends; and thus the function has a global minimum. Likewise, if a is negative, it decreases to negative infinity and has a global maximum. In both cases it may or may not have another local maximum and another local minimum.

The degree four (*quartic* case) is the highest degree such that every polynomial equation can be solved by radicals.

History

Lodovico Ferrari is credited with the discovery of the solution to the quartic in 1540, but since this solution, like all algebraic solutions of the quartic, requires the solution of a cubic to be found, it could not be published immediately. The solution of the quartic was published together with that of the cubic by Ferrari's mentor Gerolamo Cardano in the book *Ars Magna*.

The Soviet historian I. Y. Depman claimed that even earlier, in 1486, Spanish mathematician Valmes was burned at the stake for claiming to have solved the quartic equation. Inquisitor General Tomás de Torquemada allegedly told Valmes that it was the will of God that such a solution be inaccessible to human understanding. However Beckmann, who popularized this story of Depman in the West, said that it was unreliable and hinted that it may have been invented as Soviet antireligious propaganda. Beckmann's version of this story has been widely copied in several books and internet sites, usually without his reservations and sometimes with fanciful embellishments. Several attempts to find corroborating evidence for this story, or even for the existence of Valmes, have failed.

The proof that four is the highest degree of a general polynomial for which such solutions can be found was first given in the Abel–Ruffini theorem in 1824, proving that all attempts at solving the higher order polynomials would be futile. The notes left by Évariste Galois prior to dying in a duel in 1832 later led to an elegant complete theory of the roots of polynomials, of which this theorem was one result.

Applications

Each coordinate of the intersection points of two conic sections is a solution of a quartic equation. The same is true for the intersection of a line and a torus. It follows that quartic equations often arise in computational geometry and all related fields such as computer graphics, computer-aided design, computer-aided manufacturing and optics. Here are example of other geometric problems whose solution amounts of solving a quartic equation.

In computer-aided manufacturing, the torus is a shape that is commonly associated with the end-mill cutter. To calculate its location relative to a triangulated surface, the position of a horizontal torus on the z-axis must be found where it is tangent to a fixed line, and this requires the solution of a general quartic equation to be calculated.

A quartic equation arises also in the process of solving the crossed ladders problem, in which the lengths of two crossed ladders, each based against one wall and leaning against another, are given along with the height at which they cross, and the distance between the walls is to be found.

In optics, Alhazen's problem is *"Given a light source and a spherical mirror, find the point on the mirror where the light will be reflected to the eye of an observer."* This leads to a quartic equation.

Finding the distance of closest approach of two ellipses involves solving a quartic equation.

The eigenvalues of a 4×4 matrix are the roots of a quartic polynomial which is the characteristic polynomial of the matrix.

The characteristic equation of a fourth-order linear difference equation or differential equation is a quartic equation. An example arises in the Timoshenko-Rayleigh theory of beam bending.

Intersections between spheres, cylinders, or other quadrics can be found using quartic equations.

Inflection Points and Golden Ratio

Letting F and G be the distinct inflection points of a quartic, and letting H be the intersection of the inflection secant line FG and the quartic, nearer to G than to F, then G divides FH into the golden section:

$$\frac{FG}{GH} = \frac{1+\sqrt{5}}{2} = \text{the golden ratio.}$$

Moreover, the area of the region between the secant line and the quartic below the secant line equals the area of the region between the secant line and the quartic above the secant line. One of those regions is disjointed into sub-regions of equal area.

Solving a Quartic Equation

Nature of the Roots

Given the general quartic equation

$$ax^4 + bx^3 + cx^2 + dx + e = 0$$

with real coefficients and $a \neq 0$ the nature of its roots is mainly determined by the sign of its discriminant

$$\Delta = 256a^3e^3 - 192a^2bde^2 - 128a^2c^2e^2 + 144a^2cd^2e - 27a^2d^4$$
$$+144ab^2ce^2 - 6ab^2d^2e - 80abc^2de + 18abcd^3 + 16ac^4e$$
$$-4ac^3d^2 - 27b^4e^2 + 18b^3cde - 4b^3d^3 - 4b^2c^3e + b^2c^2d^2$$

This may be refined by considering the signs of four other polynomials:

$$P = 8ac - 3b^2$$

such that $P/8a^2$ is the second degree coefficient of the associated depressed quartic;

$$Q = b^3 + 8da^2 - 4abc,$$

such that $Q/8a^3$ is the first degree coefficient of the associated depressed quartic;

$$\Delta_0 = c^2 - 3bd + 12ae,$$

which is 0 if the quartic has a triple root; and

$$D = 64a^3e - 16a^2c^2 + 16ab^2c - 16a^2bd - 3b^4$$

which is 0 if the quartic has two double roots.

The possible cases for the nature of the roots are as follows:

- If $\Delta < 0$ then the equation has two distinct real roots and two complex conjugate non-real roots.

- If $\Delta > 0$ then either the equation's four roots are all real or none is.

  - If $P < 0$ and $D < 0$ then all four roots are real and distinct.

  - If $P > 0$ or $D > 0$ then there are two pairs of non-real complex conjugate roots.

- If $\Delta = 0$ then (and only then) the polynomial has a multiple root. Here are the different cases that can occur:

  - If $P < 0$ and $D < 0$ and $\Delta_0 \neq 0$, there is a real double root and two real simple roots.

  - If $D > 0$ or ($P > 0$ and ($D \neq 0$ or $Q \neq 0$)), there is a real double root and two complex conjugate roots.

  - If $\Delta_0 = 0$ and $D \neq 0$, there is a triple root and a simple root, all real.

  - If $D = 0$, then:

    - If $P < 0$, there are two real double roots.

    - If $P > 0$ and $Q = 0$, there are two complex conjugate double roots.

    - If $\Delta_0 = 0$, all four roots are equal to $-b/4a$

There are some cases that do not seem to be covered, but they cannot occur. For example, $\Delta_0 > 0$, $P = 0$ and $D \leq 0$ is not one of the cases. However, if $\Delta_0 > 0$ and $P = 0$ then $D > 0$ so this combination is not possible.

General Formula for Roots

Quartic formula written out in full. This formula is too unwieldy for general use; hence other methods, or simpler formulas for special cases, are generally used.

The four roots x_1, x_2, x_3, and x_4 for the general quartic equation

$$ax^4 + bx^3 + cx^2 + dx + e = 0$$

with $a \neq 0$ are given in the following formula, which is deduced from the one in the section on Ferrari's method by back changing the variables and using the formulas for the quadratic and cubic equations.

$$x_{1,2} = -\frac{b}{4a} - S \pm \frac{1}{2}\sqrt{-4S^2 - 2p + \frac{q}{S}}$$

$$x_{3,4} = -\frac{b}{4a} + S \pm \frac{1}{2}\sqrt{-4S^2 - 2p - \frac{q}{S}}$$

where p and q are the coefficients of the second and of the first degree respectively in the associated depressed quartic

$$p = \frac{8ac - 3b^2}{8a^2}$$

$$q = \frac{b^3 - 4abc + 8a^2d}{8a^3}$$

and where

$$S = \frac{1}{2}\sqrt{-\frac{2}{3}p + \frac{1}{3a}\left(Q + \frac{\Delta_0}{Q}\right)}$$

$$Q = \sqrt[3]{\frac{\Delta_1 + \sqrt{\Delta_1^2 - 4\Delta_0^3}}{2}}$$

(if $S = 0$ or $Q = 0$)

with

$$\Delta_0 = c^2 - 3bd + 12ae$$

$$\Delta_1 = 2c^3 - 9bcd + 27b^2e + 27ad^2 - 72ace$$

and

$\Delta_1^2 - 4\Delta_0^3 = -27\Delta$, where Δ is the aforementioned discriminant. For the cube root expression for Q, any of the three cube roots in the complex plane can be used, although if one of them is real that is the natural and simplest one to choose. The mathematical expressions of these last four terms are very similar to those of their cubic counterparts.

Special Cases of the Formula

- If $\Delta > 0$, the value of Q is a non-real complex number. In this case, either all roots are non-real or they are all real. In the latter case, the value of S is also real, and one may prefer to express it in a purely real way, by using trigonometric functions, as follows:

$$S = \frac{1}{2}\sqrt{-\frac{2}{3}p + \frac{2}{3a}\sqrt{\Delta_0}\cos\frac{\phi}{3}}$$

where

$$\phi = \arccos\left(\frac{\Delta_1}{2\sqrt{\Delta_0^3}}\right).$$

- If $\Delta \neq 0$ and $\Delta_0 = 0$, the sign of $\sqrt{\Delta_1^2 - 4\Delta_0^3} = \sqrt{\Delta_1^2}$ has to be chosen to have $Q \neq 0$, that is one should define $\sqrt{\Delta_1^2}$ as Δ_1, maintaining the sign of Δ_1.

- If $S = 0$, then one must change the choice of the cube root in Q in order to have This is always possible except if the quartic may be factored into $\left(x + \frac{b}{4a}\right)^4$. The result is then correct, but misleading because it hides the fact that no cube root is needed in this case. In fact this case may occur only if the numerator of q is zero, in which case the associated depressed quartic is biquadratic; it may thus be solved by the method described below.

- If $\Delta = 0$ and $\Delta_0 = 0$, and thus also $\Delta_1 = 0$, at least three roots are equal, and the roots are rational functions of the coefficients.

- If $\Delta = 0$ and $\ddot{A}_0 \neq 0$, the above expression for the roots is correct but misleading, hiding the fact that the polynomial is reducible and no cube root is needed to represent the roots.

Simpler Cases

Reducible Quartics

Consider the general quartic

$$Q(x) = a_4 x^4 + a_3 x^3 + a_2 x^2 + a_1 x + a_0.$$

It is reducible if $Q(x) = R(x) \times S(x)$, where $R(x)$ and $S(x)$ are non-constant polynomials with rational coefficients (or more generally with coefficients in the same field as the coefficients of $Q(x)$). There are two ways to write such a factorization: either

$$Q(x) = (x - x_1)(b_3 x^3 + b_2 x^2 + b_1 x + b_0)$$

or

$$Q(x) = (c_2 x^2 + c_1 x + c_0)(d_2 x^2 + d_1 x + d_0).$$

In either case, the roots of $Q(x)$ are the roots of the factors, which may be computed by solving quadratic or cubic equations.

Detecting the existence of such factorizations can be done using the resolvent cubic of $Q(x)$. It turns out that:

- if we are working over R (or, more generally, over some real closed field) then there is always such a factorization;

- if we are working over Q, then there is an algorithm to determine whether or not $Q(x)$ is reducible and, if it is, how to express it as a product of polynomials of smaller degree.

In fact, several methods of solving quartic equations (Ferrari's method, Descartes' method, and, to a lesser extent, Euler's method) are based upon finding such factorizations.

Biquadratic Equation

If $a_3 = a_1 = 0$ then the biquadratic function

$$Q(x) = a_4x^4 + a_2x^2 + a_0$$

defines a biquadratic equation, which is easy to solve.

Let $z = x^2$. Then $Q(x)$ becomes a quadratic q in z: $q(z) = a_4z^2 + a_2z + a_0$. Let $z+$ and $z-$ be the roots of $q(z)$. Then the roots of our quartic $Q(x)$ are

$$x_1 = +\sqrt{z_+},$$
$$x_2 = -\sqrt{z_+},$$
$$x_3 = +\sqrt{z_-},$$
$$x_4 = -\sqrt{z_-}.$$

Quasi-palindromic Equation

The polynomial

$$P(x) = a_0x^4 + a_1x^3 + a_2x^2 + a_1mx + a_0m^2$$

is almost palindromic, as $P(mx) = x^4/m^2P(m/x)$ (it is palindromic if $m = 1$). The change of variables $z = x + m/x$ in $P(x)/x^2 = 0$ produces the quadratic equation $a_0z^2 + a_1z + a_2 - 2ma_0 = 0$. Since $x^2 - xz + m = 0$, the quartic equation $P(x) = 0$ may be solved by applying the quadratic formula twice.

Solution Methods

Converting to a Depressed Quartic

For solving purposes, it is generally better to convert the quartic into a depressed quartic by the following simple change of variable. All formulas are simpler and some methods work only in this case. The roots of the original quartic are easily recovered from that of the depressed quartic by the reverse change of variable.

Let

$$a_4x^4 + a_3x^3 + a_2x^2 + a_1x + a_0 = 0$$

be the general quartic equation we want to solve.

Dividing by a_4, provides the equivalent equation $x^4 + bx^3 + cx^2 + dx + e = 0$, with $b = a_3/a_4$, $c = a_2/a_4$, $d = a_1/a_4$, and $e = a_0/a_4$. Substituting $y - b/4$ for x gives, after regrouping the terms, the equation $y^4 + py^2 + qy + r = 0$, where

$$p = \frac{8c - 3b^2}{8} = \frac{8a_2 a_4 - 3a_3^2}{8a_4^2}$$

$$q = \frac{b^3 - 4bc + 8d}{8} = \frac{a_3^3 - 4a_2 a_3 a_4 + 8a_1 a_4^2}{8a_4^3}$$

$$r = \frac{-3b^4 + 256e - 64bd + 16b^2 c}{256} = \frac{-3a_3^4 + 256a_0 a_4^3 - 64a_1 a_3 a_4^2 + 16a_2 a_3^2 a_4}{256a_4^4}.$$

If y_0 is a root of this depressed quartic, then $y_0 - b/4$ (that is $y_0 - a_3/4a_4$) is a root of the original quartic and every root of the original quartic can be obtained by this process.

Ferrari's Solution

As explained in the preceding section, we may start with the *depressed quartic equation*

$$y^4 + py^2 + qy + r = 0.$$

This depressed quartic can be solved by means of a method discovered by Lodovico Ferrari. The depressed equation may be rewritten (this is easily verified by expanding the square and regrouping all terms in the left-hand side) as

$$\left(y^2 + \frac{p}{2} \right)^2 = -qy - r + \frac{p^2}{4}.$$

Then, we introduce a variable m into the factor on the left-hand side by adding $2y^2 m + pm + m^2$ to both sides. After regrouping the coefficients of the power of y in the right-hand side, this gives the equation

$$\left(y^2 + \frac{p}{2} + m \right)^2 = 2my^2 - qy + m^2 + mp + \frac{p^2}{4} - r, \tag{1}$$

which is equivalent to the original equation, whichever value is given to m.

As the value of m may be arbitrarily chosen, we will choose it in order to get a perfect square in the right-hand side. This implies that the discriminant in y of this quadratic equation is zero, that is m is a root of the equation

$$(-q)^2 - 4(2m)\left(m^2 + pm + \frac{p^2}{4} - r \right) = 0,$$

which may be rewritten as

$$8m^3 + 8pm^2 + (2p^2 - 8r)m - q^2 = 0.$$

This is the resolvent cubic of the quartic equation. The value of m may thus be obtained from Cardano's formula. When m is a root of this equation, the right-hand side of equation (1) is the square

$$\left(\sqrt{2m}\,y - \frac{q}{2\sqrt{2m}} \right)^2 .$$

However, this induces a division by zero if $m = 0$. This implies $q = 0$, and thus that the depressed equation is bi-quadratic, and may be solved by an easier method. This was not a problem at the time of Ferrari, when one solved only explicitly given equations with numeric coefficients. For a general formula that is always true, one thus needs to choose a root of the cubic equation such that $m \neq 0$. This is always possible except for the depressed equation $x^4 = 0$.

Now, if m is a root of the cubic equation such that $m \neq 0$, equation (1) becomes

$$\left(y^2 + \frac{p}{2} + m \right)^2 = \left(y\sqrt{2m} - \frac{q}{2\sqrt{2m}} \right)^2 .$$

This equation is of the form $M^2 = N^2$, which can be rearranged as $M^2 - N^2 = 0$ or $(M + N)(M - N) = 0$. Therefore, equation (1) may be rewritten as

$$\left(y^2 + \frac{p}{2} + m + \sqrt{2m}\,y - \frac{q}{2\sqrt{2m}} \right)\left(y^2 + \frac{p}{2} + m - \sqrt{2m}\,y + \frac{q}{2\sqrt{2m}} \right) = 0.$$

This equation is easily solved by applying to each factor the quadratic formula. Solving them we may write the four roots as

$$y = \frac{\pm_1 \sqrt{2m} \pm_2 \sqrt{-\left(2p + 2m \pm_1 \dfrac{\sqrt{2}q}{\sqrt{m}} \right)}}{2},$$

where $\pm_1$ and $\pm_2$ denote either $+$ or $-$. As the two occurrences of $\pm_1$ must denote the same sign, this leaves four possibilities, one for each root.

Therefore, the solutions of the original quartic equation are

$$x = -\frac{a_3}{4a_4} + \frac{\pm_1 \sqrt{2m} \pm_2 \sqrt{-\left(2p + 2m \pm_1 \dfrac{\sqrt{2}q}{\sqrt{m}} \right)}}{2}.$$

A comparison with the general formula above shows that $\sqrt{2m} = 2S$.

Descartes' Solution

Descartes introduced in 1637 the method of finding the roots of a quartic polynomial by factoring it into two quadratic ones. Let

$$x^4 + bx^3 + cx^2 + dx + e = (x^2 + sx + t)(x^2 + ux + v)$$
$$= x^4 + (s + u)x^3 + (t + v + su)x^2 + (sv + tu)x + tv$$

By equating coefficients, this results in the following system of equations:

$$\begin{cases} b = s + u \\ c = t + v + su \\ d = sv + tu \\ e = tv \end{cases}$$

This can be simplified by starting again with the depressed quartic $y^4 + py^2 + qy + r$, which can be obtained by substituting $y - b/4$ for x. Since the coefficient of y^3 is 0, we get $s = -u$, and:

$$\begin{cases} p + u^2 = t + v \\ q = u(t - v) \\ r = tv \end{cases}$$

One can now eliminate both t and v by doing the following:

$$u^2(p + u^2)^2 - q^2 = u^2(t + v)^2 - u^2(t - v)^2$$
$$= u^2[(t + v + (t - v))(t + v - (t - v))]$$
$$= u^2(2t)(2v)$$
$$= 4u^2 tv$$
$$= 4u^2 r$$

If we set $U = u^2$, then solving this equation becomes finding the roots of the resolvent cubic

$$U^3 + 2pU^2 + (p^2 - 4r)U - q^2, \qquad (2)$$

which is done elsewhere. Then, if u is a square root of a non-zero root of this resolvent (such a non-zero root exists except for the quartic x^4, which is trivially factored),

$$\begin{cases} s = -u \\ 2t = p + u^2 + q/u \\ 2v = p + u^2 - q/u \end{cases}$$

The symmetries in this solution are as follows. There are three roots of the cubic, corresponding to the three ways that a quartic can be factored into two quadratics, and choosing positive or negative values of u for the square root of U merely exchanges the two quadratics with one another.

The above solution shows that a quartic polynomial with rational coefficients and a zero coefficient on the cubic term is factorable into quadratics with rational coefficients if and only if either the resolvent cubic (2) has a non-zero root which is the square of a rational, or $p^2 - 4r$ is the square of rational and $q = 0$; this can readily be checked using the rational root test.

Euler's Solution

A variant of the previous method is due to Euler. Unlike the previous methods, both of which use *some* root of the resolvent cubic, Euler's method uses all of them. Let us consider a depressed quartic $x^4 + px^2 + qx + r$. Observe that, if

- $x^4 + px^2 + qx + r = (x^2 + sx + t)(x^2 - sx + v)$,

- r_1 and r_2 are the roots of $x^2 + sx + t$,

- r_3 and r_4 are the roots of $x^2 - sx + v$,

then

- the roots of $x^4 + px^2 + qx + r$ are r_1, r_2, r_3, and r_4,

- $r_1 + r_2 = -s$,

- $r_3 + r_4 = s$.

Therefore, $(r_1 + r_2)(r_3 + r_4) = -s^2$. In other words, $-(r_1 + r_2)(r_3 + r_4)$ is one of the roots of the resolvent cubic (2) and this suggests that the roots of that cubic are equal to $-(r_1 + r_2)(r_3 + r_4)$, $-(r_1 + r_3)(r_2 + r_4)$, and $-(r_1 + r_4)(r_2 + r_3)$. This is indeed true and it follows from Vieta's formulas. It also follows from Vieta's formulas, together with the fact that we are working with a depressed quartic, that $r_1 + r_2 + r_3 + r_4 = 0$. (Of course, this also follows from the fact that $r_1 + r_2 + r_3 + r_4 = -s + s$.) Therefore, if α, β, and γ are the roots of the resolvent cubic, then the numbers r_1, r_2, r_3, and r_4 are such that

$$\begin{cases} r_1 + r_2 + r_3 + r_4 = 0 \\ (r_1 + r_2)(r_3 + r_4) = -\alpha \\ (r_1 + r_3)(r_2 + r_4) = -\beta \\ (r_1 + r_4)(r_2 + r_3) = -\gamma. \end{cases}$$

It is a consequence of the first two equations that $r_1 + r_2$ is a square root of $-\alpha$ and that $r_3 + r_4$ is the other square root of $-\alpha$. For the same reason,

- $r_1 + r_3$ is a square root of $-\beta$,

- $r_2 + r_4$ is the other square root of $-\beta$,

- $r_1 + r_4$ is a square root of $-\gamma$,

- $r_2 + r_3$ is the other square root of $-\gamma$.

Therefore, the numbers r_1, r_2, r_3, and r_4 are such that

$$\begin{cases} r_1 + r_2 + r_3 + r_4 = 0 \\ r_1 + r_2 = \sqrt{-\alpha} \\ r_1 + r_3 = \sqrt{-\beta} \\ r_1 + r_4 = \sqrt{-\gamma}; \end{cases}$$

the sign of the square roots will be dealt with below. The only solution of this system is:

$$\begin{cases} r_1 = \dfrac{\sqrt{-\alpha} + \sqrt{-\beta} + \sqrt{-\gamma}}{2} \\[2mm] r_2 = \dfrac{\sqrt{-\alpha} - \sqrt{-\beta} - \sqrt{-\gamma}}{2} \\[2mm] r_3 = \dfrac{-\sqrt{-\alpha} + \sqrt{-\beta} - \sqrt{-\gamma}}{2} \\[2mm] r_4 = \dfrac{-\sqrt{-\alpha} - \sqrt{-\beta} + \sqrt{-\gamma}}{2}. \end{cases}$$

Since, in general, there are two choices for each square root, it might look as if this provides 8 (= 2^3) choices for the set $\{r_1, r_2, r_3, r_4\}$, but, in fact, it provides no more than 2 such choices, because the consequence of replacing one of the square roots by the symmetric one is that the set $\{r_1, r_2, r_3, r_4\}$ becomes the set $\{-r_1, -r_2, -r_3, -r_4\}$.

In order to determine the right sign of the square roots, one simply chooses some square root for each of the numbers $-\alpha$, $-\beta$, and $-\gamma$ and uses them to compute the numbers r_1, r_2, r_3, and r_4 from the previous equalities. Then, one computes the number $\sqrt{-\alpha}\sqrt{-\beta}\sqrt{-\gamma}$. Note that, since α, β, and γ are the roots of (2), it is a consequence of Vieta's formulas that their product is equal to q^2 and therefore that $\sqrt{-\alpha}\sqrt{-\beta}\sqrt{-\gamma} = \pm q$. But a straightforward computation shows that

$$\sqrt{-\alpha}\sqrt{-\beta}\sqrt{-\gamma} = r_1 r_2 r_3 + r_1 r_2 r_4 + r_1 r_3 r_4 + r_2 r_3 r_4.$$

If this number is $-q$, then the choice of the square roots was a good one (again, by Vieta's formulas); otherwise, the roots of the polynomial will be $-r_1$, $-r_2$, $-r_3$, and $-r_4$, which are the numbers obtained if one of the square roots is replaced by the symmetric one (or, what amounts to the same thing, if each of the three square roots is replaced by the symmetric one).

This argument suggests another way of choosing the square roots:

- pick *any* square root $\sqrt{-\alpha}$ of $-\alpha$ and *any* square root $\sqrt{-\beta}$ of $-\beta$;

- *define* $\sqrt{-\gamma}$ as $-\dfrac{q}{\sqrt{-\alpha}\sqrt{-\beta}}$.

Of course, this will make no sense if α or β is equal to 0, but 0 is a root of (2) only when $q = 0$, that is, only when we are dealing with a biquadratic equation, in which case there is a much simpler approach.

Solving by Lagrange Resolvent

The symmetric group S_4 on four elements has the Klein four-group as a normal subgroup. This suggests using a resolvent cubic whose roots may be variously described as a discrete Fourier transform or a Hadamard matrix transform of the roots; see Lagrange resolvents for the general method. Denote by x_i, for i from 0 to 3, the four roots of $x^4 + bx^3 + cx^2 + dx + e$. If we set

$$s_0 = \tfrac{1}{2}(x_0 + x_1 + x_2 + x_3),$$
$$s_1 = \tfrac{1}{2}(x_0 - x_1 + x_2 - x_3),$$
$$s_2 = \tfrac{1}{2}(x_0 + x_1 - x_2 - x_3),$$
$$s_3 = \tfrac{1}{2}(x_0 - x_1 - x_2 + x_3),$$

then since the transformation is an involution we may express the roots in terms of the four s_i in exactly the same way. Since we know the value $s_0 = -b/2$, we only need the values for s_1, s_2 and s_3. These are the roots of the polynomial

$$(s^2 - s_1^{\,2})(s^2 - s_2^{\,2})(s^2 - s_3^{\,2}).$$

Substituting the s_i by their values in term of the x_i, this polynomial may be expanded in a polynomial in s whose coefficients are symmetric polynomials in the x_i. By the fundamental theorem of symmetric polynomials, these coefficients may be expressed as polynomials in the coefficients of the monic quartic. If, for simplification, we suppose that the quartic is depressed, that is $b = 0$, this results in the polynomial

$$s^6 + 2cs^4 + (c^2 - 4e)s^2 - d^2 \qquad\qquad\qquad\qquad (3)$$

This polynomial is of degree six, but only of degree three in s^2, and so the corresponding equation is solvable by the method described in the article about cubic function. By substituting the roots in the expression of the x_i in terms of the s_i, we obtain expression for the roots. In fact we obtain, apparently, several expressions, depending on the numbering of the roots of the cubic polynomial and of the signs given to their square roots. All these different expressions may be deduced from one of them by simply changing the numbering of the x_i.

These expressions are unnecessarily complicated, involving the cubic roots of unity, which can be avoided as follows. If s is any non-zero root of (3), and if we set

$$F_1(x) = x^2 + sx + \frac{c}{2} + \frac{s^2}{2} - \frac{d}{2s}$$

$$F_2(x) = x^2 - sx + \frac{c}{2} + \frac{s^2}{2} + \frac{d}{2s}$$

then

$$F_1(x) \times F_2(x) = x^4 + cx^2 + dx + e.$$

We therefore can solve the quartic by solving for s and then solving for the roots of the two factors using the quadratic formula.

Note that this gives exactly the same formula for the roots as the one provided by Descartes' method.

Solving with Algebraic Geometry

There is an alternative solution using algebraic geometry In brief, one interprets the roots as the intersection of two quadratic curves, then finds the three reducible quadratic curves (pairs of lines) that pass through these points (this corresponds to the resolvent cubic, the pairs of lines being the Lagrange resolvents), and then use these linear equations to solve the quadratic.

The four roots of the depressed quartic $x^4 + px^2 + qx + r = 0$ may also be expressed as the x coordinates of the intersections of the two quadratic equations $y^2 + py + qx + r = 0$ and $y - x^2 = 0$ i.e., using the substitution $y = x^2$ that two quadratics intersect in four points is an instance of Bézout's theorem. Explicitly, the four points are $P_i := (x_i, x_i^2)$ for the four roots x_i of the quartic.

These four points are not collinear because they lie on the irreducible quadratic $y = x^2$ and thus there is a 1-parameter family of quadratics (a pencil of curves) passing through these points. Writing the projectivization of the two quadratics as quadratic forms in three variables:

$$F_1(X,Y,Z) := Y^2 + pYZ + qXZ + rZ^2,$$
$$F_2(X,Y,Z) := YZ - X^2$$

the pencil is given by the forms $\lambda F_1 + \mu F_2$ for any point $[\lambda, \mu]$ in the projective line — in other words, where λ and μ are not both zero, and multiplying a quadratic form by a constant does not change its quadratic curve of zeros.

This pencil contains three reducible quadratics, each corresponding to a pair of lines, each passing through two of the four points, which can be done $\binom{4}{2} = 6$ different ways. Denote these $Q_1 = L_{12} + L_{34}$, $Q_2 = L_{13} + L_{24}$, and $Q_3 = L_{14} + L_{23}$. Given any two of these, their intersection has exactly the four points.

The reducible quadratics, in turn, may be determined by expressing the quadratic form $\lambda F_1 + \mu F_2$ as a 3×3 matrix: reducible quadratics correspond to this matrix being singular, which is equivalent to its determinant being zero, and the determinant is a homogeneous degree three polynomial in λ and μ and corresponds to the resolvent cubic.

References

- Anglin, W. S.; Lambek, Joachim (1995), "Mathematics in the Renaissance", The Heritage of Thales, Springers, pp. 125–131, ISBN 978-0-387-94544-6 Ch. 24.

- Press, WH; Teukolsky, SA; Vetterling, WT; Flannery, BP (2007), "Section 5.6 Quadratic and Cubic Equations", Numerical Recipes: The Art of Scientific Computing (3rd ed.), New York: Cambridge University Press, ISBN 978-0-521-88068-8

- Felix Klein, Lectures on the Icosahedron and the Solution of Equations of the Fifth Degree, trans. George Gavin Morrice, Trübner & Co., 1888. ISBN 0-486-49528-0.

- Ian Stewart, Galois Theory 2nd Edition, Chapman and Hall, 1989. ISBN 0-412-34550-1. Discusses Galois Theory in general including a proof of insolvability of the general quintic.

- Descartes, René (1954) [1637], "Book III: On the construction of solid and supersolid problems", The Geometry of Rene Descartes with a facsimile of the first edition, Dover, ISBN 0-486-60068-8.

- van der Waerden, Bartel Leendert (1991), "The Galois theory: Equations of the second, third, and fourth degrees", Algebra, 1 (7th ed.), Springer-Verlag, ISBN 0-387-97424-5.

- Euler, Leonhard (1984) [1765], "Of a new method of resolving equations of the fourth degree", Elements of Algebra, Springer-Verlag, ISBN 978-1-4613-8511-0.

Algebraic Structure: An Integrated Study

Algebraic structures are sets of finitary operations defined on it that satisfies a list of axioms. Associative algebra is an algebraic structure with compatible functions of addition and multiplication. This chapter helps the reader in developing an in-depth understanding of algebraic structures.

Algebraic Structure

In mathematics, and more specifically in abstract algebra, an algebraic structure is a set (called carrier set or underlying set) with one or more finitary operations defined on it that satisfies a list of axioms.

Examples of algebraic structures include groups, rings, fields, and lattices. More complex structures can be defined by introducing multiple operations, different underlying sets, or by altering the defining axioms. Examples of more complex algebraic structures include vector spaces, modules, and algebras.

The properties of specific algebraic structures are studied in abstract algebra. The general theory of algebraic structures has been formalized in universal algebra. Category theory is used to study the relationships between two or more classes of algebraic structures, often of different kinds. For example, Galois theory studies the connection between certain fields and groups, algebraic structures of two different kinds.

Introduction

Addition and multiplication on numbers are the prototypical example of an operation that combines two elements of a set to produce a third. These operations obey several algebraic laws. For example, $a + (b + c) = (a + b) + c$ and $a(bc) = (ab)c$, both examples of the *associative law*. Also $a + b = b + a$, and $ab = ba$, the *commutative law*. Many systems studied by mathematicians have operations that obey some, but not necessarily all, of the laws of ordinary arithmetic. For example, rotations of objects in three-dimensional space can be combined by performing the first rotation and then applying the second rotation to the object in its new orientation. This operation on rotations obeys the associative law, but can fail the commutative law.

Mathematicians give names to sets with one or more operations that obey a particular collection of laws, and study them in the abstract as algebraic structures. When a new problem can be shown to follow the laws of one of these algebraic structures, all the work that has been done on that category in the past can be applied to the new problem.

In full generality, algebraic structures may involve an arbitrary number of sets and operations that can combine more than two elements (higher arity), but this article focuses on binary operations on one or two sets. The examples here are by no means a complete list, but they are meant to be a representative list and include the most common structures. Longer lists of algebraic structures may be found in the external links and within *Category:Algebraic structures*. Structures are listed in approximate order of increasing complexity.

Examples

One Set with Operations

Simple structures: no binary operation:

- Set: a degenerate algebraic structure S having no operations.

- Pointed set: S has one or more distinguished elements, often 0, 1, or both.

- Unary system: S and a single unary operation over S.

- Pointed unary system: a unary system with S a pointed set.

Group-like structures: one binary operation. The binary operation can be indicated by any symbol, or with no symbol (juxtaposition) as is done for ordinary multiplication of real numbers.

Magma or groupoid: S and a single binary operation over S.

- Semigroup: an associative magma.

- Monoid: a semigroup with identity.

- Group: a monoid with a unary operation (inverse), giving rise to inverse elements.

- Abelian group: a group whose binary operation is commutative.

- Semilattice: a semigroup whose operation is idempotent and commutative. The binary operation can be called either meet or join.

- Quasigroup: a magma obeying the latin square property. A quasigroup may also be represented using three binary operations.

- Loop: a quasigroup with identity.

- Ring-like structures or Ringoids: two binary operations, often called addition and multiplication, with multiplication distributing over addition.

- Semiring: a ringoid such that S is a monoid under each operation. Addition is typically assumed to be commutative and associative, and the monoid product is assumed to distribute over the addition on both sides, and the additive identity satisfies $0\,x = 0$ for all x.

- Near-ring: a semiring whose additive monoid is a (not necessarily abelian) group.

- Ring: a semiring whose additive monoid is an abelian group.

- Lie ring: a ringoid whose additive monoid is an abelian group, but whose multiplicative operation satisfies the Jacobi identity rather than associativity.

- Boolean ring: a commutative ring with idempotent multiplication operation.

- Field: a commutative ring which contains a multiplicative inverse for every nonzero element

- Kleene algebras: a semiring with idempotent addition and a unary operation, the Kleene star, satisfying additional properties.

- *-algebra: a ring with an additional unary operation (*) satisfying additional properties.

Lattice structures: two or more binary operations, including operations called meet and join, connected by the absorption law.

- Complete lattice: a lattice in which arbitrary meet and joins exist.

- Bounded lattice: a lattice with a greatest element and least element.

- Complemented lattice: a bounded lattice with a unary operation, complementation, denoted by postfix $^{\perp}$. The join of an element with its complement is the greatest element, and the meet of the two elements is the least element.

- Modular lattice: a lattice whose elements satisfy the additional *modular identity*.

- Distributive lattice: a lattice in which each of meet and join distributes over the other. Distributive lattices are modular, but the converse does not hold.

- Boolean algebra: a complemented distributive lattice. Either of meet or join can be defined in terms of the other and complementation. This can be shown to be equivalent with the ring-like structure of the same name above.

- Heyting algebra: a bounded distributive lattice with an added binary operation, relative pseudo-complement, denoted by infix $\rightarrow$, and governed by the axioms $x \rightarrow x = 1$, $x (x \rightarrow y) = x y$, $y (x \rightarrow y) = y$, $x \rightarrow (y z) = (x \rightarrow y) (x \rightarrow z)$.

Arithmetics: two binary operations, addition and multiplication. S is an infinite set. Arithmetics are pointed unary systems, whose unary operation is injective successor, and with distinguished element 0.

- Robinson arithmetic. Addition and multiplication are recursively defined by means of successor. 0 is the identity element for addition, and annihilates multiplication. Robinson arithmetic is listed here even though it is a variety, because of its closeness to Peano arithmetic.

- Peano arithmetic. Robinson arithmetic with an axiom schema of induction. Most ring and field axioms bearing on the properties of addition and multiplication are theorems of Peano arithmetic or of proper extensions thereof.

Two Sets with Operations

Module-like structures: composite systems involving two sets and employing at least two binary operations.

- Group with operators: a group G with a set Ω and a binary operation $\Omega \times G \rightarrow G$ satisfying certain axioms.

- Module: an Abelian group M and a ring R acting as operators on M. The members of R are sometimes called scalars, and the binary operation of *scalar multiplication* is a function $R \times M \to M$, which satisfies several axioms. Counting the ring operations these systems have at least three operations.

- Vector space: a module where the ring R is a division ring or field.

- Graded vector space: a vector space with a direct sum decomposition breaking the space into "grades".

- Quadratic space: a vector space V over a field F with a function from V into F satisfying certain properties. Every quadratic space is also an inner product space.

Algebra-like structures: composite system defined over two sets, a ring R and a R module M equipped with an operation called multiplication. This can be viewed as a system with five binary operations: two operations on R, two on M and one involving both R and M.

- Algebra over a ring (also *R-algebra*): a module over a commutative ring R, which also carries a multiplication operation that is compatible with the module structure. This includes distributivity over addition and linearity with respect to multiplication by elements of R. The theory of an algebra over a field is especially well developed.

- Associative algebra: an algebra over a ring such that the multiplication is associative.

- Nonassociative algebra: a module over a commutative ring, equipped with a ring multiplication operation that is not necessarily associative. Often associativity is replaced with a different identity, such as alternation, the Jacobi identity, or the Jordan identity.

- Coalgebra: a vector space with a "comultiplication" defined dually to that of associative algebras.

- Lie algebra: a special type of nonassociative algebra whose product satisfies the Jacobi identity.

- Lie coalgebra: a vector space with a "comultiplication" defined dually to that of Lie algebras.

- Graded algebra: a graded vector space with an algebra structure compatible with the grading. The idea is that if the grades of two elements a and b are known, then the grade of ab is known, and so the location of the product ab is determined in the decomposition.

- Inner product space: an F vector space V with a bilinear binary operation from $V \times V \to F$.

Four or more binary operations:

- Bialgebra: an associative algebra with a compatible coalgebra structure.

- Lie bialgebra: a Lie algebra with a compatible bialgebra structure.

- Hopf algebra: a bialgebra with a connection axiom (antipode).

- Clifford algebra: a graded associative algebra equipped with an exterior product from which may be derived several possible inner products. Exterior algebras and geometric algebras are special cases of this construction.

Hybrid Structures

Algebraic structures can also coexist with added structure of non-algebraic nature, such as partial order or a topology. The added structure must be compatible, in some sense, with the algebraic structure.

- Topological group: a group with a topology compatible with the group operation.

- Lie group: a topological group with a compatible smooth manifold structure.

- Ordered groups, ordered rings and ordered fields: each type of structure with a compatible partial order.

- Archimedean group: a linearly ordered group for which the Archimedean property holds.

- Topological vector space: a vector space whose M has a compatible topology.

- Normed vector space: a vector space with a compatible norm. If such a space is complete (as a metric space) then it is called a Banach space.

- Hilbert space: an inner product space over the real or complex numbers whose inner product gives rise to a Banach space structure.

- Vertex operator algebra

- Von Neumann algebra: a *-algebra of operators on a Hilbert space equipped with the weak operator topology.

Universal Algebra

Algebraic structures are defined through different configurations of axioms. Universal algebra abstractly studies such objects. One major dichotomy is between structures that are axiomatized entirely by *identities* and structures that are not. If all axioms defining a class of algebras are identities, then the class of objects is a variety.

Identities are equations formulated using only the operations the structure allows, and variables that are tacitly universally quantified over the relevant universe. Identities contain no connectives, existentially quantified variables, or relations of any kind other than the allowed operations. The study of varieties is an important part of universal algebra. An algebraic structure in a variety may be understood as the quotient algebra of term algebra (also called "absolutely free algebra") divided by the equivalence relations generated by a set of identities. So, a collection of functions with given signatures generate a free algebra, the term algebra T. Given a set of equational identities (the axioms), one may consider their symmetric, transitive closure E. The quotient algebra T/E is then the algebraic structure or variety. Thus, for example, groups have a signature containing two operators: the multiplication operator m, taking two arguments, and the inverse operator i, taking one argument, and the identity element e, a constant, which may be considered an operator that takes zero arguments. Given a (countable) set of variables x, y, z, etc. the term algebra is the collection of all possible terms involving m, i, e and the variables; so for example, $m(i(x), m(x,m(y,e)))$ would be an element of the term algebra. One of the axioms defining a group is the identity $m(x, i(x)) = e$; another is $m(x,e) = x$. The axioms can be represented as trees. These equations induce equivalence classes on the free algebra; the quotient algebra then has the algebraic structure of a group.

Several non-variety structures fail to be varieties, because either:

- It is necessary that 0 ≠ 1, 0 being the additive identity element and 1 being a multiplicative identity element, but this is a nonidentity;

- Structures such as fields have some axioms that hold only for nonzero members of S. For an algebraic structure to be a variety, its operations must be defined for *all* members of S; there can be no partial operations.

Structures whose axioms unavoidably include nonidentities are among the most important ones in mathematics, e.g., fields and division rings. Although structures with nonidentities retain an undoubted algebraic flavor, they suffer from defects varieties do not have. For example, the product of two fields is not a field.

Category Theory

Category theory is another tool for studying algebraic structures. A category is a collection of *objects* with associated *morphisms*. Every algebraic structure has its own notion of homomorphism, namely any function compatible with the operation(s) defining the structure. In this way, every algebraic structure gives rise to a category. For example, the category of groups has all groups as objects and all group homomorphisms as morphisms. This concrete category may be seen as a category of sets with added category-theoretic structure. Likewise, the category of topological groups (whose morphisms are the continuous group homomorphisms) is a category of topological spaces with extra structure. A forgetful functor between categories of algebraic structures "forgets" a part of a structure.

There are various concepts in category theory that try to capture the algebraic character of a context, for instance

- algebraic category

- essentially algebraic category

- presentable category

- locally presentable category

- monadic functors and categories

- universal property.

Different Meanings of "Structure"

In a slight abuse of notation, the word "structure" can also refer to just the operations on a structure, instead of the underlying set itself. For example, the sentence, "We have defined a ring *structure* on the set A," means that we have defined ring *operations* on the set A. For another example, the group $(\mathbb{Z}, +)$ can be seen as a set $\mathbb{Z}$ that is equipped with an *algebraic structure,* namely the *operation* $+$.

Associative Algebra

In mathematics, an associative algebra is an algebraic structure with compatible operations of addition, multiplication (assumed to be associative), and a scalar multiplication by elements in some field. The addition and multiplication operations together give A the structure of a ring; the addition and scalar multiplication operations together give A the structure of a vector space over K. In this article we will also use the term K-algebra to mean an associative algebra over the field K. A standard first example of a K-algebra is a ring of square matrices over a field K, with the usual matrix multiplication.

In this article associative algebras are assumed to have a multiplicative unit, denoted 1; they are sometimes called unital associative algebras for clarification. In some areas of mathematics this assumption is not made, and we will call such structures non-unital associative algebras. We will also assume that all rings are unital, and all ring homomorphisms are unital.

Many authors consider the more general concept of an associative algebra over a commutative ring R, instead of a field: An **R**-algebra is an R-module with an associative R-bilinear binary operation, which also contains a multiplicative identity. For examples of this concept, if S is any ring with center C, then S is an associative C-algebra.

Definition

Let R be a fixed commutative ring (so R could be a field). An associative **R**-algebra (or more simply, an **R**-algebra) is an additive abelian group A which has the structure of both a ring and an R-module in such a way that the scalar multiplication satisfies

$$r \cdot (xy) = (r \cdot x)y = x(r \cdot y)$$

for all $r \in R$ and $x, y \in A$. Furthermore, A is assumed to be unital, which is to say it contains an element 1 such that

$$1x = x = x1$$

for all $x \in A$. Note that such an element 1 must be unique.

In other words, A is an R-module together with (1) an R-bilinear map $A \times A \to A$, called the multiplication, and (2) the multiplicative identity, such that the multiplication is associative:

$$x(yz) = (xy)z$$

for all x, y, and z in A. (Technical note: the multiplicative identity is a datum, while associativity is a property. By the uniqueness of the multiplicative identity, "unitarity" is often treated like a property.) If one drops the requirement for the associativity, then one obtains a non-associative algebra.

If A itself is commutative (as a ring) then it is called a commutative **R**-algebra.

As a Monoid Object in the Category of Modules

The definition is equivalent to saying that a unital associative R-algebra is a monoid object in **R**-Mod (the monoidal category of R-modules). By definition, a ring is a monoid object in the category of abelian groups; thus, an associative algebra is obtained by replacing the category of abelian groups with the category of modules.

This reinterpretation of the definition is convenient for further generalization since one does not need to refer to elements of an algebra A explicitly. For example, the associativity can be expressed as follows. By the universal property of a tensor product of modules, the multiplication (the R-bilinear map) corresponds to a unique R-linear map

$$m : A \otimes_R A \to A.$$

The associativity then refers to the identity:

$$m \circ (\mathrm{id} \otimes m) = m \circ (m \otimes \mathrm{id}).$$

From Ring Homomorphisms

An associative algebra amounts to a ring homomorphism whose image lies in the center. Indeed, starting with a ring A and a ring homomorphism $\eta : R \to A$ whose image lies in the center of A, we can make A an R-algebra by defining

$$r \cdot x = \eta(r)x$$

for all $r \in R$ and $x \in A$. If A is an R-algebra, taking $x = 1$, the same formula in turn defines a ring homomorphism $\eta : R \to A$ whose image lies in the center.

If A is commutative then the center of A is equal to A, so that a commutative R-algebra can be defined simply as a homomorphism $\eta : R \to A$ of commutative rings.

The ring homomorphism η appearing in the above is often called a structure map. In the commutative case, one can consider the category whose objects are ring homomorphisms $R \to A$; i.e., commutative R-algebras and whose morphisms are ring homomorphisms $A \to A'$ that are under R; i.e., $R \to A \to A'$ is $R \to A'$ (i.e., the coslice category of the category of commutative rings under R.) The prime spectrum functor Spec then determines an anti-equivalence of this category to the category of affine schemes over Spec R. (How to weaken the commutativity assumption is a subject matter of noncommutative algebraic geometry and, more recently, of derived algebraic geometry.)

Algebra Homomorphisms

A homomorphism between two R-algebras is an R-linear ring homomorphism. Explicitly, $\phi : A_1 \to A_2$ is an associative algebra homomorphism if

$$\phi(r \cdot x) = r \cdot \phi(x)$$

$$\phi(x + y) = \phi(x) + \phi(y)$$

$$\phi(xy) = \phi(x)\phi(y)$$

$$\phi(1) = 1$$

The class of all R-algebras together with algebra homomorphisms between them form a category, sometimes denoted $\boldsymbol{R}$-Alg.

The subcategory of commutative R-algebras can be characterized as the coslice category $R/$CRing where CRing is the category of commutative rings.

Examples

The most basic example is a ring itself; it is an algebra over its center or any subring lying in the center. In particular, any commutative ring is an algebra over any of its subrings. Other examples abound both from algebra and other fields of mathematics.

Algebra

- Any ring A can be considered as a Z-algebra. The unique ring homomorphism from Z to A is determined by the fact that it must send 1 to the identity in A. Therefore, rings and Z-algebras are equivalent concepts, in the same way that abelian groups and Z-modules are equivalent.

- Any ring of characteristic n is a (Z/nZ)-algebra in the same way.

- Given an R-module M, the endomorphism ring of M, denoted $\text{End}_R(M)$ is an R-algebra by defining $(r \cdot \varphi)(x) = r \cdot \varphi(x)$.

- Any ring of matrices with coefficients in a commutative ring R forms an R-algebra under matrix addition and multiplication. This coincides with the previous example when M is a finitely-generated, free R-module.

- The square n-by-n matrices with entries from the field K form an associative algebra over K. In particular, the 2×2 real matrices form an associative algebra useful in plane mapping.

- The complex numbers form a 2-dimensional associative algebra over the real numbers.

- The quaternions form a 4-dimensional associative algebra over the reals (but not an algebra over the complex numbers, since the complex numbers are not in the center of the quaternions).

- The polynomials with real coefficients form an associative algebra over the reals.

- Every polynomial ring $R[x_1, ..., x_n]$ is a commutative R-algebra. In fact, this is the free commutative R-algebra on the set $\{x_1, ..., x_n\}$.

- The free R-algebra on a set E is an algebra of polynomials with coefficients in R and non-commuting indeterminates taken from the set E.

- The tensor algebra of an R-module is naturally an R-algebra. The same is true for quotients such as the exterior and symmetric algebras. Categorically speaking, the functor which maps an R-module to its tensor algebra is left adjoint to the functor which sends an R-algebra to its underlying R-module (forgetting the ring structure).

- Given a commutative ring R and any ring A the tensor product $R \otimes_Z A$ can be given the structure of an R-algebra by defining $r \cdot (s \otimes a) = (rs \otimes a)$. The functor which sends A to

$R \otimes_Z A$ is left adjoint to the functor which sends an R-algebra to its underlying ring (forgetting the module structure).

Representation theory

- The universal enveloping algebra of a Lie algebra is an associative algebra that can be used to study the given Lie algebra.

- If G is a group and R is a commutative ring, the set of all functions from G to R with finite support form an R-algebra with the convolution as multiplication. It is called the group algebra of G. The construction is the starting point for the application to the study of (discrete) groups.

- If G is an algebraic group (e.g., semisimple complex Lie group), then the coordinate ring of G is the Hopf algebra A corresponding to G. Many structures of G translate to those of A.

Analysis

- Given any Banach space X, the continuous linear operators $A : X \to X$ form an associative algebra (using composition of operators as multiplication); this is a Banach algebra.

- Given any topological space X, the continuous real- or complex-valued functions on X form a real or complex associative algebra; here the functions are added and multiplied pointwise.

- The set of semimartingales defined on the filtered probability space $(\Omega, F, (F_t)_{t \geq 0}, P)$ forms a ring under stochastic integration.

- The Weyl algebra

Geometry and combinatorics

- The Clifford algebras, which are useful in geometry and physics.

- Incidence algebras of locally finite partially ordered sets are associative algebras considered in combinatorics.

Constructions

Subalgebras

A subalgebra of an R-algebra A is a subset of A which is both a subring and a submodule of A. That is, it must be closed under addition, ring multiplication, scalar multiplication, and it must contain the identity element of A.

Quotient algebras

Let A be an R-algebra. Any ring-theoretic ideal I in A is automatically an R-module since $r \cdot x = (r1_A)x$. This gives the quotient ring A/I the structure of an R-module and, in fact, an R-algebra. It follows that any ring homomorphic image of A is also an R-algebra.

Direct products

The direct product of a family of R-algebras is the ring-theoretic direct product. This becomes an R-algebra with the obvious scalar multiplication.

Free products

One can form a free product of R-algebras in a manner similar to the free product of groups. The free product is the coproduct in the category of R-algebras.

Tensor products

The tensor product of two R-algebras is also an R-algebra in a natural way.

Coalgebras

An associative algebra over K is given by a K-vector space A endowed with a bilinear map $A \times A \to A$ having 2 inputs (multiplicator and multiplicand) and one output (product), as well as a morphism $K \to A$ identifying the scalar multiples of the multiplicative identity. If the bilinear map $A \times A \to A$ is reinterpreted as a linear map (i. e., morphism in the category of K-vector spaces) $A \otimes A \to A$ (by the universal property of the tensor product), then we can view an associative algebra over K as a K-vector space A endowed with two morphisms (one of the form $A \otimes A \to A$ and one of the form $K \to A$) satisfying certain conditions which boil down to the algebra axioms. These two morphisms can be dualized using categorial duality by reversing all arrows in the commutative diagrams which describe the algebra axioms; this defines the structure of a coalgebra.

There is also an abstract notion of F-coalgebra, where F is a functor. This is vaguely related to the notion of coalgebra discussed above.

Representations

A representation of an algebra A is an algebra homomorphism $\rho: A \to \text{End}(V)$ from A to the endomorphism algebra of some vector space (or module) V. The property of ρ being an algebra homomorphism means that ρ preserves the multiplicative operation (that is, $\rho(xy)=\rho(x)\rho(y)$ for all x and y in A), and that ρ sends the unity of A to the unity of $\text{End}(V)$ (that is, to the identity endomorphism of V).

If A and B are two algebras, and $\rho: A \to \text{End}(V)$ and $\tau: B \to \text{End}(W)$ are two representations, then there is a (canonical) representation $A \otimes B \to \text{End}(V \otimes W)$ of the tensor product algebra $A\,B$ on the vector space $V \otimes W$. However, there is no natural way of defining a tensor product of two representations of a single associative algebra in such a way that the result is still a representation of that same algebra (not of its tensor product with itself), without somehow imposing additional conditions. Here, by *tensor product of representations*, the usual meaning is intended: the result should be a linear representation of the same algebra on the product vector space. Imposing such additional structure typically leads to the idea of a Hopf algebra or a Lie algebra, as demonstrated below.

Motivation for a Hopf algebra

Consider, for example, two representations $\sigma: A \to \text{End}(V)$ and $\tau: A \to \text{End}(W)$. One might try to form a tensor product representation $\rho: x \mapsto \sigma(x) \otimes \tau(x)$ according to how it acts on the product vector space, so that

$$\rho(x)(v \otimes w) = (\sigma(x)(v)) \otimes (\tau(x)(w)).$$

However, such a map would not be linear, since one would have

$$\rho = (\sigma \otimes \tau)^\circ \Delta.$$

for $k \in K$. One can rescue this attempt and restore linearity by imposing additional structure, by defining an algebra homomorphism $\Delta: A \to A \otimes A$, and defining the tensor product representation as

$$x \mapsto \rho(x) = \sigma(x) \otimes \mathrm{Id}_W + \mathrm{Id}_V \otimes \tau(x)$$

Such a homomorphism Δ is called a comultiplication if it satisfies certain axioms. The resulting structure is called a bialgebra. To be consistent with the definitions of the associative algebra, the coalgebra must be co-associative, and, if the algebra is unital, then the co-algebra must be co-unital as well. A Hopf algebra is a bialgebra with an additional piece of structure (the so-called antipode), which allows not only to define the tensor product of two representations, but also the Hom module of two representations (again, similarly to how it is done in the representation theory of groups).

Motivation for a Lie Algebra

One can try to be more clever in defining a tensor product. Consider, for example,

$$x \mapsto \rho(x) = \sigma(x) \otimes \mathrm{Id}_W + \mathrm{Id}_V \otimes \tau(x)$$

so that the action on the tensor product space is given by

$$\rho(x)(v \otimes w) = (\sigma(x)v) \otimes w + v \otimes (\tau(x)w).$$

This map is clearly linear in x, and so it does not have the problem of the earlier definition. However, it fails to preserve multiplication:

$$\rho(xy) = \sigma(x)\sigma(y) \otimes \mathrm{Id}_W + \mathrm{Id}_V \otimes \tau(x)\tau(y).$$

But, in general, this does not equal

$$\rho(x)\rho(y) = \sigma(x)\sigma(y) \otimes \mathrm{Id}_W + \sigma(x) \otimes \tau(y) + \sigma(y) \otimes \tau(x) + \mathrm{Id}_V \otimes \tau(x)\tau(y).$$

This shows that this definition of a tensor product is too naive; the obvious fix is to define it such that it is antisymmetric, so that the middle two terms cancel. This leads to the concept of a Lie algebra.

Non-unital Algebras

Some authors use the term "associative algebra" to refer to structures with do not necessarily have a multiplicative identity, and hence consider homomorphisms which are not necessarily unital.

An example of a non-unital associative algebra is given by the set of all functions $f\colon \mathrm{R} \to \mathrm{R}$ whose limit as x nears infinity is zero.

Non-associative Algebra

A non-associative algebra (or distributive algebra) is an algebra over a field where the binary multiplication operation is not assumed to be associative. That is, an algebraic structure A is a non-associative algebra over a field K if it is a vector space over K and is equipped with a K-bilinear binary multiplication operation $A \times A \to A$ which may or may not be associative. Examples include Lie algebras, Jordan algebras, the octonions, and three-dimensional Euclidian space equipped with the cross product operation. Since it is not assumed that the multiplication is associative, using parentheses to indicate the order of multiplications is necessary. For example, the expressions $(ab)(cd)$, $(a(bc))d$ and $a(b(cd))$ may all yield different answers.

While this use of *non-associative* means that associativity is not assumed, it does not mean that associativity is disallowed. In other words, "non-associative" means "not necessarily associative", just as "noncommutative" means "not necessarily commutative" for noncommutative rings.

An algebra is *unital* or *unitary* if it has an identity element I with $Ix = x = xI$ for all x in the algebra. For example, the octonions are unital, but Lie algebras never are.

The nonassociative algebra structure of A may be studied by associating it with other associative algebras which are subalgebra of the full algebra of K-endomorphisms of A as a K-vector space. Two such are the derivation algebra and the (associative) enveloping algebra, the latter being in a sense "the smallest associative algebra containing A".

More generally, some authors consider the concept of a non-associative algebra over a commutative ring R: An R-module equipped with an R-bilinear binary multiplication operation. If a structure obeys all of the ring axioms apart from associativity (for example, any R-algebra), then it is naturally a $\mathbb{Z}$-algebra, so some authors refer to non-associative $\mathbb{Z}$-algebras as non-associative rings.

Algebras Satisfying Identities

Ring-like structures with two binary operations and no other restrictions are a broad class, one which is too general to study. For this reason, the best-known kinds of non-associative algebras satisfy identities which simplify multiplication somewhat. These include the following identities.

In the list, x, y and z denote arbitrary elements of an algebra.

- Associative: $(xy)z = x(yz)$.

- Commutative: $xy = yx$.

- Anticommutative: $xy = -yx$.

- Jacobi identity: $(xy)z + (yz)x + (zx)y = 0$.

- Jordan identity: $(xy)x^2 = x(yx^2)$.

- Power associative: For all x, any three nonnegative powers of x associate. That is if a, b and c are nonnegative powers of x, then $a(bc) = (ab)c$. This is equivalent to saying that $x^m x^n = x^{n+m}$ for all non-negative integers m and n.

- Alternative: $(xx)y = x(xy)$ and $(yx)x = y(xx)$.

- Flexible: $x(yx) = (xy)x$.

- Elastic: Flexible and $(xy)(xx) = x(y(xx))$, $x(xx)y = (xx)(xy)$.

These properties are related by

1. *associative* implies *alternative* implies *power associative*;

2. *associative* implies *Jordan identity* implies *power associative*;

3. Each of the properties *associative, commutative, anticommutative, Jordan identity*, and *Jacobi identity* individually imply *flexible*.

4. For a field with characteristic not two, being both commutative and anticommutative implies the algebra is just {0}.

Associator

The associator on A is the K-multilinear map $[\cdot,\cdot,\cdot]: A \times A \times A \rightarrow A$ given by

$$[x,y,z] = (xy)z - x(yz).$$

It measures the degree of nonassociativity of A, and can be used to conveniently express some possible identities satisfied by A.

- Associative: the associator is identically zero;

- Alternative: the associator is alternating, interchange of any two terms changes the sign;

- Flexible: $[x,y,x] = 0$;

- Jordan: $[x,y,x^2] = 0$..

The nucleus is the set of elements that associate with all others: that is, the n in A such that

$$[n,A,A] = [A,n,A] = [A,A,n] = \{0\}.$$

Examples

- Euclidean space $\mathbf{R}^3$ with multiplication given by the vector cross product is an example of an algebra which is anticommutative and not associative. The cross product also satisfies the Jacobi identity.

- Lie algebras are algebras satisfying anticommutativity and the Jacobi identity.

- Algebras of vector fields on a differentiable manifold (if K is R or the complex numbers C) or an algebraic variety (for general K);

- Jordan algebras are algebras which satisfy the commutative law and the Jordan identity.

- Every associative algebra gives rise to a Lie algebra by using the commutator as Lie bracket. In fact every Lie algebra can either be constructed this way, or is a subalgebra of a Lie algebra so constructed.

- Every associative algebra over a field of characteristic other than 2 gives rise to a Jordan algebra by defining a new multiplication $x*y = (1/2)(xy + yx)$. In contrast to the Lie algebra case, not every Jordan algebra can be constructed this way. Those that can are called *special*.

- Alternative algebras are algebras satisfying the alternative property. The most important examples of alternative algebras are the octonions (an algebra over the reals), and generalizations of the octonions over other fields. All associative algebras are alternative. Up to isomorphism, the only finite-dimensional real alternative, division algebras are the reals, complexes, quaternions and octonions.

- Power-associative algebras, are those algebras satisfying the power-associative identity. Examples include all associative algebras, all alternative algebras, Jordan algebras, and the sedenions.

- The hyperbolic quaternion algebra over R, which was an experimental algebra before the adoption of Minkowski space for special relativity.

More classes of algebras:

- Graded algebras. These include most of the algebras of interest to multilinear algebra, such as the tensor algebra, symmetric algebra, and exterior algebra over a given vector space. Graded algebras can be generalized to filtered algebras.

- Division algebras, in which multiplicative inverses exist. The finite-dimensional alternative division algebras over the field of real numbers have been classified. They are the real numbers (dimension 1), the complex numbers (dimension 2), the quaternions (dimension 4), and the octonions (dimension 8). The quaternions and octonions are not commutative. Of these algebras, all are associative except for the octonions.

- Quadratic algebras, which require that $xx = re + sx$, for some elements r and s in the ground field, and e a unit for the algebra. Examples include all finite-dimensional alternative algebras, and the algebra of real 2-by-2 matrices. Up to isomorphism the only alternative, quadratic real algebras without divisors of zero are the reals, complexes, quaternions, and octonions.

- The Cayley–Dickson algebras (where K is R), which begin with:

  - C (a commutative and associative algebra);

  - the quaternions H (an associative algebra);

  - the octonions (an alternative algebra);

  - the sedenions, and the infinite sequence of Cayley-Dickson algebras (power-associative algebras).

- The Poisson algebras are considered in geometric quantization. They carry two multiplications, turning them into commutative algebras and Lie algebras in different ways.

- Genetic algebras are non-associative algebras used in mathematical genetics.

- Triple systems

Properties

There are several properties that may be familiar from ring theory, or from associative algebras, which are not always true for non-associative algebras. Unlike the associative case, elements with a (two-sided) multiplicative inverse might also be a zero divisor. For example, all non-zero elements of the sedenions have a two-sided inverse, but some of them are also zero divisors.

Free Non-associative Algebra

The free non-associative algebra on a set X over a field K is defined as the algebra with basis consisting of all non-associative monomials, finite formal products of elements of X retaining parentheses. The product of monomials u, v is just $(u)(v)$. The algebra is unital if one takes the empty product as a monomial.

Kurosh proved that every subalgebra of a free non-associative algebra is free.

Associated Algebras

An algebra A over a field K is in particular a K-vector space and so one can consider the associative algebra $\text{End}_K(A)$ of K-linear vector space endomorphism of A. We can associate to the algebra structure on A two subalgebras of $\text{End}_K(A)$, the derivation algebra and the (associative) enveloping algebra.

Derivation Algebra

A *derivation* on A is a map D with the property

$$D(x \cdot y) = D(x) \cdot y + x \cdot D(y).$$

The derivations on A form a subspace $\text{Der}_K(A)$ in $\text{End}_K(A)$. The commutator of two derivations is again a derivation, so that the Lie bracket gives $\text{Der}_K(A)$ a structure of Lie algebra.

Enveloping Algebra

There are linear maps L and R attached to each element a of an algebra A:

$L(a) : x \mapsto ax; \ \ R(a) : x \mapsto xa$.

The *associative enveloping algebra* or *multiplication algebra* of A is the associative algebra generated by the left and right linear maps. The *centroid* of A is the centraliser of the enveloping algebra in the endomorphism algebra $\text{End}_K(A)$. An algebra is *central* if its centroid consists of the K-scalar multiples of the identity.

Some of the possible identities satisfied by non-associative algebras may be conveniently expressed in terms of the linear maps:

- Commutative: each $L(a)$ is equal to the corresponding $R(a)$;

- Associative: any L commutes with any R;

- Flexible: every $L(a)$ commutes with the corresponding $R(a)$;

- Jordan: every $L(a)$ commutes with $R(a^2)$;

- Alternative: every $L(a)^2 = L(a^2)$ and similarly for the right.

The *quadratic representation* Q is defined by

$$Q(a): x \mapsto 2a \cdot (a \cdot x) - (a \cdot a) \cdot x$$

or equivalently

$$Q(a) = 2L^2(a) - L(a^2).$$

The article on universal enveloping algebras describes the canonical construction of enveloping algebras, as well as the PBW-type theorems for them. For Lie algebras, such enveloping algebras have a universal property, which does not hold, in general, for non-associative algebras. The best-known example is, perhaps the Albert algebra, an exceptional Jordan algebra that is not enveloped by the canonical construction of the enveloping algebra for Jordan algebras.

Magma (Algebra)

In abstract algebra, a magma is a basic kind of algebraic structure. Specifically, a magma consists of a set, M, equipped with a single binary operation, $M \times M \to M$. The binary operation must be closed by definition but no other properties are imposed.

History and Terminology

The term *groupoid* was introduced in 1926 by Heinrich Brandt describing his Brandt groupoid (translated from the German *Gruppoid*). The term was then appropriated by B. A. Hausmann and Øystein Ore (1937) in the sense (of a set with a binary operation) used in this article. In a couple of reviews of subsequent papers in Zentralblatt, Brandt strongly disagreed with this overloading of terminology. The Brandt groupoid is a groupoid in the sense used in category theory, but not in the sense used by Hausmann and Ore. Nevertheless, influential books in semigroup theory, including Clifford and Preston (1961) and Howie (1995) use groupoid in the sense of Hausmann and Ore. Hollings (2014) writes that the term *groupoid* is "perhaps most often used in modern mathematics" in the sense given to it in category theory.

According to Bergman and Hausknecht (1996): "There is no generally accepted word for a set with a not necessarily associative binary operation. The word *groupoid* is used by many universal algebraists, but workers in category theory and related areas object strongly to this usage because they use the same word to mean "category in which all morphisms are invertible". The term *magma* was used by Serre [Lie Algebras and Lie Groups, 1965]." It also appears in Bourbaki's *Éléments de mathématique*, Algèbre, chapitres 1 à 3, 1970.

Definition

A magma is a set M matched with an operation, •, that sends any two elements a, $b \in M$ to another element, $a • b$. The symbol, •, is a general placeholder for a properly defined operation. To qualify as a magma, the set and operation $(M, •)$ must satisfy the following requirement (known as the *magma or closure axiom*):

For all a, b in M, the result of the operation $a • b$ is also in M.

And in mathematical notation:

$$\forall\, a, b \in M: a • b \in M.$$

If • is instead a partial operation, then S is called a partial magma or more often a partial groupoid.

Morphism of Magmas

A morphism of magmas is a function, $f : M \to N$, mapping magma M to magma N, that preserves the binary operation:

$$f(x •_M y) = f(x) •_N f(y)$$

where $•_M$ and $•_N$ denote the binary operation on M and N respectively.

Notation and Combinatorics

The magma operation may be applied repeatedly, and in the general, non-associative case, the order matters, which is notated with parentheses. Also, the operation, •, is often omitted and notated by juxtaposition:

$$(a • (b • c)) • d = (a(bc))d$$

A shorthand is often used to reduce the number of parentheses, in which the innermost operations and pairs of parentheses are omitted, being replaced just with juxtaposition, $xy • z = (x • y) • z$. For example, the above is abbreviated to the following expression, still containing parentheses:

$$(a • bc)d.$$

A way to avoid completely the use of parentheses is prefix notation, in which the same expression would be written ••$a•bcd$.

The set of all possible strings consisting of symbols denoting elements of the magma, and sets of balanced parentheses is called the Dyck language. The total number of different ways of writing n applications of the magma operator is given by the Catalan number, C_n. Thus, for example, $C_2 = 2$, which is just the statement that $(ab)c$ and $a(bc)$ are the only two ways of pairing three elements of a magma with two operations. Less trivially, $C_3 = 5$: $((ab)c)d$, $(a(bc))d$, $(ab)(cd)$, $a((bc)d)$, and $a(b(cd))$.

The number of non-isomorphic magmas having 0, 1, 2, 3, 4, ... elements are 1, 1, 10, 3330, 178981952, ... (sequence A001329 in the OEIS). The corresponding numbers of non-isomorphic and non-antiisomorphic magmas are 1, 1, 7, 1734, 89521056, ... (sequence A001424 in the OEIS).

Free Magma

A free magma, M_X, on a set, X, is the "most general possible" magma generated by X (i.e., there are no relations or axioms imposed on the generators; see free object). It can be described as the set of non-associative words on X with parentheses retained.

It can also be viewed, in terms familiar in computer science, as the magma of binary trees with leaves labelled by elements of X. The operation is that of joining trees at the root. It therefore has a foundational role in syntax.

A free magma has the universal property such that, if $f : X \rightarrow N$ is a function from X to any magma, N, then there is a unique extension of f to a morphism of magmas, f'

$$f' : M_X \rightarrow N.$$

Types of Magmas

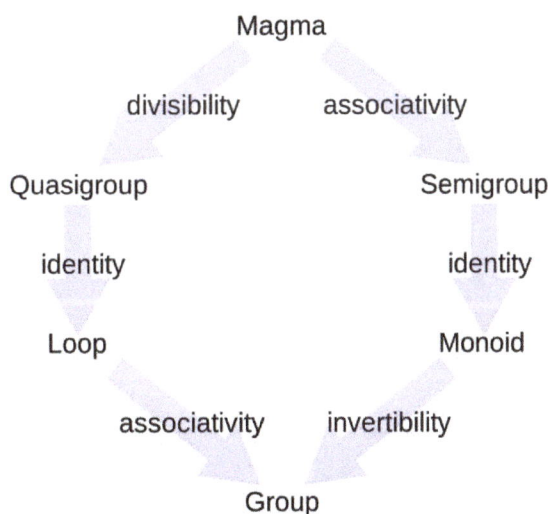

Magma

divisibility   associativity

Quasigroup   Semigroup

identity   identity

Loop   Monoid

associativity   invertibility

Group

Magmas are not often studied as such; instead there are several different kinds of magmas, depending on what axioms one might require of the operation. Commonly studied types of magmas include:

Quasigroups

Magmas where division is always possible

Loops

Quasigroups with identity elements

AG-groupoids: Magmas satisfy left invertive law

AG-monoids

> AG-groupoid with left identity

AG-groups

> AG-monoid with inverse elements

Semigroups

> Magmas where the operation is associative

Semilattices

> Semigroups where the operation is commutative and idempotent

Monoids

> Semigroups with identity elements

Groups

> Monoids with inverse elements, or equivalently, associative loops or non-empty associative quasigroups

Abelian groups

> Groups where the operation is commutative

Note that each of divisibility and invertibility imply the cancellation property.

Classification by Properties

A magma $(S, \bullet)$, with $x, y, u, z \in S$, is called

Medial

> If it satisfies the identity, $xy \bullet uz \equiv xu \bullet yz$

Left semimedial

> If it satisfies the identity, $xx \bullet yz \equiv xy \bullet xz$

Right semimedial

> If it satisfies the identity, $yz \bullet xx \equiv yx \bullet zx$

Semimedial

> If it is both left and right semimedial

Left distributive

If it satisfies the identity, $x \cdot yz \equiv xy \cdot xz$

Right distributive

If it satisfies the identity, $yz \cdot x \equiv yx \cdot zx$

Autodistributive

If it is both left and right distributive

Commutative

If it satisfies the identity, $xy \equiv yx$

Idempotent

If it satisfies the identity, $xx \equiv x$

Unipotent

If it satisfies the identity, $xx \equiv yy$

Zeropotent

If it satisfies the identities, $xx \cdot y \equiv xx \equiv y \cdot xx$

Alternative

If it satisfies the identities $xx \cdot y \equiv x \cdot xy$ and $x \cdot yy \equiv xy \cdot y$

Power-associative

If the submagma generated by any element is associative

A semigroup, or associative

If it satisfies the identity, $x \cdot yz \equiv xy \cdot z$

A left unar

If it satisfies the identity, $xy \equiv xz$

A right unar

If it satisfies the identity, $yx \equiv zx$

Semigroup with zero multiplication, or null semigroup

If it satisfies the identity, $xy \equiv uv$

Unital

If it has an identity element

Left-cancellative

If, for all x, y, and, z, $xy = xz$ implies $y = z$

Right-cancellative

If, for all x, y, and, z, $yx = zx$ implies $y = z$

Cancellative

If it is both right-cancellative and left-cancellative

A semigroup with left zeros

If it is a semigroup and, for all x, the identity, $x \equiv xy$, holds

A semigroup with right zeros

If it is a semigroup and, for all x, the identity, $x \equiv yx$, holds

Trimedial

If any triple of (not necessarily distinct) elements generates a medial submagma

Entropic

If it is a homomorphic image of a medial cancellation magma.

Group (Mathematics)

The manipulations of this Rubik's Cube form the Rubik's Cube group.

In mathematics, a group is an algebraic structure consisting of a set of elements equipped with an operation that combines any two elements to form a third element. The operation satisfies four conditions called the group axioms, namely closure, associativity, identity and invertibility. One of the most familiar examples of a group is the set of integers together with the addition operation, but the abstract formalization of the group axioms, detached as it is from the concrete nature of any particular group and its operation, applies much more widely. It allows entities with highly diverse mathematical origins in abstract algebra and beyond to be handled in a flexible way while retaining their essential structural aspects. The ubiquity of groups in numerous areas within and outside mathematics makes them a central organizing principle of contemporary mathematics.

Groups share a fundamental kinship with the notion of symmetry. For example, a symmetry group encodes symmetry features of a geometrical object: the group consists of the set of transformations that leave the object unchanged and the operation of combining two such transformations by performing one after the other. Lie groups are the symmetry groups used in the Standard Model of particle physics; Poincaré groups, which are also Lie groups, can express the physical symmetry underlying special relativity; and point groups are used to help understand symmetry phenomena in molecular chemistry.

The concept of a group arose from the study of polynomial equations, starting with Évariste Galois in the 1830s. After contributions from other fields such as number theory and geometry, the group notion was generalized and firmly established around 1870. Modern group theory—an active mathematical discipline—studies groups in their own right.[a][>] To explore groups, mathematicians have devised various notions to break groups into smaller, better-understandable pieces, such as subgroups, quotient groups and simple groups. In addition to their abstract properties, group theorists also study the different ways in which a group can be expressed concretely, both from a point of view of representation theory (that is, through the representations of the group) and of computational group theory. A theory has been developed for finite groups, which culminated with the classification of finite simple groups, completed in 2004.[aa][>] Since the mid-1980s, geometric group theory, which studies finitely generated groups as geometric objects, has become a particularly active area in group theory.

Definition and Illustration

First Example: the Integers

One of the most familiar groups is the set of integers Z which consists of the numbers

..., −4, −3, −2, −1, 0, 1, 2, 3, 4, ..., together with addition.

The following properties of integer addition serve as a model for the abstract group axioms given in the definition below.

- For any two integers a and b, the sum $a + b$ is also an integer. That is, addition of integers always yields an integer. This property is known as *closure* under addition.

- For all integers a, b and c, $(a + b) + c = a + (b + c)$. Expressed in words, adding a to b first, and then adding the result to c gives the same final result as adding a to the sum of b and c, a property known as *associativity*.

- If a is any integer, then $0 + a = a + 0 = a$. Zero is called the *identity element* of addition because adding it to any integer returns the same integer.

- For every integer a, there is an integer b such that $a + b = b + a = 0$. The integer b is called the *inverse element* of the integer a and is denoted $-a$.

The integers, together with the operation +, form a mathematical object belonging to a broad class sharing similar structural aspects. To appropriately understand these structures as a collective, the following abstract definition is developed.

Definition

[T]he axioms for a group are short and natural... Yet somehow hidden behind these axioms is the monster simple group, a huge and extraordinary mathematical object, which appears to rely on numerous bizarre coincidences to exist. The axioms for groups give no obvious hint that anything like this exists.

Richard Borcherds in Mathematicians: An Outer View of the Inner World

A group is a set, G, together with an operation • (called the *group law* of G) that combines any two elements a and b to form another element, denoted $a \cdot b$ or ab. To qualify as a group, the set and operation, $(G, \cdot)$, must satisfy four requirements known as the *group axioms*:

Closure

> For all a, b in G, the result of the operation, $a \cdot b$, is also in G.[b][>]

Associativity

> For all a, b and c in G, $(a \cdot b) \cdot c = a \cdot (b \cdot c)$.

Identity element

> There exists an element e in G, such that for every element a in G, the equation $e \cdot a = a \cdot e = a$ holds. Such an element is unique, and thus one speaks of *the* identity element.

Inverse element

> For each a in G, there exists an element b in G, commonly denoted a^{-1} (or $-a$, if the operation is denoted "+"), such that $a \cdot b = b \cdot a = e$, where e is the identity element.

The result of an operation may depend on the order of the operands. In other words, the result of combining element a with element b need not yield the same result as combining element b with element a; the equation

$$a \cdot b = b \cdot a$$

may not always be true. This equation always holds in the group of integers under addition, because $a + b = b + a$ for any two integers (commutativity of addition). Groups for which the commutativity equation $a \cdot b = b \cdot a$ always holds are called *abelian groups* (in honor of Niels Henrik Abel). The symmetry group described in the following section is an example of a group that is not abelian.

The identity element of a group G is often written as 1 or 1_G, a notation inherited from the multiplicative identity. If a group is abelian, then one may choose to denote the group operation by + and the identity element by 0; in that case, the group is called an additive group. The identity element can also be written as *id*.

The set G is called the *underlying set* of the group $(G, \bullet)$. Often the group's underlying set G is used as a short name for the group $(G, \bullet)$. Along the same lines, shorthand expressions such as "a subset of the group G" or "an element of group G" are used when what is actually meant is "a subset of the underlying set G of the group $(G, \bullet)$" or "an element of the underlying set G of the group $(G, \bullet)$". Usually, it is clear from the context whether a symbol like G refers to a group or to an underlying set.

Second Example: a Symmetry Group

Two figures in the plane are congruent if one can be changed into the other using a combination of rotations, reflections, and translations. Any figure is congruent to itself. However, some figures are congruent to themselves in more than one way, and these extra congruences are called symmetries. A square has eight symmetries. These are:

The elements of the symmetry group of the square (D_4). Vertices are identified by color or number.

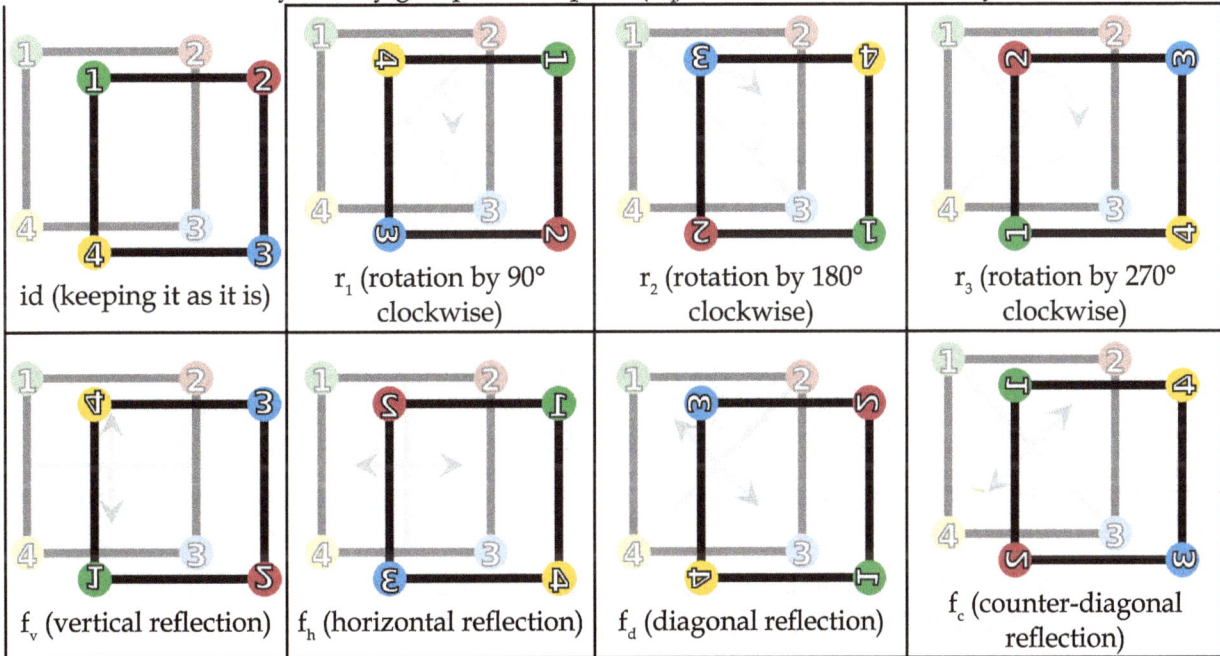

id (keeping it as it is)	r_1 (rotation by 90° clockwise)	r_2 (rotation by 180° clockwise)	r_3 (rotation by 270° clockwise)
f_v (vertical reflection)	f_h (horizontal reflection)	f_d (diagonal reflection)	f_c (counter-diagonal reflection)

- the identity operation leaving everything unchanged, denoted id;

- rotations of the square around its center by 90° clockwise, 180° clockwise, and 270° clockwise, denoted by r_1, r_2 and r_3, respectively;

- reflections about the vertical and horizontal middle line (f_h and f_v), or through the two diagonals (f_d and f_c).

These symmetries are represented by functions. Each of these functions sends a point in the square to the corresponding point under the symmetry. For example, r_1 sends a point to its rotation 90° clockwise around the square's center, and f_h sends a point to its reflection across the square's vertical middle line. Composing two of these symmetry functions gives another symmetry function. These symmetries determine a group called the dihedral group of degree 4 and denoted D_4. The underlying set of the group is the above set of symmetry functions, and the group operation is function composition. Two symmetries are combined by composing them as functions, that is, applying the first one to the square, and the second one to the result of the first application. The result of performing first a and then b is written symbolically *from right to left* as

$b \cdot a$ ("apply the symmetry b after performing the symmetry a").

The right-to-left notation is the same notation that is used for composition of functions.

The group table on the right lists the results of all such compositions possible. For example, rotating by 270° clockwise (r_3) and then reflecting horizontally (f_h) is the same as performing a reflection along the diagonal (f_d). Using the above symbols, highlighted in blue in the group table:

$f_h \cdot r_3 = f_d.$

Group table of D_4								
$\cdot$	id	r_1	r_2	r_3	f_v	f_h	f_d	f_c
id	id	r_1	r_2	r_3	f_v	f_h	f_d	f_c
r_1	r_1	r_2	r_3	id	f_c	f_d	f_v	f_h
r_2	r_2	r_3	id	r_1	f_h	f_v	f_c	f_d
r_3	r_3	id	r_1	r_2	f_d	f_c	f_h	f_v
f_v	f_v	f_d	f_h	f_c	id	r_2	r_1	r_3
f_h	f_h	f_c	f_v	f_d	r_2	id	r_3	r_1
f_d	f_d	f_h	f_c	f_v	r_3	r_1	id	r_2
f_c	f_c	f_v	f_d	f_h	r_1	r_3	r_2	id

The elements id, r_1, r_2, and r_3 form a subgroup, highlighted in red (upper left region). A left and right coset of this subgroup is highlighted in green (in the last row) and yellow (last column), respectively.

Given this set of symmetries and the described operation, the group axioms can be understood as follows:

1. The closure axiom demands that the composition $b \cdot a$ of any two symmetries a and b is also a symmetry. Another example for the group operation is

$r_3 \cdot f_h = f_c,$

i.e. rotating 270° clockwise after reflecting horizontally equals reflecting along the counter-diagonal (f_c). Indeed every other combination of two symmetries still gives a symmetry, as can be checked using the group table.

2. The associativity constraint deals with composing more than two symmetries: Starting with three elements a, b and c of D_4, there are two possible ways of using these three symmetries in this order to determine a symmetry of the square. One of these ways is to first compose a and b into a single symmetry, then to compose that symmetry with c. The other way is to first compose b and c, then to compose the resulting symmetry with a. The associativity condition

$(a \cdot b) \cdot c = a \cdot (b \cdot c)$

means that these two ways are the same, i.e., a product of many group elements can be simplified in any grouping. For example, $(f_d \bullet f_v) \bullet r_2 = f_d \bullet (f_v \bullet r_2)$ can be checked using the group table at the right

| $(f_d \bullet f_v) \bullet r_2$ | = | $r_3 \bullet r_2$ | = | r_1, which equals |
| $f_d \bullet (f_v \bullet r_2)$ | = | $f_d \bullet f_h$ | = | r_1. |

While associativity is true for the symmetries of the square and addition of numbers, it is not true for all operations. For instance, subtraction of numbers is not associative: $(7 - 3) - 2 = 2$ is not the same as $7 - (3 - 2) = 6$.

3. The identity element is the symmetry id leaving everything unchanged: for any symmetry a, performing id after a (or a after id) equals a, in symbolic form,

    $$\text{id} \bullet a = a,$$

    $$a \bullet \text{id} = a.$$

4. An inverse element undoes the transformation of some other element. Every symmetry can be undone: each of the following transformations—identity id, the reflections f_h, f_v, f_d, f_c and the 180° rotation r_2—is its own inverse, because performing it twice brings the square back to its original orientation. The rotations r_3 and r_1 are each other's inverses, because rotating 90° and then rotation 270° (or vice versa) yields a rotation over 360° which leaves the square unchanged. In symbols,

    $$f_h \bullet f_h = \text{id},$$

    $$r_3 \bullet r_1 = r_1 \bullet r_3 = \text{id}.$$

In contrast to the group of integers above, where the order of the operation is irrelevant, it does matter in D_4: $f_h \bullet r_1 = f_c$ but $r_1 \bullet f_h = f_d$. In other words, D_4 is not abelian, which makes the group structure more difficult than the integers introduced first.

History

The modern concept of an abstract group developed out of several fields of mathematics. The original motivation for group theory was the quest for solutions of polynomial equations of degree higher than 4. The 19th-century French mathematician Évariste Galois, extending prior work of Paolo Ruffini and Joseph-Louis Lagrange, gave a criterion for the solvability of a particular polynomial equation in terms of the symmetry group of its roots (solutions). The elements of such a Galois group correspond to certain permutations of the roots. At first, Galois' ideas were rejected by his contemporaries, and published only posthumously. More general permutation groups were investigated in particular by Augustin Louis Cauchy. Arthur Cayley's *On the theory of groups, as depending on the symbolic equation $\theta^n = 1$* (1854) gives the first abstract definition of a finite group.

Geometry was a second field in which groups were used systematically, especially symmetry groups as part of Felix Klein's 1872 Erlangen program. After novel geometries such as hyperbolic and projective geometry had emerged, Klein used group theory to organize them in a more coherent way. Further advancing these ideas, Sophus Lie founded the study of Lie groups in 1884.

The third field contributing to group theory was number theory. Certain abelian group structures had been used implicitly in Carl Friedrich Gauss' number-theoretical work *Disquisitiones Arithmeticae* (1798), and more explicitly by Leopold Kronecker. In 1847, Ernst Kummer made early attempts to prove Fermat's Last Theorem by developing groups describing factorization into prime numbers.

The convergence of these various sources into a uniform theory of groups started with Camille Jordan's *Traité des substitutions et des équations algébriques* (1870). Walther von Dyck (1882) introduced the idea of specifying a group by means of generators and relations, and was also the first to give an axiomatic definition of an "abstract group", in the terminology of the time. As of the 20th century, groups gained wide recognition by the pioneering work of Ferdinand Georg Frobenius and William Burnside, who worked on representation theory of finite groups, Richard Brauer's modular representation theory and Issai Schur's papers. The theory of Lie groups, and more generally locally compact groups was studied by Hermann Weyl, Élie Cartan and many others. Its algebraic counterpart, the theory of algebraic groups, was first shaped by Claude Chevalley (from the late 1930s) and later by the work of Armand Borel and Jacques Tits.

The University of Chicago's 1960–61 Group Theory Year brought together group theorists such as Daniel Gorenstein, John G. Thompson and Walter Feit, laying the foundation of a collaboration that, with input from numerous other mathematicians, led to the classification of finite simple groups, with the final step taken by Aschbacher and Smith in 2004. This project exceeded previous mathematical endeavours by its sheer size, in both length of proof and number of researchers. Research is ongoing to simplify the proof of this classification. These days, group theory is still a highly active mathematical branch, impacting many other fields.

Elementary Consequences of the Group Axioms

Basic facts about all groups that can be obtained directly from the group axioms are commonly subsumed under *elementary group theory*. For example, repeated applications of the associativity axiom show that the unambiguity of

$$a \bullet b \bullet c = (a \bullet b) \bullet c = a \bullet (b \bullet c)$$

generalizes to more than three factors. Because this implies that parentheses can be inserted anywhere within such a series of terms, parentheses are usually omitted.

The axioms may be weakened to assert only the existence of a left identity and left inverses. Both can be shown to be actually two-sided, so the resulting definition is equivalent to the one given above.

Uniqueness of Identity Element and Inverses

Two important consequences of the group axioms are the uniqueness of the identity element and the uniqueness of inverse elements. There can be only one identity element in a group, and each element in a group has exactly one inverse element. Thus, it is customary to speak of *the* identity, and *the* inverse of an element.

To prove the uniqueness of an inverse element of a, suppose that a has two inverses, denoted b and c, in a group $(G, \bullet)$. Then

b	=	b • e		as e is the identity element
	=	b • (a • c)		because c is an inverse of a, so e = a • c
	=	(b • a) • c		by associativity, which allows to rearrange the parentheses
	=	e • c		since b is an inverse of a, i.e. b • a = e
	=	c		for e is the identity element

The term b on the first line above and the c on the last are equal, since they are connected by a chain of equalities. In other words, there is only one inverse element of a. Similarly, to prove that the identity element of a group is unique, assume G is a group with two identity elements e and f. Then $e = e \bullet f = f$, hence e and f are equal.

Division

In groups, the existence of inverse elements implies that division is possible: given elements a and b of the group G, there is exactly one solution x in G to the equation $x \bullet a = b$, namely $b \bullet a^{-1}$. In fact, we have

$$(b \bullet a^{-1}) \bullet a = b \bullet (a^{-1} \bullet a) = b \bullet e = b.$$

Uniqueness results by multiplying the two sides of the equation $x \bullet a = b$ by a^{-1}. The element $b \bullet a^{-1}$, often denoted b / a, is called the *right quotient* of b by a, or the result of the *right division* of b by a.

Similarly there is exactly one solution y in G to the equation $a \bullet y = b$, namely $y = a^{-1} \bullet b$. This solution is the *left quotient* of b by a, and is sometimes denoted $a \setminus b$.

In general a / b and $a \setminus b$ may be different, but, if the group operation is commutative (that is, if the group is abelian), they are equal. In this case, the group operation is often denoted as an addition, and one talks of *subtraction* and *difference* instead of division and quotient.

A consequence of this is that multiplication by a group element g is a bijection. Specifically, if g is an element of the group G, the function (mathematics) from G to itself that maps $h \in G$ to $g \bullet h$ is a bijection. This function is called the *left translation* by g. Similarly, the *right translation* by g is the bijection from G to itself, that maps h to $h \bullet g$. If G is abelian, the left and the right translation by a group element are the same.

Basic Concepts

The following sections use mathematical symbols such as $X = \{x, y, z\}$ to denote a set X containing elements x, y, and z, or alternatively $x \in X$ to restate that x is an element of X. The notation $f: X \to Y$ means f is a function assigning to every element of X an element of Y.

To understand groups beyond the level of mere symbolic manipulations as above, more structural concepts have to be employed.[>] There is a conceptual principle underlying all of the following notions: to take advantage of the structure offered by groups (which sets, being "structureless", do not have), constructions related to groups have to be *compatible* with the group operation. This compatibility manifests itself in the following notions in various ways. For example, groups can be related to each other via functions called group homomorphisms. By the mentioned principle, they are required to respect the group structures in a precise sense. The structure of groups can also be understood by breaking them into pieces called subgroups and quotient groups. The principle of "preserving structures"—a recurring topic in mathematics throughout—is an instance of working in a category, in this case the category of groups.

Group Homomorphisms

Group homomorphisms[g[·]] are functions that preserve group structure. A function $a: G \to H$ between two groups $(G, \bullet)$ and $(H, *)$ is called a *homomorphism* if the equation

$$a(g \bullet k) = a(g) * a(k)$$

holds for all elements g, k in G. In other words, the result is the same when performing the group operation after or before applying the map a. This requirement ensures that $a(1_G) = 1_H$, and also $a(g)^{-1} = a(g^{-1})$ for all g in G. Thus a group homomorphism respects all the structure of G provided by the group axioms.

Two groups G and H are called *isomorphic* if there exist group homomorphisms $a: G \to H$ and $b: H \to G$, such that applying the two functions one after another in each of the two possible orders gives the identity functions of G and H. That is, $a(b(h)) = h$ and $b(a(g)) = g$ for any g in G and h in H. From an abstract point of view, isomorphic groups carry the same information. For example, proving that $g \bullet g = 1_G$ for some element g of G is equivalent to proving that $a(g) * a(g) = 1_H$, because applying a to the first equality yields the second, and applying b to the second gives back the first.

Subgroups

Informally, a *subgroup* is a group H contained within a bigger one, G. Concretely, the identity element of G is contained in H, and whenever h_1 and h_2 are in H, then so are $h_1 \bullet h_2$ and h_1^{-1}, so the elements of H, equipped with the group operation on G restricted to H, indeed form a group.

In the example above, the identity and the rotations constitute a subgroup $R = \{id, r_1, r_2, r_3\}$, highlighted in red in the group table above: any two rotations composed are still a rotation, and a rotation can be undone by (i.e. is inverse to) the complementary rotations 270° for 90°, 180° for 180°, and 90° for 270° (note that rotation in the opposite direction is not defined). The subgroup test is a necessary and sufficient condition for a nonempty subset H of a group G to be a subgroup: it is sufficient to check that $g^{-1}h \in H$ for all elements $g, h \in H$. Knowing the subgroups is important in understanding the group as a whole.

Given any subset S of a group G, the subgroup generated by S consists of products of elements of S and their inverses. It is the smallest subgroup of G containing S. In the introductory example above, the subgroup generated by r_2 and f_v consists of these two elements, the identity element id and $f_h = f_v \bullet r_2$. Again, this is a subgroup, because combining any two of these four elements or their inverses (which are, in this particular case, these same elements) yields an element of this subgroup.

Cosets

In many situations it is desirable to consider two group elements the same if they differ by an element of a given subgroup. For example, in D_4 above, once a reflection is performed, the square never gets back to the r_2 configuration by just applying the rotation operations (and no further reflections), i.e. the rotation operations are irrelevant to the question whether a reflection has been performed. Cosets are used to formalize this insight: a subgroup H defines left and right cosets, which can be thought of as translations of H by arbitrary group elements g. In symbolic terms, the *left* and *right* cosets of H containing g are

$$gH = \{g \bullet h : h \in H\} \text{ and } Hg = \{h \bullet g : h \in H\}, \text{ respectively.}$$

The left cosets of any subgroup H form a partition of G; that is, the union of all left cosets is equal to G and two left cosets are either equal or have an empty intersection. The first case $g_1 H = g_2 H$ happens precisely when $g_1^{-1} \bullet g_2 \in H$, i.e. if the two elements differ by an element of H. Similar considerations apply to the right cosets of H. The left and right cosets of H may or may not be equal. If they are, i.e. for all g in G, $gH = Hg$, then H is said to be a *normal subgroup*.

In D_4, the introductory symmetry group, the left cosets gR of the subgroup R consisting of the rotations are either equal to R, if g is an element of R itself, or otherwise equal to $U = f_c R = \{f_c, f_v, f_d, f_h\}$ (highlighted in green). The subgroup R is also normal, because $f_c R = U = Rf_c$ and similarly for any element other than f_c. (In fact, in the case of D_4, observe that all such cosets are equal, such that $f_h R = f_v R = f_d R = f_c R$.)

Quotient Groups

In some situations the set of cosets of a subgroup can be endowed with a group law, giving a *quotient group* or *factor group*. For this to be possible, the subgroup has to be normal. Given any normal subgroup N, the quotient group is defined by

$$G / N = \{gN, g \in G\}, \text{``}G \text{ modulo } N\text{''.}$$

This set inherits a group operation (sometimes called coset multiplication, or coset addition) from the original group G: $(gN) \bullet (hN) = (gh)N$ for all g and h in G. This definition is motivated by the idea (itself an instance of general structural considerations outlined above) that the map $G \to G / N$ that associates to any element g its coset gN be a group homomorphism, or by general abstract considerations called universal properties. The coset $eN = N$ serves as the identity in this group, and the inverse of gN in the quotient group is $(gN)^{-1} = (g^{-1})N$.[>]

•	R	U
R	R	U
U	U	R

Group table of the quotient group D_4 / R

The elements of the quotient group D_4 / R are R itself, which represents the identity, and $U = f_v R$. The group operation on the quotient is shown at the right. For example, $U \bullet U = f_v R \bullet f_v R = (f_v \bullet f_v)R = R$. Both the subgroup $R = \{\text{id}, r_1, r_2, r_3\}$, as well as the corresponding quotient are abelian, whereas D_4 is not abelian. Building bigger groups by smaller ones, such as D_4 from its subgroup R and the quotient D_4 / R is abstracted by a notion called semidirect product.

Quotient groups and subgroups together form a way of describing every group by its *presentation*: any group is the quotient of the free group over the *generators* of the group, quotiented by the subgroup of *relations*. The dihedral group D_4, for example, can be generated by two elements r and f (for example, $r = r_1$, the right rotation and $f = f_v$ the vertical (or any other) reflection), which means that every symmetry of the square is a finite composition of these two symmetries or their inverses. Together with the relations

$$r^4 = f^2 = (r \bullet f)^2 = 1,$$

the group is completely described. A presentation of a group can also be used to construct the Cayley graph, a device used to graphically capture discrete groups.

Sub- and quotient groups are related in the following way: a subset H of G can be seen as an injective map $H \to G$, i.e. any element of the target has at most one element that maps to it. The counterpart to injective maps are surjective maps (every element of the target is mapped onto), such as the canonical map $G \to G / N$.[y][>] Interpreting subgroup and quotients in light of these homomorphisms emphasizes the structural concept inherent to these definitions alluded to in the introduction. In general, homomorphisms are neither injective nor surjective. Kernel and image of group homomorphisms and the first isomorphism theorem address this phenomenon.

Examples and Applications

A periodic wallpaper pattern gives rise to a wallpaper group.

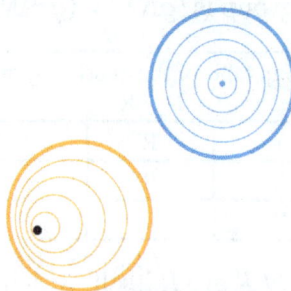

The fundamental group of a plane minus a point (bold) consists of loops around the missing point. This group is isomorphic to the integers.

Examples and applications of groups abound. A starting point is the group Z of integers with addition as group operation, introduced above. If instead of addition multiplication is considered, one obtains multiplicative groups. These groups are predecessors of important constructions in abstract algebra.

Groups are also applied in many other mathematical areas. Mathematical objects are often examined by associating groups to them and studying the properties of the corresponding groups. For example, Henri Poincaré founded what is now called algebraic topology by introducing the fundamental group. By means of this connection, topological properties such as proximity and continuity translate into properties of groups.[i] For example, elements of the fundamental group are represented by loops. The second image at the right shows some loops in a plane minus a point. The blue loop is considered null-homotopic (and thus irrelevant), because it can be continuously shrunk to a point. The presence of the hole prevents the orange loop from being shrunk to a point. The fundamental group of the plane with a point deleted turns out to be infinite cyclic, generated by the orange loop (or any other loop winding once around the hole). This way, the fundamental group detects the hole.

In more recent applications, the influence has also been reversed to motivate geometric constructions by a group-theoretical background.[j] In a similar vein, geometric group theory employs geometric concepts, for example in the study of hyperbolic groups. Further branches crucially applying groups include algebraic geometry and number theory.

In addition to the above theoretical applications, many practical applications of groups exist. Cryptography relies on the combination of the abstract group theory approach together with algorithmical knowledge obtained in computational group theory, in particular when implemented for finite groups. Applications of group theory are not restricted to mathematics; sciences such as physics, chemistry and computer science benefit from the concept.

Numbers

Many number systems, such as the integers and the rationals enjoy a naturally given group structure. In some cases, such as with the rationals, both addition and multiplication operations give rise to group structures. Such number systems are predecessors to more general algebraic structures known as rings and fields. Further abstract algebraic concepts such as modules, vector spaces and algebras also form groups.

Integers

The group of integers Z under addition, denoted (Z, +), has been described above. The integers, with the operation of multiplication instead of addition, (Z, ·) do *not* form a group. The closure, associativity and identity axioms are satisfied, but inverses do not exist: for example, $a = 2$ is an integer, but the only solution to the equation $a \cdot b = 1$ in this case is $b = 1/2$, which is a rational number, but not an integer. Hence not every element of Z has a (multiplicative) inverse.[k]

Rationals

The desire for the existence of multiplicative inverses suggests considering fractions

$$\frac{a}{b}.$$

Fractions of integers (with b nonzero) are known as rational numbers.[l] The set of all such fractions is commonly denoted Q. There is still a minor obstacle for (Q, ·), the rationals with multiplication, being a group: because the rational number 0 does not have a multiplicative inverse (i.e., there is no x such that $x \cdot 0 = 1$), (Q, ·) is still not a group.

However, the set of all *nonzero* rational numbers $Q \setminus \{0\} = \{q \in Q \mid q \neq 0\}$ does form an abelian group under multiplication, denoted $(Q \setminus \{0\}, \cdot)$.[m] Associativity and identity element axioms follow from the properties of integers. The closure requirement still holds true after removing zero, because the product of two nonzero rationals is never zero. Finally, the inverse of a/b is b/a, therefore the axiom of the inverse element is satisfied.

The rational numbers (including 0) also form a group under addition. Intertwining addition and multiplication operations yields more complicated structures called rings and—if division is possible, such as in Q—fields, which occupy a central position in abstract algebra. Group theoretic arguments therefore underlie parts of the theory of those entities.[n]

Modular Arithmetic

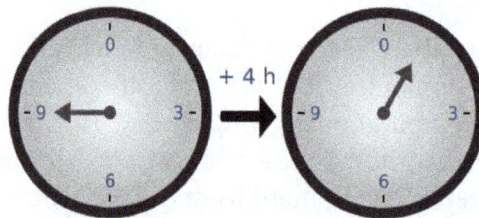

The hours on a clock form a group that uses addition modulo 12. Here 9 + 4 = 1.

In modular arithmetic, two integers are added and then the sum is divided by a positive integer called the *modulus*. The result of modular addition is the remainder of that division. For any modulus, n, the set of integers from 0 to $n - 1$ forms a group under modular addition: the inverse of any element a is $n - a$, and 0 is the identity element. This is familiar from the addition of hours on the face of a clock: if the hour hand is on 9 and is advanced 4 hours, it ends up on 1, as shown at the right. This is expressed by saying that 9 + 4 equals 1 "modulo 12" or, in symbols,

$$9 + 4 \equiv 1 \text{ modulo } 12.$$

The group of integers modulo n is written Z_n or Z/nZ.

For any prime number p, there is also the multiplicative group of integers modulo p. Its elements are the integers 1 to $p - 1$. The group operation is multiplication modulo p. That is, the usual product is divided by p and the remainder of this division is the result of modular multiplication. For example, if $p = 5$, there are four group elements 1, 2, 3, 4. In this group, $4 \cdot 4 = 1$, because the usual product 16 is equivalent to 1, which divided by 5 yields a remainder of 1. for 5 divides $16 - 1 = 15$, denoted

$$16 \equiv 1 \ (\text{mod } 5).$$

The primality of p ensures that the product of two integers neither of which is divisible by p is not divisible by p either, hence the indicated set of classes is closed under multiplication.[o[>]] The identity element is 1, as usual for a multiplicative group, and the associativity follows from the corresponding property of integers. Finally, the inverse element axiom requires that given an integer a not divisible by p, there exists an integer b such that

$a \cdot b \equiv 1 \pmod{p}$, i.e. p divides the difference $a \cdot b - 1$.

The inverse b can be found by using Bézout's identity and the fact that the greatest common divisor $\gcd(a, p)$ equals 1. In the case $p = 5$ above, the inverse of 4 is 4, and the inverse of 3 is 2, as $3 \cdot 2 = 6 \equiv 1 \pmod 5$. Hence all group axioms are fulfilled. Actually, this example is similar to $(\mathbb{Q} \setminus \{0\}, \cdot)$ above: it consists of exactly those elements in $\mathbb{Z}/p\mathbb{Z}$ that have a multiplicative inverse. These groups are denoted $F_p^{\times}$. They are crucial to public-key cryptography.[p[>]]

Cyclic Groups

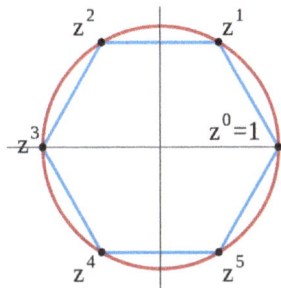

The 6th complex roots of unity form a cyclic group. z is a primitive element, but z^2 is not, because the odd powers of z are not a power of z^2.

A *cyclic group* is a group all of whose elements are powers of a particular element a. In multiplicative notation, the elements of the group are:

$\ldots, a^{-3}, a^{-2}, a^{-1}, a^0 = e, a, a^2, a^3, \ldots,$

where a^2 means $a \bullet a$, and a^{-3} stands for $a^{-1} \bullet a^{-1} \bullet a^{-1} = (a \bullet a \bullet a)^{-1}$ etc.[h[>]] Such an element a is called a generator or a primitive element of the group. In additive notation, the requirement for an element to be primitive is that each element of the group can be written as

$\ldots, -a-a, -a, 0, a, a+a, \ldots$

In the groups $\mathbb{Z}/n\mathbb{Z}$ introduced above, the element 1 is primitive, so these groups are cyclic. Indeed, each element is expressible as a sum all of whose terms are 1. Any cyclic group with n elements is isomorphic to this group. A second example for cyclic groups is the group of n-th complex roots of unity, given by complex numbers z satisfying $z^n = 1$. These numbers can be visualized as the vertices on a regular n-gon, as shown in blue at the right for $n = 6$. The group operation is multiplication of complex numbers. In the picture, multiplying with z corresponds to a counter-clockwise rotation by 60°. Using some field theory, the group $F_p^{\times}$ can be shown to be cyclic: for example, if $p = 5$, 3 is a generator since $3^1 = 3$, $3^2 = 9 \equiv 4$, $3^3 \equiv 2$, and $3^4 \equiv 1$.

Some cyclic groups have an infinite number of elements. In these groups, for every non-zero element a, all the powers of a are distinct; despite the name "cyclic group", the powers of the elements do not cycle. An infinite cyclic group is isomorphic to (Z, +), the group of integers under addition introduced above. As these two prototypes are both abelian, so is any cyclic group.

The study of finitely generated abelian groups is quite mature, including the fundamental theorem of finitely generated abelian groups; and reflecting this state of affairs, many group-related notions, such as center and commutator, describe the extent to which a given group is not abelian.

Symmetry Groups

Symmetry groups are groups consisting of symmetries of given mathematical objects—be they of geometric nature, such as the introductory symmetry group of the square, or of algebraic nature, such as polynomial equations and their solutions. Conceptually, group theory can be thought of as the study of symmetry.[*] Symmetries in mathematics greatly simplify the study of geometrical or analytical objects. A group is said to act on another mathematical object X if every group element performs some operation on X compatibly to the group law. In the rightmost example below, an element of order 7 of the (2,3,7) triangle group acts on the tiling by permuting the highlighted warped triangles (and the other ones, too). By a group action, the group pattern is connected to the structure of the object being acted on.

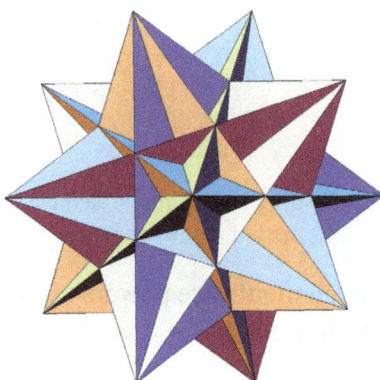

Rotations and reflections form the symmetry group of a great icosahedron.

In chemical fields, such as crystallography, space groups and point groups describe molecular symmetries and crystal symmetries. These symmetries underlie the chemical and physical behavior of these systems, and group theory enables simplification of quantum mechanical analysis of these properties. For example, group theory is used to show that optical transitions between certain quantum levels cannot occur simply because of the symmetry of the states involved.

Not only are groups useful to assess the implications of symmetries in molecules, but surprisingly they also predict that molecules sometimes can change symmetry. The Jahn-Teller effect is a distortion of a molecule of high symmetry when it adopts a particular ground state of lower symmetry from a set of possible ground states that are related to each other by the symmetry operations of the molecule.

Likewise, group theory helps predict the changes in physical properties that occur when a material undergoes a phase transition, for example, from a cubic to a tetrahedral crystalline form. An example is ferroelectric materials, where the change from a paraelectric to a ferroelectric state occurs at the Curie temperature and is related to a change from the high-symmetry paraelectric state to the lower symmetry ferroelectric state, accompanied by a so-called soft phonon mode, a vibrational lattice mode that goes to zero frequency at the transition.

Such spontaneous symmetry breaking has found further application in elementary particle physics, where its occurrence is related to the appearance of Goldstone bosons.

Buckminsterfullerene displays icosahedral symmetry, though the double bonds reduce this to pyritohedral symmetry.	Ammonia, NH_3. Its symmetry group is of order 6, generated by a 120° rotation and a reflection.	Cubane C_8H_8 features octahedral symmetry.	Hexaaquacopper(II) complex ion, $[Cu(OH_2)_6]^{2+}$. Compared to a perfectly symmetrical shape, the molecule is vertically dilated by about 22% (Jahn-Teller effect).	The (2,3,7) triangle group, a hyperbolic group, acts on this tiling of the hyperbolic plane.

Finite symmetry groups such as the Mathieu groups are used in coding theory, which is in turn applied in error correction of transmitted data, and in CD players. Another application is differential Galois theory, which characterizes functions having antiderivatives of a prescribed form, giving group-theoretic criteria for when solutions of certain differential equations are well-behaved.[u][>] Geometric properties that remain stable under group actions are investigated in (geometric) invariant theory.

General Linear Group and Representation Theory

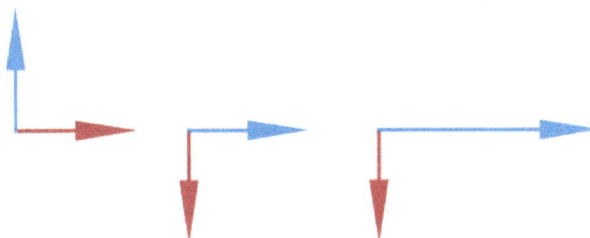

Two vectors (the left illustration) multiplied by matrices (the middle and right illustrations). The middle illustration represents a clockwise rotation by 90°, while the right-most one stretches the x-coordinate by factor 2.

Matrix groups consist of matrices together with matrix multiplication. The *general linear group* GL(n, R) consists of all invertible n-by-n matrices with real entries. Its subgroups are referred to as *matrix groups* or *linear groups*. The dihedral group example mentioned above can be viewed as a (very small) matrix group. Another important matrix group is the special orthogonal group SO(n). It describes all possible rotations in n dimensions. Via Euler angles, rotation matrices are used in computer graphics.

Representation theory is both an application of the group concept and important for a deeper understanding of groups. It studies the group by its group actions on other spaces. A broad class of group representations are linear representations, i.e. the group is acting on a vector space, such as the three-dimensional Euclidean space R³. A representation of G on an n-dimensional real vector space is simply a group homomorphism

$$\rho: G \to GL(n, R)$$

from the group to the general linear group. This way, the group operation, which may be abstractly given, translates to the multiplication of matrices making it accessible to explicit computations.

Given a group action, this gives further means to study the object being acted on. On the other hand, it also yields information about the group. Group representations are an organizing principle in the theory of finite groups, Lie groups, algebraic groups and topological groups, especially (locally) compact groups.

Galois Groups

Galois groups were developed to help solve polynomial equations by capturing their symmetry features. For example, the solutions of the quadratic equation $ax^2 + bx + c = 0$ are given by

$$x = \frac{-b \pm \sqrt{b^2 - 4ac}}{2a}.$$

Exchanging "+" and "−" in the expression, i.e. permuting the two solutions of the equation can be viewed as a (very simple) group operation. Similar formulae are known for cubic and quartic equations, but do *not* exist in general for degree 5 and higher. Abstract properties of Galois groups associated with polynomials (in particular their solvability) give a criterion for polynomials that have all their solutions expressible by radicals, i.e. solutions expressible using solely addition, multiplication, and roots similar to the formula above.

The problem can be dealt with by shifting to field theory and considering the splitting field of a polynomial. Modern Galois theory generalizes the above type of Galois groups to field extensions and establishes—via the fundamental theorem of Galois theory—a precise relationship between fields and groups, underlining once again the ubiquity of groups in mathematics.

Finite Groups

A group is called *finite* if it has a finite number of elements. The number of elements is called the order of the group. An important class is the *symmetric groups* S_N, the groups of permutations of N letters. For example, the symmetric group on 3 letters S_3 is the group consisting of all possible orderings of the three letters *ABC*, i.e. contains the elements *ABC, ACB, BAC, BCA, CAB, CBA*, in total 6 (factorial of 3) elements. This class is fundamental insofar as any finite group can be expressed as a subgroup of a symmetric group S_N for a suitable integer N, according to Cayley's theorem. Parallel to the group of symmetries of the square above, S_3 can also be interpreted as the group of symmetries of an equilateral triangle.

The order of an element a in a group G is the least positive integer n such that $a^n = e$, where a^n represents

$$\underbrace{a \cdots a}_{n \text{ factors}},$$

i.e. application of the operation • to n copies of a. (If • represents multiplication, then a^n corresponds to the nth power of a.) In infinite groups, such an n may not exist, in which case the order of a is said to be infinity. The order of an element equals the order of the cyclic subgroup generated by this element.

More sophisticated counting techniques, for example counting cosets, yield more precise statements about finite groups: Lagrange's Theorem states that for a finite group G the order of any finite subgroup H divides the order of G. The Sylow theorems give a partial converse.

The dihedral group (discussed above) is a finite group of order 8. The order of r_1 is 4, as is the order of the subgroup R it generates. The order of the reflection elements f_v etc. is 2. Both orders divide 8, as predicted by Lagrange's theorem. The groups $F_p^\times$ above have order $p - 1$.

Classification of Finite Simple Groups

Mathematicians often strive for a complete classification (or list) of a mathematical notion. In the context of finite groups, this aim leads to difficult mathematics. According to Lagrange's theorem, finite groups of order p, a prime number, are necessarily cyclic (abelian) groups Z_p. Groups of order p^2 can also be shown to be abelian, a statement which does not generalize to order p^3, as the non-abelian group D_4 of order $8 = 2^3$ above shows. Computer algebra systems can be used to list small groups, but there is no classification of all finite groups. An intermediate step is the classification of finite simple groups. A nontrivial group is called *simple* if its only normal subgroups are the trivial group and the group itself. The Jordan–Hölder theorem exhibits finite simple groups as the building blocks for all finite groups. Listing all finite simple groups was a major achievement in contemporary group theory. 1998 Fields Medal winner Richard Borcherds succeeded in proving the monstrous moonshine conjectures, a surprising and deep relation between the largest finite simple sporadic group—the "monster group"—and certain modular functions, a piece of classical complex analysis, and string theory, a theory supposed to unify the description of many physical phenomena.

Groups with Additional Structure

Many groups are simultaneously groups and examples of other mathematical structures. In the language of category theory, they are group objects in a category, meaning that they are objects (that is, examples of another mathematical structure) which come with transformations (called morphisms) that mimic the group axioms. For example, every group (as defined above) is also a set, so a group is a group object in the category of sets.

Topological Groups

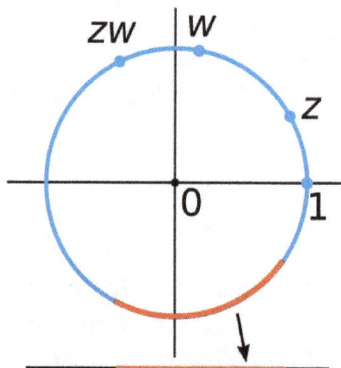

The unit circle in the complex plane under complex multiplication is a Lie group and, therefore, a topological group. It is topological since complex multiplication and division are continuous. It is a manifold and thus a Lie group, because every small piece, such as the red arc in the figure, looks like a part of the real line.

Some topological spaces may be endowed with a group law. In order for the group law and the topology to interweave well, the group operations must be continuous functions, that is, $g \bullet h$, and g^{-1} must not vary wildly if g and h vary only little. Such groups are called *topological groups,* and they are the group objects in the category of topological spaces. The most basic examples are the reals R under addition, $(R \setminus \{0\}, \cdot)$, and similarly with any other topological field such as the complex numbers or p-adic numbers. All of these groups are locally compact, so they have Haar measures and can be studied via harmonic analysis. The former offer an abstract formalism of invariant integrals. Invariance means, in the case of real numbers for example:

$$\int f(x)dx = \int f(x+c)dx$$

for any constant c. Matrix groups over these fields fall under this regime, as do adele rings and adelic algebraic groups, which are basic to number theory. Galois groups of infinite field extensions such as the absolute Galois group can also be equipped with a topology, the so-called Krull topology, which in turn is central to generalize the above sketched connection of fields and groups to infinite field extensions. An advanced generalization of this idea, adapted to the needs of algebraic geometry, is the étale fundamental group.

Lie Groups

Lie groups (in honor of Sophus Lie) are groups which also have a manifold structure, i.e. they are spaces looking locally like some Euclidean space of the appropriate dimension. Again, the additional structure, here the manifold structure, has to be compatible, i.e. the maps corresponding to multiplication and the inverse have to be smooth.

A standard example is the general linear group introduced above: it is an open subset of the space of all n-by-n matrices, because it is given by the inequality

$$\det (A) \neq 0,$$

where A denotes an n-by-n matrix.

Lie groups are of fundamental importance in modern physics: Noether's theorem links continuous symmetries to conserved quantities. Rotation, as well as translations in space and time are basic symmetries of the laws of mechanics. They can, for instance, be used to construct simple models—imposing, say, axial symmetry on a situation will typically lead to significant simplification in the equations one needs to solve to provide a physical description.[>] Another example are the Lorentz transformations, which relate measurements of time and velocity of two observers in motion relative to each other. They can be deduced in a purely group-theoretical way, by expressing the transformations as a rotational symmetry of Minkowski space. The latter serves—in the absence of significant gravitation—as a model of space time in special relativity. The full symmetry group of Minkowski space, i.e. including translations, is known as the Poincaré group. By the above, it plays a pivotal role in special relativity and, by implication, for quantum field theories. Symmetries that vary with location are central to the modern description of physical interactions with the help of gauge theory.

Generalizations

In abstract algebra, more general structures are defined by relaxing some of the axioms defining a group. For example, if the requirement that every element has an inverse is eliminated, the resulting algebraic structure is called a monoid. The natural numbers N (including 0) under addition form a monoid, as do the nonzero integers under multiplication $(Z \setminus \{0\}, \cdot)$, see above. There is a general method to formally add inverses to elements to any (abelian) monoid, much the same way as $(Q \setminus \{0\}, \cdot)$ is derived from $(Z \setminus \{0\}, \cdot)$, known as the Grothendieck group. Groupoids are similar to groups except that the composition $a \bullet b$ need not be defined for all a and b. They arise in the study of more complicated forms of symmetry, often in topological and analytical structures, such as the fundamental groupoid or stacks. Finally, it is possible to generalize any of these concepts by replacing the binary operation with an arbitrary n-ary one (i.e. an operation taking n arguments). With the proper generalization of the group axioms this gives rise to an n-ary group. The table gives a list of several structures generalizing groups.

Semigroup

In mathematics, a semigroup is an algebraic structure consisting of a set together with an associative binary operation. The binary operation of a semigroup is most often denoted multiplicatively: $x \cdot y$, or simply xy, denotes the result of applying the semigroup operation to the ordered pair (x, y). Associativity is formally expressed as that $(x \cdot y) \cdot z = x \cdot (y \cdot z)$ for all x, y and z in the semigroup.

The name "semigroup" originates in the fact that a semigroup generalizes a group by preserving only associativity and closure under the binary operation from the axioms defining a group.[note 1] From the opposite point of view (of adding rather than removing axioms), a semigroup is an associative magma. As in the case of groups or magmas, the semigroup operation need not be commutative, so $x \cdot y$ is not necessarily equal to $y \cdot x$; a typical example of associative but non-commutative operation is matrix multiplication. If the semigroup operation is commutative, then the semigroup is called a *commutative semigroup* or (less often than in the analogous case of groups) it may be called an *abelian semigroup*.

A monoid is an algebraic structure intermediate between groups and semigroups, and is a semigroup having an identity element, thus obeying all but one of the axioms of a group; existence of inverses is not required of a monoid. A natural example is strings with concatenation as the binary operation, and the empty string as the identity element. Restricting to non-empty strings gives an example of a semigroup that is not a monoid. Positive integers with addition form a commutative semigroup that is not a monoid, whereas the non-negative integers do form a monoid. A semigroup without an identity element can be easily turned into a monoid by just adding an identity element. Consequently, monoids are studied in the theory of semigroups rather than in group theory. Semigroups should not be confused with quasigroups, which are a generalization of groups in a different direction; the operation in a quasigroup need not be associative but quasigroups preserve from groups a notion of division. Division in semigroups (or in monoids) is not possible in general.

The formal study of semigroups began in the early 20th century. Early results include a Cayley theorem for semigroups realizing any semigroup as transformation semigroup, in which arbitrary functions replace the role of bijections from group theory. Other fundamental techniques of studying semigroups like Green's relations do not imitate anything in group theory though. A deep result in the classification of finite semigroups is Krohn–Rhodes theory. The theory of finite semigroups has been of particular importance in theoretical computer science since the 1950s because of the natural link between finite semigroups and finite automata via the syntactic monoid. In probability theory, semigroups are associated with Markov processes. In other areas of applied mathematics, semigroups are fundamental models for linear time-invariant systems. In partial differential equations, a semigroup is associated to any equation whose spatial evolution is independent of time. There are numerous special classes of semigroups, semigroups with additional properties, which appear in particular applications. Some of these classes are even closer to groups by exhibiting some additional but not all properties of a group. Of these we mention: regular semigroups, orthodox semigroups, semigroups with involution, inverse semigroups and cancellative semigroups. There also interesting classes of semigroups that do not contain any groups except the trivial group; examples of the latter kind are bands and their commutative subclass—semilattices, which are also ordered algebraic structures.

Definition

A semigroup is a set S together with a binary operation "$\cdot$" (that is, a function $\cdot : S \times S \to S$) that satisfies the associative property:

> For all $a, b, c \in S$, the equation $(a \cdot b) \cdot c = a \cdot (b \cdot c)$ holds.

More succinctly, a semigroup is an associative magma.

Examples of Semigroups

- Empty semigroup: the empty set forms a semigroup with the empty function as the binary operation.

- Semigroup with one element: there is essentially only one, the singleton $\{a\}$ with operation $a \cdot a = a$.

- Semigroup with two elements: there are five which are essentially different.

- The set of positive integers with addition. (With 0 included, this becomes a monoid.)

- The set of integers with minimum or maximum. (With positive/negative infinity included, this becomes a monoid.)

- Square nonnegative matrices of a given size with matrix multiplication.

- Any ideal of a ring with the multiplication of the ring.

- The set of all finite strings over a fixed alphabet Σ with concatenation of strings as the semigroup operation — the so-called "free semigroup over Σ". With the empty string included, this semigroup becomes the free monoid over Σ.

- A probability distribution F together with all convolution powers of F, with convolution as the operation. This is called a convolution semigroup.

- A monoid is a semigroup with an identity element.

- A group is a monoid in which every element has an inverse element.

- Transformation semigroups and monoids

- The set of continuous functions from a topological space to itself

Basic Concepts

Identity and Zero

A left identity of a semigroup S (or more generally, magma) is an element e such that for all x in S, $ex = x$. Similarly, a right identity is an element f such that for all x in S, $xf = x$. Left and right identities are both called one-sided identities. A semigroup may have one or more left identities but no right identity, and vice versa.

A two-sided identity (or just identity) is an element which is both a left and right identity. Semigroups with a two-sided identity are called monoids. A semigroup may have at most one two-sided identity. If a semigroup has a two-sided identity, then the two-sided identity is the only one-sided identity in the semigroup. If a semigroup has both a left identity and a right identity, then it has a two-sided identity (which is therefore the unique one-sided identity).

A semigroup S without identity may be embedded in a monoid formed by adjoining an element $e \notin S$ to S and defining $e \cdot s = s \cdot e = s$ for all $s \in S \cup \{e\}$. The notation S^1 denotes a monoid obtained from S by adjoining an identity *if necessary* ($S^1 = S$ for a monoid).

Similarly, every magma has at most one absorbing element, which in semigroup theory is called a zero. Analogous to the above construction, for every semigroup S, one can define S^0, a semigroup with 0 that embeds S.

Subsemigroups and Ideals

The semigroup operation induces an operation on the collection of its subsets: given subsets A and B of a semigroup S, their product $A \cdot B$, written commonly as AB, is the set $\{ ab \mid a$ in A and b in $B \}$. (This notion is defined identically as it is for groups.) In terms of this operation, a subset A is called

- a subsemigroup if AA is a subset of A,

- a right ideal if AS is a subset of A, and

- a left ideal if SA is a subset of A.

If A is both a left ideal and a right ideal then it is called an ideal (or a two-sided ideal).

If S is a semigroup, then the intersection of any collection of subsemigroups of S is also a subsemigroup of S. So the subsemigroups of S form a complete lattice.

An example of semigroup with no minimal ideal is the set of positive integers under addition. The minimal ideal of a commutative semigroup, when it exists, is a group.

Green's relations, a set of five equivalence relations that characterise the elements in terms of the principal ideals they generate, are important tools for analysing the ideals of a semigroup and related notions of structure.

The subset with the property that its every element commutes with any other element of the semigroup is called the center of the semigroup. The center of a semigroup is actually a subsemigroup.

Homomorphisms and Congruences

A semigroup homomorphism is a function that preserves semigroup structure. A function $f\colon S \to T$ between two semigroups is a homomorphism if the equation

$$f(ab) = f(a)f(b).$$

holds for all elements a, b in S, i.e. the result is the same when performing the semigroup operation after or before applying the map f.

A semigroup homomorphism between monoids preserves identity if it is a monoid homomorphism. But there are semigroup homomorphisms which are not monoid homomorphisms, e.g. the canonical embedding of a semigroup S without identity into S^1. Conditions characterizing monoid homomorphisms are discussed further. Let $f : S_0 \to S_1$ be a semigroup homomorphism. The image of f is also a semigroup. If S_0 is a monoid with an identity element e_0, then $f(e_0)$ is the identity element in the image of f. If S_1 is also a monoid with an identity element e_1 and e_1 belongs to the image of f, then $f(e_0) = e_1$, i.e. f is a monoid homomorphism. Particularly, if f is surjective, then it is a monoid homomorphism.

Two semigroups S and T are said to be isomorphic if there is a bijection $f : S \leftrightarrow T$ with the property that, for any elements a, b in S, $f(ab) = f(a)f(b)$. Isomorphic semigroups have the same structure.

A semigroup congruence $\sim$ is an equivalence relation that is compatible with the semigroup operation. That is, a subset $\sim\, \subseteq S \times S$ that is an equivalence relation and $x \sim y$ and $u \sim v$ implies $xu \sim yv$ for every x, y, u, v in S. Like any equivalence relation, a semigroup congruence induces congruence classes

$$[a]_\sim = \{x \in S \mid x \sim a\}$$

and the semigroup operation induces a binary operation $\circ$ on the congruence classes:

$$[u]_\sim \circ [v]_\sim = [uv]_\sim$$

Because $\sim$ is a congruence, the set of all congruence classes of $\sim$ forms a semigroup with $\circ$, called the quotient semigroup or factor semigroup, and denoted $S/\sim$. The mapping $x \mapsto [x]_\sim$ is a semigroup homomorphism, called the quotient map, canonical surjection or projection; if S is a monoid then quotient semigroup is a monoid with identity. Conversely, the kernel of any semigroup homomorphism is a semigroup congruence. These results are nothing more than a particularization of the first isomorphism theorem in universal algebra. Congruence classes and factor monoids are the objects of study in string rewriting systems.

A nuclear congruence on S is one which is the kernel of an endomorphism of S.

A semigroup S satisfies the maximal condition on congruences if any family of congruences on S, ordered by inclusion, has a maximal element. By Zorn's lemma, this is equivalent to saying that the ascending chain condition holds: there is no infinite strictly ascending chain of congruences on S.

Every ideal I of a semigroup induces a subsemigroup, the Rees factor semigroup via the congruence $x \rho y \Leftrightarrow$ either $x = y$ or both x and y are in I.

Structure of Semigroups

For any subset A of S there is a smallest subsemigroup T of S which contains A, and we say that A generates T. A single element x of S generates the subsemigroup $\{ x^n \mid n$ is a positive integer $\}$. If this is finite, then x is said to be of finite order, otherwise it is of infinite order. A semigroup is said to be periodic if all of its elements are of finite order. A semigroup generated by a single element is said to be monogenic (or cyclic). If a monogenic semigroup is infinite then it is isomorphic to the semigroup of positive integers with the operation of addition. If it is finite and nonempty, then it must contain at least one idempotent. It follows that every nonempty periodic semigroup has at least one idempotent.

A subsemigroup which is also a group is called a subgroup. There is a close relationship between the subgroups of a semigroup and its idempotents. Each subgroup contains exactly one idempotent, namely the identity element of the subgroup. For each idempotent e of the semigroup there is a unique maximal subgroup containing e. Each maximal subgroup arises in this way, so there is a one-to-one correspondence between idempotents and maximal subgroups. Here the term *maximal subgroup* differs from its standard use in group theory.

More can often be said when the order is finite. For example, every nonempty finite semigroup is periodic, and has a minimal ideal and at least one idempotent. The number of finite semigroups of a given size (greater than 1) is (obviously) larger than the number of groups of the same size. For example, of the sixteen possible "multiplication tables" for a set of two elements {a, b}, eight form semigroups[whereas only four of these are monoids and only two form groups. For more on the structure of finite semigroups.

Special Classes of Semigroups

- A monoid is a semigroup with identity.

- A subsemigroup is a subset of a semigroup that is closed under the semigroup operation.

- A band is a semigroup the operation of which is idempotent.

- A cancellative semigroup is one having the cancellation property: $a \cdot b = a \cdot c$ implies $b = c$ and similarly for $b \cdot a = c \cdot a$.

- A semilattice is a semigroup whose operation is idempotent and commutative.

- 0-simple semigroups.

- Transformation semigroups: any finite semigroup S can be represented by transformations of a (state-) set Q of at most $|S| + 1$ states. Each element x of S then maps Q into itself x: $Q \to Q$ and sequence xy is defined by $q(xy) = (qx)y$ for each q in Q. Sequencing clearly is an associative operation, here equivalent to function composition. This representation is basic for any automaton or finite state machine (FSM).

- The bicyclic semigroup is in fact a monoid, which can be described as the free semigroup on two generators p and q, under the relation $pq = 1$.

- C_0-semigroups.

- Regular semigroups. Every element x has at least one inverse y satisfying $xyx=x$ and $yxy=y$; the elements x and y are sometimes called "mutually inverse".

- Inverse semigroups are regular semigroups where every element has exactly one inverse. Alternatively, a regular semigroup is inverse if and only if any two idempotents commute.

- Affine semigroup: a semigroup that is isomorphic to a finitely-generated subsemigroup of Z^d. These semigroups have applications to commutative algebra.

Structure Theorem for Commutative Semigroups

There is a structure theorem for commutative semigroups in terms of semilattices. A semilattice (or more precisely a meet-semilattice) $(L, \leq)$ is a partially ordered set where every pair of elements $a, b \in L$ has a greatest lower bound, denoted $a \wedge b$. The operation $\wedge$ makes L into a semigroup satisfying the additional idempotence law $a \wedge a = a..$

Given a homomorphism $f : S \to L$ from an arbitrary semigroup to a semilattice, each inverse image $a, b \in L$ is a (possibly empty) semigroup. Moreover, $a \wedge b.$ becomes graded by L, in the sense that

$$S_a S_b \subseteq S_{a \wedge b}$$

If f is onto, the semilattice L is isomorphic to the quotient of S by the equivalence relation $\sim$ such that $x \sim y$ iff $f(x) = f(y)$. This equivalence relation is a semigroup congruence, as defined above.

Whenever we take the quotient of a commutative semigroup by a congruence, we get another commutative semigroup. The structure theorem says that for any commutative semigroup S, there is a finest congruence L such that the quotient of S by this equivalence relation is a semilattice. Denoting this semilattice by L, we get a homomorphism f from S onto L. As mentioned, S becomes graded by this semilattice.

Furthermore, the components S_a are all Archimedean semigroups. An Archimedean semigroup is one where given any pair of elements x, y, there exists an element z and $n > 0$ such that $x^n = yz$.

The Archimedean property follows immediately from the ordering in the semilattice L, since with this ordering we have $f(x) \leq f(y)$ if and only if $x^n = yz$ for some z and $n > 0$.

Group of Fractions

The group of fractions or group completion of a semigroup S is the group $G = G(S)$ generated by the elements of S as generators and all equations $xy = z$ which hold true in S as relations. There is an obvious semigroup homomorphism $j : S \to G(S)$ which sends each element of S to the corresponding generator. This has a universal property for morphisms from S to a group: given any group H and any semigroup homomorphism $k : S \to H$, there exists a unique group homomorphism $f : G \to H$ with $k=fj$. We may think of G as the "most general" group that contains a homomorphic image of S.

An important question is to characterize those semigroups for which this map is an embedding. This need not always be the case: for example, take S to be the semigroup of subsets of some set X with set-theoretic intersection as the binary operation (this is an example of a semilattice). Since $A.A = A$ holds for all elements of S, this must be true for all generators of $G(S)$ as well: which is therefore the trivial group. It is clearly necessary for embeddability that S have the cancellation property. When S is commutative this condition is also sufficient and the Grothendieck group of the semigroup provides a construction of the group of fractions. The problem for non-commutative semigroups can be traced to the first substantial paper on semigroups. Anatoly Maltsev gave necessary and sufficient conditions for embeddability in 1937.

Semigroup Methods in Partial Differential Equations

Semigroup theory can be used to study some problems in the field of partial differential equations. Roughly speaking, the semigroup approach is to regard a time-dependent partial differential equation as an ordinary differential equation on a function space. For example, consider the following initial/boundary value problem for the heat equation on the spatial interval $(0, 1) \subset R$ and times $t \geq 0$:

$$\begin{cases} \partial_t u(t,x) = \partial_x^2 u(t,x), & x \in (0,1), t > 0; \\ u(t,x) = 0, & x \in \{0,1\}, t > 0; \\ u(t,x) = u_0(x), & x \in (0,1), t = 0. \end{cases}$$

Let $X = L^2((0, 1) R)$ be the L^p space of square-integrable real-valued functions with domain the interval $(0, 1)$ and let A be the second-derivative operator with domain

$$D(A) = \left\{ u \in H^2((0,1); \mathbf{R}) \,\middle|\, u(0) = u(1) = 0 \right\},$$

where H^2 is a Hardy space. Then the above initial/boundary value problem can be interpreted as an initial value problem for an ordinary differential equation on the space X:

$$\begin{cases} \dot{u}(t) = Au(t); \\ u(0) = u_0. \end{cases}$$

On an heuristic level, the solution to this problem "ought" to be $u(t) = \exp(tA)u_0$. However, for a rigorous treatment, a meaning must be given to the exponential of tA. As a function of t, $\exp(tA)$ is a semigroup of operators from X to itself, taking the initial state u_0 at time $t = 0$ to the state $u(t) = \exp(tA)u_0$ at time t. The operator A is said to be the infinitesimal generator of the semigroup.

History

The study of semigroups trailed behind that of other algebraic structures with more complex axioms such as groups or rings. A number of sources attribute the first use of the term (in French) to J.-A. de Séguier in *Élements de la Théorie des Groupes Abstraits* (Elements of the Theory of Abstract Groups) in 1904. The term is used in English in 1908 in Harold Hinton's *Theory of Groups of Finite Order*.

Anton Suschkewitsch obtained the first non-trivial results about semigroups. His 1928 paper *Über die endlichen Gruppen ohne das Gesetz der eindeutigen Umkehrbarkeit* (*On finite groups without the rule of unique invertibility*) determined the structure of finite simple semigroups and showed that the minimal ideal (or Green's relations J-class) of a finite semigroup is simple. From that point on, the foundations of semigroup theory were further laid by David Rees, James Alexander Green, Evgenii Sergeevich Lyapin, Alfred H. Clifford and Gordon Preston. The latter two published a two-volume monograph on semigroup theory in 1961 and 1967 respectively. In 1970, a new periodical called *Semigroup Forum* (currently edited by Springer Verlag) became one of the few mathematical journals devoted entirely to semigroup theory.

In recent years researchers in the field have become more specialized with dedicated monographs appearing on important classes of semigroups, like inverse semigroups, as well as monographs focusing on applications in algebraic automata theory, particularly for finite automata, and also in functional analysis.

Generalizations

If the associativity axiom of a semigroup is dropped, the result is a magma, which is nothing more than a set M equipped with a binary operation $M \times M \to M$.

Generalizing in a different direction, an *n*-ary semigroup (also *n*-semigroup, polyadic semigroup or multiary semigroup) is a generalization of a semigroup to a set G with a n-ary operation instead of a binary operation. The associative law is generalized as follows: ternary associativity is $(abc)de = a(bcd)e = ab(cde)$, i.e. the string $abcde$ with any three adjacent elements bracketed. N-ary associativity is a string of length $n + (n - 1)$ with any n adjacent elements bracketed. A 2-ary semigroup is just a semigroup. Further axioms lead to an n-ary group.

A third generalization is the semigroupoid, in which the requirement that the binary relation be total is lifted. As categories generalize monoids in the same way, a semigroupoid behaves much like a category but lacks identities.

Infinitary generalizations of commutative semigroups have sometimes been considered by various authors.

Vertex Operator Algebra

In mathematics, a vertex operator algebra (VOA) is an algebraic structure that plays an important role in conformal field theory and string theory. In addition to physical applications, vertex operator algebras have proven useful in purely mathematical contexts such as monstrous moonshine and the geometric Langlands correspondence.

The related notion of vertex algebra was introduced by Richard Borcherds in 1986, motivated by a construction of an infinite-dimensional Lie algebra due to Frenkel. In the course of this construction, one employs a Fock space that admits an action of vertex operators attached to lattice vectors. Borcherds formulated the notion of vertex algebra by axiomatizing the relations between the lattice vertex operators, producing an algebraic structure that allows one to construct new Lie algebras by following Frenkel's method.

The notion of vertex operator algebra was introduced as a modification of the notion of vertex algebra, by Frenkel, Lepowsky, and Meurman in 1988, as part of their project to construct the moonshine module. They observed that many vertex algebras that appear in nature have a useful additional structure (an action of the Virasoro algebra), and satisfy a bounded-below property with respect to an energy operator. Motivated by this observation, they added the Virasoro action and bounded-below property as axioms.

We now have post-hoc motivation for these notions from physics, together with several interpretations of the axioms that were not initially known. Physically, the vertex operators arising from holomorphic field insertions at points (i.e., vertices) in two dimensional conformal field theory admit operator product expansions when insertions collide, and these satisfy precisely the relations specified in the definition of vertex operator algebra. Indeed, the axioms of a vertex operator algebra are a formal algebraic interpretation of what physicists call chiral algebras, or "algebras of chiral symmetries", where these symmetries describe the Ward identities satisfied by a given conformal field theory, including conformal invariance. Other formulations of the vertex algebra axioms include Borcherds's later work on singular commutative rings, algebras over certain operads on curves introduced by Huang, Kriz, and others, and D-module-theoretic objects called chiral algebras introduced by Alexander Beilinson and Vladimir Drinfeld. While related, these chiral algebras are not precisely the same as the objects with the same name that physicists use.

Important basic examples of vertex operator algebras include lattice VOAs (modeling lattice conformal field theories), VOAs given by representations of affine Kac–Moody algebras (from the WZW model), the Virasoro VOAs (i.e., VOAs corresponding to representations of the Virasoro algebra) and the moonshine module $V^\natural$, which is distinguished by its monster symmetry. More sophisticated examples such as affine W-algebras and the chiral de Rham complex on a complex manifold arise in geometric representation theory and mathematical physics.

Formal Definition

Vertex Algebra

A vertex algebra is a collection of data that satisfy certain axioms.

Data

- a vector space V, called the space of states. The underlying field is typically taken to be the complex numbers, although Borcherds's original formulation allowed for an arbitrary commutative ring.

- an identity element $1 \in V$, sometimes written $|0\rangle$ or Ω to indicate a vacuum state.

- an endomorphism $T: V \to V$, called "translation". (Borcherds's original formulation included a system of divided powers of T, because he did not assume the ground ring was divisible.)

- a linear multiplication map $Y: V \otimes V \to V((z))$, where $V((z))$ is the space of all formal Laurent series with coefficients in V. This structure is alternatively presented as an infinite collection of bilinear products $u_n v$, or as a left-multiplication map $V \to \mathrm{End}(V)[[z^{\pm 1}]]$, called the state-field correspondence. For each $u \in V$, the operator-valued formal distribution $Y(u, z)$ is called a vertex operator or a field (inserted at zero), and the coefficient of z^{-n-1} is the operator u_n. The standard notation for the multiplication is

$$u \otimes v \mapsto Y(u, z)v = \sum_{n \in \mathbf{Z}} u_n v z^{-n-1}.$$

Axioms

These data are required to satisfy the following axioms:

- Identity. For any $u \in V$, $Y(1, z)u = u = uz^0$ and $Y(u, z)1 \in u + zV[[z]]$.

- Translation. $T(1) = 0$, and for any $u, v \in V$,

$$[T, Y(u, z)]v = TY(u, z)v - Y(u, z)Tv = \frac{d}{dz} Y(u, z)v$$

- Locality (Jacobi identity, or Borcherds identity). For any $u, v \in V$, there exists a positive integer N such that:

$$(z - x)^N Y(u, z)Y(v, x) = (z - x)^N Y(v, x)Y(u, z).$$

Equivalent Formulations of Locality Axiom

The Locality axiom has several equivalent formulations in the literature, e.g., Frenkel-Lepowsky-Meurman introduced the Jacobi identity:

$$\forall u,v,w \in V: \quad z^{-1}\delta\left(\frac{y-x}{z}\right)Y(u,x)Y(v,y)w - z^{-1}\delta\left(\frac{-y+x}{z}\right)Y(v,y)Y(u,x)w = y^{-1}\delta\left(\frac{x+z}{y}\right)Y(Y(u,z)v,y)w,$$

where we define the formal delta series by:

$$\delta\left(\frac{y-x}{z}\right) := \sum_{s\geq 0, r\in \mathbf{Z}} \binom{r}{s}(-1)^s y^{r-s}x^s z^{-r}.$$

Borcherds initially used the following identity: for any vectors u, v, and w, and integers m and n we have

$$(u_m(v))_n(w) = \sum_{i\geq 0}(-1)^i\binom{m}{i}\left(u_{m-i}(v_{n+i}(w)) - (-1)^m v_{m+n-i}(u_i(w))\right)$$

He later gave a more expansive version that is equivalent but easier to use: for any vectors u, v, and w, and integers m, n, and q we have

$$\sum_{i\in\mathbf{Z}}\left(u_{q+i}(v)\right)_{m+n-i}(w) = \sum_{i\in\mathbf{Z}}(-1)^i\binom{q}{i}\left(u_{m+q-i}\left(v_{n+i}(w)\right) - (-1)^q v_{n+q-i}\left(u_{m+i}(w)\right)\right)$$

Finally, there is a formal function version of locality: For any $u, v, w \in V$, there is an element

$$X(u,v,w;z,x) \in V[[z,x]]\left[z^{-1}, x^{-1}, (z-x)^{-1}\right]$$

such that $Y(u,z)Y(v,x)w$ and $Y(v,x)Y(u,z)w$ are the corresponding expansions of $X(u,v,w;z,x)$ in $V((z))((x))$ and $V((x))((z))$.

Vertex Operator Algebra

A vertex operator algebra is a vertex algebra equipped with a conformal element ω, such that the vertex operator $Y(\omega, z)$ is the weight two Virasoro field $L(z)$:

$$Y(\omega,z) = \sum_{n\in\mathbf{Z}}\omega_n z^{-n-1} = L(z) = \sum_{n\in\mathbf{Z}}L_n z^{-n-2}$$

and satisfies the following properties:

- $[L_m, L_n] = (m-n)L_{m+n} + (\delta_{m+n,0}/12)(m^3-m)c\,\mathrm{Id}_V$, where c is a constant called the central charge, or rank of V. In particular, the coefficients of this vertex operator endow V with an action of the Virasoro algebra with central charge c.

- L_0 acts semisimply on V with integer eigenvalues that are bounded below.

- Under the grading provided by the eigenvalues of L_0, the multiplication on V is homogeneous in the sense that if u and v are homogeneous, then $u_n v$ is homogeneous of degree $\deg(u) + \deg(v) - n - 1$.

- The identity 1 has degree 0, and the conformal element ω has degree 2.

- $L_{-1} = T$.

A homomorphism of vertex algebras is a map of the underlying vector spaces that respects the additional identity, translation, and multiplication structure. Homomorphisms of vertex operator algebras have "weak" and "strong" forms, depending on whether they respect conformal vectors.

Commutative Vertex Algebras

A vertex algebra V is commutative if all vertex operators commute with each other. This is equivalent to the property that all products $Y(u,z)v$ lie in $V[[z]]$. Given a commutative vertex algebra, the constant terms of multiplication endow the vector space with a commutative ring structure, and T is a derivation. Conversely, any commutative ring V with derivation T has a canonical vertex algebra structure, where we set $Y(u,z)v = u_{-1}v z^0 = uv$. If the derivation T vanishes, we may set $\omega = 0$ to obtain a vertex operator algebra concentrated in degree zero.

Any finite-dimensional vertex algebra is commutative. In particular, even the smallest examples of noncommutative vertex algebras require significant introduction.

Basic Properties

The translation operator T in a vertex algebra induces infinitesimal symmetries on the product structure, and satisfies the following properties:

- $Y(u,z)1 = e^{zT}u$

- $Tu = u_{-2}1$, so T is determined by Y.

- $Y(Tu,z) = d(Y(u,z))/dz$

- $e^{xT}Y(u,z)e^{-xT} = Y(e^{xT}u,z) = Y(u,z+x)$

- (skew-symmetry) $Y(u,z)v = e^{zT}Y(v,-z)u$

For a vertex operator algebra, the other Virasoro operators satisfy similar properties:

- $x^{L_0}Y(u,z)x^{-L}_0 = Y(x^{L_0}u,xz)$

- $e^{xL_1}Y(u,z)e^{-xL_1} = Y(e^{x(1-xz)L_1}(1-xz)^{-2L_0}u,z(1-xz)^{-1})$

- (quasi-conformality) $[L_m, Y(u,z)] = \sum_{k=0}^{m+1} \binom{m+1}{k} z^k Y(L_{m-k}u, z)$ for all $m \geq -1$.

- (Associativity, or Cousin property): For any $u, v, w \in V$, the element

$$X(u,v,w;z,x) \in V[[z,x]][z^{-1}, x^{-1}, (z-x)^{-1}]$$

given in the definition also expands to $Y(Y(u,z-x)v,x)w$ in $V((x))((z-x))$.

The associativity property of a vertex algebra follows from the fact that the commutator of $Y(u,z)$ and $Y(v,x)$ is annihilated by a finite power of $z-x$, i.e., one can expand it as a finite linear combination of derivatives of the formal delta function in $(z-x)$, with coefficients in End(V).

Reconstruction: Let V be a vertex algebra, and let $\{J^a\}$ be a set of vectors, with corresponding fields $J^a(z) \in \text{End}(V)[[z^{\pm 1}]]$. If V is spanned by monomials in the positive weight coefficients of the fields (i.e., finite products of operators J^a_n applied to 1, where n is negative), then we may write the operator product of such a monomial as a normally ordered product of divided power derivatives of fields (here, normal ordering means polar terms on the left are moved to the right). Specifically,

$$Y(J^{a_1}_{n_1+1}J^{a_2}_{n_2+1}...J^{a_k}_{n_k+1}1,z) =: \frac{\partial^{n_1}}{\partial_z^{n_1}}\frac{J^{a_1}(z)}{n_1!}\frac{\partial^{n_2}}{\partial_z^{n_2}}\frac{J^{a_2}(z)}{n_2!}\cdots\frac{\partial^{n_k}}{\partial_z^{n_k}}\frac{J^{a_k}(z)}{n_k!}:$$

More generally, if one is given a vector space V with an endomorphism T and vector 1, and one assigns to a set of vectors J^a a set of fields $J^a(z) \in \text{End}(V)[[z^{\pm 1}]]$ that are mutually local, whose positive weight coefficients generate V, and that satisfy the identity and translation conditions, then the previous formula describes a vertex algebra structure.

Example: The Rank 1 Free Boson

A basic example of a noncommutative vertex algebra is the rank 1 free boson, also called the Heisenberg vertex operator algebra. It is "generated" by a single vector b, in the sense that by applying the coefficients of the field $b(z) = Y(b,z)$ to the vector 1, we obtain a spanning set. The underlying vector space is the infinite-variable polynomial ring $C[x_1,x_2,...]$, where for positive n, the coefficient b_{-n} of $Y(b,z)$ acts as multiplication by x_n, and b_n acts as n times the partial derivative in x_n. The action of b_0 is multiplication by zero, producing the "momentum zero" Fock representation V_0 of the Heisenberg Lie algebra (generated by b_n for integers n, with commutation relations $[b_n,b_m]=n\,\delta_{n,-m}$), i.e., induced by the trivial representation of the subalgebra spanned by b_n, $n \geq 0$.

The Fock space V_0 can be made into a vertex algebra by following reconstruction:

$$Y(x_{n_1+1}x_{n_2+1}x_{n_3+1}...x_{n_k+1},z) \equiv \frac{1}{n_1!n_2!..n_k!}:\partial^{n_1}b(z)\partial^{n_2}b(z)...\partial^{n_k}b(z):$$

where $:..:$ denotes normal ordering (i.e. moving all derivatives in x to the right). The vertex operators may also be written as a functional of a multivariable function f as:

$$Y[f,z] \equiv: f(\frac{b(z)}{0!},\frac{b'(z)}{1!},\frac{b''(z)}{2!},...):$$

if we understand that each term in the expansion of f is normal ordered.

The rank n free boson is given by taking an n-fold tensor product of the rank 1 free boson. For any vector b in n-dimensional space, one has a field $b(z)$ whose coefficients are elements of the rank n Heisenberg algebra, whose commutation relations have an extra inner product term: $[b_n,c_m]=n$ (b,c) $\delta_{n,-m}$.

Example: Virasoro Vertex Operator Algebras

Virasoro vertex operator algebras are important for two reasons: First, the conformal element in a vertex operator algebra canonically induces a homomorphism from a Virasoro vertex operator algebra, so they play a universal role in the theory. Second, they are intimately connected to the theory of unitary representations of the Virasoro algebra, and these play a major role in conformal field theory. In particular, the unitary Virasoro minimal models are simple quotients of these vertex algebras, and their tensor products provide a way to combinatorially construct more complicated vertex operator algebras.

The Virasoro vertex operator algebra is defined as an induced representation of the Virasoro algebra: If we choose a central charge c, there is a unique one-dimensional module for the subalgebra $C[z]\partial_z + K$ for which K acts by $c\mathrm{Id}$, and $C[z]\partial_z$ acts trivially, and the corresponding induced module is spanned by polynomials in $L_{-n} = -z^{-n-1}\partial_z$ as n ranges over integers greater than 1. The module then has partition function

$$Tr_V q^{L_0} = \sum_{n \in \mathbf{R}} \dim V_n q^n = \prod_{n \geq 2}(1 - q^n)^{-1}.$$

This space has a vertex operator algebra structure, where the vertex operators are defined by:

$$Y(L_{-n_1-2}L_{-n_2-2}...L_{-n_k-2}|0\rangle, z) \equiv \frac{1}{n_1!n_2!..n_k!} : \partial^{n_1}L(z)\partial^{n_2}L(z)...\partial^{n_k}L(z):$$

and $\omega = L_{-2}|0\rangle$. The fact that the Virasoro field $L(z)$ is local with respect to itself can be deduced from the formula for its self-commutator:

$$[L(z), L(x)] = \left(\frac{\partial}{\partial x}L(x)\right)w^{-1}\delta\left(\frac{z}{x}\right) - 2L(x)x^{-1}\frac{\partial}{\partial z}\delta\left(\frac{z}{x}\right) - \frac{1}{12}cx^{-1}\left(\frac{\partial}{\partial z}\right)^3\delta\left(\frac{z}{x}\right)$$

where c is the central charge.

Given a vertex algebra homomorphism from a Virasoro vertex algebra of central charge c to any other vertex algebra, the vertex operator attached to the image of ω automatically satisfies the Virasoro relations, i.e., the image of ω is a conformal vector. Conversely, any conformal vector in a vertex algebra induces a distinguished vertex algebra homomorphism from some Virasoro vertex operator algebra.

The Virasoro vertex operator algebras are simple, except when c has the form $1-6(p-q)^2/pq$ for coprime integers p,q strictly greater than 1 - this follows from Kac's determinant formula. In these exceptional cases, one has a unique maximal ideal, and the corresponding quotient is called a minimal model. When $p = q+1$, the vertex algebras are unitary representations of Virasoro, and their modules are known as discrete series representations. They play an important role in conformal field theory in part because they are unusually tractable, and for small p, they correspond to well-known statistical mechanics systems at criticality, e.g., the Ising model, the tri-critical Ising model, the three-state Potts model, etc. By work of Weiqang Wang concerning fusion rules, we have a full description of the tensor categories of unitary minimal models. For example, when $c=1/2$ (Ising), there three irreducible modules with lowest L_0-weight 0, 1/2, and 1/16, and its fusion ring is $Z[x,y]/(x^2-1, y^2-x-1, xy-y)$.

Example: WZW Vacuum Modules

By replacing the Heisenberg Lie algebra with an untwisted affine Kac–Moody Lie algebra (i.e., the universal central extension of the loop algebra on a finite-dimensional simple Lie algebra), one may construct the vacuum representation in much the same way as the free boson vertex algebra is constructed. Here, WZW refers to the Wess–Zumino–Witten model, which produces the anomaly that is interpreted as the central extension.

Concretely, pulling back the central extension

$$0 \to \mathbb{C} \to \hat{\mathfrak{g}} \to \mathfrak{g}[t, t^{-1}] \to 0$$

along the inclusion $\mathfrak{g}[t] \to \mathfrak{g}[t, t^{-1}]$ yields a split extension, and the vacuum module is induced from the one-dimensional representation of the latter on which a central basis element acts by some chosen constant called the "level". Since central elements can be identified with invariant inner products on the finite type Lie algebra $\mathfrak{g}$, one typically normalizes the level so that the Killing form has level twice the dual Coxeter number. Equivalently, level one gives the inner product for which the longest root has norm 2. This matches the loop algebra convention, where levels are discretized by third cohomology of simply connected compact Lie groups.

By choosing a basis J^a of the finite type Lie algebra, one may form a basis of the affine Lie algebra using $J^a_n = J^a t^n$ together with a central element K. By reconstruction, we can describe the vertex operators by normal ordered products of derivatives of the fields

$$J^a(z) = \sum_{n=-\infty}^{\infty} J^a_n z^{-n-1} = \sum_{n=-\infty}^{\infty} (J^a t^n) z^{-n-1}.$$

When the level is non-critical, i.e., the inner product is not minus one half of the Killing form, the vacuum representation has a conformal element, given by the Sugawara construction. For any choice of dual bases J^a, J_a with respect to the level 1 inner product, the conformal element is

$$\omega = \frac{1}{2(k + h^{\vee})} \sum_a J_{a,-1} J^a_{-1} 1$$

and yields a vertex operator algebra whose central charge is $k \cdot \dim \mathfrak{g} / (k + h^{\vee})$. At critical level, the conformal structure is destroyed, since the denominator is zero, but one may produce operators L_n for $n \geq -1$ by taking a limit as k approaches criticality.

This construction can be altered to work for the rank 1 free boson. In fact, the Virasoro vectors form a one-parameter family $\omega_s = 1/2\, x_1^2 + s\, x_2$, endowing the resulting vertex operator algebras with central charge $1 - 12 s^2$. When $s=0$, we have the following formula for the graded dimension:

$$Tr_V q^{L_0} = \sum_{n \in \mathbf{Z}} \dim V_n q^n = \prod_{n \geq 1} (1 - q^n)^{-1}$$

This is known as the generating function for partitions, and is also written as $q^{1/24}$ times the weight $-1/2$ modular form $1/\eta$ (the Dedekind eta function). The rank n free boson then has an n parameter family of Virasoro vectors, and when those parameters are zero, the character is $q^{n/24}$ times the weight $-n/2$ modular form η^{-n}.

Modules

Much like ordinary rings, vertex algebras admit a notion of module, or representation. Modules play an important role in conformal field theory, where they are often called sectors. A standard assumption in the physics literature is that the full Hilbert space of a conformal field theory decomposes into a sum of tensor products of left-moving and right-moving sectors:

$$\mathcal{H} \cong \bigoplus_{i \in I} M_i \otimes \overline{M}_i$$

That is, a conformal field theory has a vertex operator algebra of left-moving chiral symmetries, a vertex operator algebra of right-moving chiral symmetries, and the sectors moving in a given direction are modules for the corresponding vertex operator algebra.

Given a vertex algebra V with multiplication Y, a V-module is a vector space M equipped with an action $Y^M \colon V \otimes M \to M((z))$, satisfying the following conditions:

(Identity) $Y^M(1,z) = \mathrm{Id}_M$

(Associativity, or Jacobi identity) For any $u, v \in V$, $w \in M$, there is an element

$$X(u,v,w;z,x) \in M[[z,x]][z^{-1},x^{-1},(z-x)^{-1}]$$

such that $Y^M(u,z)Y^M(v,x)w$ and $Y^M(Y(u,z-x)v,x)w$ are the corresponding expansions of $X(u,v,w;z,x)$ in $M((z))((x))$ and $M((x))((z-x))$. Equivalently, the following "Jacobi identity" holds:

$$z^{-1}\delta\left(\frac{y-x}{z}\right)Y^M(u,x)Y^M(v,y)w - z^{-1}\delta\left(\frac{-y+x}{z}\right)Y^M(v,y)Y^M(u,x)w = y^{-1}\delta\left(\frac{x+z}{y}\right)Y^M(Y(u,z)v,y)w.$$

The modules of a vertex algebra form an abelian category. When working with vertex operator algebras, the previous definition is given the name "weak module", and V-modules are required to satisfy the additional condition that L_0 acts semisimply with finite-dimensional eigenspaces and eigenvalues bounded below in each coset of Z. Work of Huang, Lepowsky, Miyamoto, and Zhang has shown at various levels of generality that modules of a vertex operator algebra admit a fusion tensor product operation, and form a braided tensor category.

When the category of V-modules is semisimple with finitely many irreducible objects, the vertex operator algebra V is called rational. Rational vertex operator algebras satisfying an additional finiteness hypothesis (known as Zhu's C_2-cofiniteness condition) are known to be particularly well-behaved, and are called "regular". For example, Zhu's 1996 modular invariance theorem asserts that the characters of modules of a regular VOA form a vector-valued representation of $SL_2(\mathbb{Z})$. In particular, if a VOA is *holomorphic*, i.e., its representation category is equivalent to that of vector spaces, then its partition function is $SL_2(\mathbb{Z})$-invariant up to a constant. Huang showed that the category of modules of a regular VOA is a modular tensor category, and its fusion rules satisfy the Verlinde formula.

To connect with our first example, the irreducible modules of the rank 1 free boson are given by Fock spaces V_λ with some fixed momentum λ, i.e., induced representations of the Heisenberg Lie algebra, where the element b_0 acts by scalar multiplication by λ. The space can be written as C[x-

$_1$,x_2,...]v_λ, where v_λ is a distinguished ground-state vector. The module category is not semisimple, since one may induce a representation of the abelian Lie algebra where b_0 acts by a nontrivial Jordan block. For the rank n free boson, one has an irreducible module V_λ for each vector λ in complex n-dimensional space. Each vector $b \in \mathbb{C}^n$ yields the operator b_0, and the Fock space V_λ is distinguished by the property that each such b_0 acts as scalar multiplication by the inner product (b,λ).

Unlike ordinary rings, vertex algebras admit a notion of twisted module attached to an automorphism. For an automorphism σ of order N, the action has the form $V \otimes M \to M((z^{1/N}))$, with the following monodromy condition: if $u \in V$ satisfies $\sigma u = \exp(2\pi i k/N)u$, then $u_n = 0$ unless n satisfies $n+k/N \in \mathbb{Z}$ (there is some disagreement about signs among specialists). Geometrically, twisted modules can be attached to branch points on an algebraic curve with a ramified Galois cover. In the conformal field theory literature, twisted modules are called twisted sectors, and are intimately connected with string theory on orbifolds.

Vertex Operator Algebra Defined by an Even Lattice

The lattice vertex algebra construction was the original motivation for defining vertex algebras. It is constructed by taking a sum of irreducible modules for the free boson corresponding to lattice vectors, and defining a multiplication operation by specifying intertwining operators between them. That is, if Λ is an even lattice, the lattice vertex algebra V_Λ decomposes into free bosonic modules as:

$$V_\Lambda \cong \bigoplus_{\lambda \in \Lambda} V_\lambda$$

Lattice vertex algebras are canonically attached to double covers of even integral lattices, rather than the lattices themselves. While each such lattice has a unique lattice vertex algebra up to isomorphism, the vertex algebra construction is not functorial, because lattice automorphisms have an ambiguity in lifting.

The double covers in question are uniquely determined up to isomorphism by the following rule: elements have the form $\pm e_\alpha$ for lattice vectors $\alpha \in \Lambda$ (i.e., there is a map to Λ sending e_α to α that forgets signs), and multiplication satisfies the relations $e_\alpha e_\beta = (-1)^{(\alpha,\beta)}e_\beta e_\alpha$. Another way to describe this is that given an even lattice Λ, there is a unique (up to coboundary) normalised cocycle $\varepsilon(\alpha, \beta)$ with values ± 1 such that $(-1)^{(\alpha,\beta)} = \varepsilon(\alpha, \beta)\,\varepsilon(\beta, \alpha)$, where the normalization condition is that $\varepsilon(\alpha,0) = \varepsilon(0,\alpha) = 1$ for all $\alpha \in \Lambda$. This cocycle induces a central extension of Λ by a group of order 2, and we obtain a twisted group ring $\mathbb{C}_\varepsilon[\Lambda]$ with basis e_α ($\alpha \in \Lambda$), and multiplication rule $e_\alpha e_\beta = \varepsilon(\alpha, \beta)e_{\alpha+\beta}$ - the cocycle condition on ε ensures associativity of the ring.

The vertex operator attached to lowest weight vector v_λ in the Fock space V_λ is

$$Y(v_\lambda, z) = e_\lambda : \exp\int \lambda(z) := e_\lambda z^\lambda \exp\left(\sum_{n<0} \lambda_n \frac{z^{-n}}{n}\right)\exp\left(\sum_{n>0} \lambda_n \frac{z^{-n}}{n}\right),$$

where z^λ is a shorthand for the linear map that takes any element of the α-Fock space V_α to the monomial $z^{(\lambda,\alpha)}$. The vertex operators for other elements of the Fock space are then determined by reconstruction.

As in the case of the free boson, one has a choice of conformal vector, given by an element s of the vector space $\Lambda \otimes C$, but the condition that the extra Fock spaces have integer L_0 eigenvalues constrains the choice of s: for an orthonormal basis x_i, the vector $1/2\, x_{i,1}{}^2 + s_2$ must satisfy $(s, \lambda) \in Z$ for all $\lambda \in \Lambda$, i.e., s lies in the dual lattice.

If the even lattice Λ is generated by its "root vectors" (those satisfying $(\alpha, \alpha) = 2$), and any two root vectors are joined by a chain of root vectors with consecutive inner products non-zero then the vertex operator algebra is the unique simple quotient of the vacuum module of the affine Kac–Moody algebra of the corresponding simply laced simple Lie algebra at level one. This is known as the Frenkel–Kac (or Frenkel–Kac–Segal) construction, and is based on the earlier construction by Sergio Fubini and Gabriele Veneziano of the tachyonic vertex operator in the dual resonance model. Among other features, the zero modes of the vertex operators corresponding to root vectors give a construction of the underlying simple Lie algebra, related to a presentation originally due to Jacques Tits. In particular, one obtains a construction of all ADE type Lie groups directly from their root lattices. And this is commonly considered the simplest way to construct the 248 dimensional group E_8.

Vertex Operator Superalgebras

By allowing the underlying vector space to be a superspace (i.e., a $Z/2Z$-graded vector space $V = V_+ \oplus V_-$) one can define a *vertex superalgebra* by the same data as a vertex algebra, with 1 in V_+ and T an even operator. The axioms are essentially the same, but one must incorporate suitable signs into the locality axiom, or one of the equivalent formulations. That is, if a and b are homogeneous, one compares $Y(a,z)Y(b,w)$ with $\varepsilon Y(b,w)Y(a,z)$, where ε is -1 if both a and b are odd and 1 otherwise. If in addition there is a Virasoro element ω in the even part of V_2, and the usual grading restrictions are satisfied, then V is called a *vertex operator superalgebra*.

One of the simplest examples is the vertex operator superalgebra generated by a single free fermion ψ. As a Virasoro representation, it has central charge $1/2$, and decomposes as a direct sum of Ising modules of lowest weight 0 and $1/2$. One may also describe it as a spin representation of the Clifford algebra on the quadratic space $t^{1/2}C[t,t^{-1}](dt)^{1/2}$ with residue pairing. The vertex operator superalgebra is holomorphic, in the sense that all modules are direct sums of itself, i.e., the module category is equivalent to the category of vector spaces.

The tensor square of the free fermion is called the free charged fermion, and by Boson-Fermion correspondence, it is isomorphic to the lattice vertex superalgebra attached to the odd lattice Z. This correspondence has been used by Date-Jimbo-Kashiwara-Miwa to construct soliton solutions to the KP hierarchy of nonlinear PDEs.

Superconformal Structures

The Virasoro algebra has some supersymmetric extensions that naturally appear in superconformal field theory and superstring theory. The $N=1$, 2, and 4 superconformal algebras are of particular importance.

Infinitesimal holomorphic superconformal transformations of a supercurve (with one even local coordinate z and N odd local coordinates $\theta_1, ..., \theta_N$) are generated by the coefficients of a super-stress–energy tensor $T(z, \theta_1, \dots, \theta_N)$.

When $N=1$, T has odd part given by a Virasoro field $L(z)$, and even part given by a field

$$G(z) = \sum_n G_n z^{-n-3/2}$$

subject to commutation relations

- $[G_m, L_n] = (m - n/2)G_{m+n}$

- $[G_m, G_n] = (m - n)L_{m+n} + \delta_{m,-n} \dfrac{4m^2 + 1}{12} c$

By examining the symmetry of the operator products, one finds that there are two possibilities for the field G: the indices n are either all integers, yielding the Ramond algebra, or all half-integers, yielding the Neveu-Schwarz algebra. These algebras have unitary discrete series representations at central charge

$$\hat{c} = \frac{2}{3} c = 1 - \frac{8}{m(m+2)} \quad m \geq 3$$

and unitary representations for all c greater than $3/2$, with lowest weight h only constrained by $h \geq 0$ for Neveu-Schwartz and $h \geq c/24$ for Ramond.

An $N=1$ superconformal vector in a vertex operator algebra V of central charge c is an odd element $\tau \in V$ of weight $3/2$, such that

$$Y(\tau, z) = G(z) = \sum_{m \in \mathbb{Z}+1/2} G_n z^{-n-3/2},$$

$G_{-1/2}\tau = \omega$, and the coefficients of $G(z)$ yield an action of the $N=1$ Neveu-Schwarz algebra at central charge c.

For $N=2$ supersymmetry, one obtains even fields $L(z)$ and $J(z)$, and odd fields $G^+(z)$ and $G^-(z)$. The field $J(z)$ generates an action of the Heisenberg algebras (described by physicists as a $U(1)$ current). There are both Ramond and Neveu-Schwarz $N=2$ superconformal algebras, depending on whether the indexing on the G fields is integral or half-integral. However, the $U(1)$ current gives rise to a one-parameter family of isomorphic superconformal algebras interpolating between Ramond and Neveu-Schwartz, and this deformation of structure is known as spectral flow. The unitary representations are given by discrete series with central charge $c = 3-6/m$ for integers m at least 3, and a continuum of lowest weights for $c > 3$.

An $N=2$ superconformal structure on a vertex operator algebra is a pair of odd elements τ^+, τ^- of weight $3/2$, and an even element μ of weight 1 such that $\tau^\pm$ generate $G^\pm(z)$, and μ generates $J(z)$.

For $N=3$ and 4, unitary representations only have central charges in a discrete family, with $c=3k/2$ and $6k$, respectively, as k ranges over positive integers.

Additional Constructions

- Fixed point subalgebras: Given an action of a symmetry group on a vertex operator algebra, the subalgebra of fixed vectors is also a vertex operator algebra. In 2013, Miyamoto proved that two important finiteness properties, namely Zhu's condition C_2 and regularity, are preserved when taking fixed points under finite solvable group actions.

- Current extensions: Given a vertex operator algebra and some modules of integral conformal weight, one may under favorable circumstances describe a vertex operator algebra structure on the direct sum. Lattice vertex algebras are a standard example of this. Another family of examples are framed VOAs, which start with tensor products of Ising models, and add modules that correspond to suitably even codes.

- Orbifolds: Given a finite cyclic group acting on a holomorphic VOA, it is conjectured that one may construct a second holomorphic VOA by adjoining irreducible twisted modules and taking fixed points under an induced automorphism, as long as those twisted modules have suitable conformal weight. This is known to be true in special cases, e.g., groups of order at most 3 acting on lattice VOAs.

- The coset construction (due to Goddard, Kent, and Olive): Given a vertex operator algebra V of central charge c and a set S of vectors, one may define the commutant $C(V,S)$ to be the subspace of vectors v strictly commute with all fields coming from S, i.e., such that $Y(s,z) v \in V[[z]]$ for all $s \in S$. This turns out to be a vertex subalgebra, with Y, T, and identity inherited from V. and if S is a VOA of central charge c_S, the commutant is a VOA of central charge $c-c_S$. For example, the embedding of $SU(2)$ at level $k+1$ into the tensor product of two $SU(2)$ algebras at levels k and 1 yields the Virasoro discrete series with $p=k+2$, $q=k+3$, and this was used to prove their existence in the 1980s. Again with $SU(2)$, the embedding of level $k+2$ into the tensor product of level k and level 2 yields the $N=1$ superconformal discrete series.

- BRST reduction: For any degree 1 vector v satisfying $v_0{}^2=0$, the cohomology of this operator has a graded vertex superalgebra structure. More generally, one may use any weight 1 field whose residue has square zero. The usual method is to tensor with fermions, as one then has a canonical differential. An important special case is quantum Drinfeld-Sokolov reduction applied to affine Kac–Moody algebras to obtain affine W-algebras as degree 0 cohomology. These W algebras also admit constructions as vertex subalgebras of free bosons given by kernels of screening operators.

Additional Examples

- The monster vertex algebra $V^\natural$ (also called the "moonshine module"), the key to Borcherds's proof of the Monstrous moonshine conjectures, was constructed by Frenkel, Lepowsky, and Meurman in 1988. It is notable because its partition function is the modular invariant $j-744$, and its automorphism group is the largest sporadic simple group, known as the monster group. It is constructed by orbifolding the Leech lattice VOA by the order 2 automorphism induced by reflecting the Leech lattice in the origin. That is, one forms the direct sum of the Leech lattice VOA with the twisted module, and takes the fixed points under an induced involution. Frenkel, Lepowsky, and Meurman conjectured in 1988 that $V^\natural$ is the unique holomorphic vertex operator algebra with central charge 24, and partition function $j-744$. This conjecture is still open.

- Chiral de Rham complex: Malikov, Schechtman, and Vaintrob showed that by a method of localization, one may canonically attach a bcβγ (boson-fermion superfield) system to a smooth complex manifold. This complex of sheaves has a distinguished differential, and the global cohomology is a vertex superalgebra. Ben-Zvi, Heluani, and Szczesny showed that a Riemannian metric on the manifold induces an $N=1$ superconformal structure, which is promoted to an $N=2$ structure if the metric is Kähler and Ricci-flat, and a hyperKähler structure induces an $N=4$ structure. Borisov and Libgober showed that one may obtain the two-variable elliptic genus of a compact complex manifold from the cohomology of Chiral de Rham - if the manifold is Calabi-Yau, then this genus is a weak Jacobi form.

Related Algebraic Structures

- If one considers only the singular part of the OPE in a vertex algebra, one arrives at the definition of a Lie conformal algebra. Since one is often only concerned with the singular part of the OPE, this makes Lie conformal algebras a natural object to study. There is a functor from vertex algebras to Lie conformal algebras that forgets the regular part of OPEs, and it has a left adjoint, called the "universal vertex algebra" functor. Vacuum modules of affine Kac–Moody algebras and Virasoro vertex algebras are universal vertex algebras, and in particular, they can be described very concisely once the background theory is developed.

- There are several generalizations of the notion of vertex algebra in the literature. Some mild generalizations involve a weakening of the locality axiom to allow monodromy, e.g., the *abelian intertwining algebras* of Dong and Lepowsky. One may view these roughly as vertex algebra objects in a braided tensor category of graded vector spaces, in much the same way that a vertex superalgebra is such an object in the category of super vector spaces. More complicated generalizations relate to q-deformations and representations of quantum groups, such as in work of Frenkel–Reshetikhin, Etingof–Kazhdan, and Li.

- Beilinson and Drinfeld introduced a sheaf-theoretic notion of *chiral algebra* that is closely related to the notion of vertex algebra, but is defined without using any visible power series. Given an algebraic curve X, a chiral algebra on X is a D_X-module A equipped with a multiplication operation $j_* j^* (A \boxtimes A) \to \Delta_* A$ on $X \times X$ that satisfies an associativity condition. They also introduced an equivalent notion of *factorization algebra* that is a system of quasicoherent sheaves on all finite products of the curve, together with a compatibility condition involving pullbacks to the complement of various diagonals. Any translation-equivariant chiral algebra on the affine line can be identified with a vertex algebra by taking the fiber at a point, and there is a natural way to attach a chiral algebra on a smooth algebraic curve to any vertex operator algebra.

References

- Frenkel, Igor; Lepowsky, James; Meurman, Arne (1988), Vertex operator algebras and the Monster, Pure and Applied Mathematics, 134, Academic Press, ISBN 0-12-267065-5

- Kac, Victor (1998), Vertex algebras for beginners, University Lecture Series, 10 (2nd ed.), American

Mathematical Society, ISBN 0-8218-1396-X

- Frenkel, Edward; Ben-Zvi, David (2001), Vertex algebras and Algebraic Curves, Mathematical Surveys and Monographs (88), American Mathematical Society, ISBN 0-8218-2894-0

- Lothaire, M. (2011) [2002], Algebraic combinatorics on words, Encyclopedia of Mathematics and Its Applications, 90, Cambridge University Press, ISBN 978-0-521-18071-9, Zbl 1221.68183

- Hille, Einar; Phillips, Ralph S. (1974), Functional analysis and semi-groups, American Mathematical Society, ISBN 0821874640, MR 0423094.

- Grillet, Pierre A. (1995), Semigroups: An Introduction to the Structure Theory, Marcel Dekker, ISBN 978-0-8247-9662-4, Zbl 0830.20079.

- Hollings, Christopher (2014), Mathematics across the Iron Curtain: A History of the Algebraic Theory of Semigroups, American Mathematical Society, ISBN 978-1-4704-1493-1, Zbl 06329297.

- Robinson, Derek John Scott (1996), A course in the theory of groups, Berlin, New York: Springer-Verlag, ISBN 978-0-387-94461-6.

- Lang, Serge (2002), Algebra, Graduate Texts in Mathematics, 211 (Revised third ed.), New York: Springer-Verlag, ISBN 978-0-387-95385-4, MR 1878556

- Herstein, Israel Nathan (1996), Abstract algebra (3rd ed.), Upper Saddle River, NJ: Prentice Hall Inc., ISBN 978-0-13-374562-7, MR 1375019.

- Fulton, William; Harris, Joe (1991). Representation theory. A first course. Graduate Texts in Mathematics, Readings in Mathematics. 129. New York: Springer-Verlag. ISBN 978-0-387-97495-8. MR 1153249, ISBN 978-0-387-97527-6..

- Rowen, Louis Halle (2008), "Definition 21B.1.", Graduate Algebra: Noncommutative View, Graduate Studies in Mathematics, American Mathematical Society, p. 321, ISBN 0-8218-8408-5

Algebraic Equation: An Overview

In mathematics, any equation which is represented in the form of P=Q is an algebraic equation. Linear equation is an algebraic equation in which either a term is constant of a product or a constant. This section is an overview of the subject matter incorporating all the major aspects of algebraic equation.

Algebraic Equation

In mathematics, an algebraic equation or polynomial equation is an equation of the form

$$P = Q$$

where P and Q are polynomials with coefficients in some field, often the field of the rational numbers. For most authors, an algebraic equation is *univariate*, which means that it involves only one variable. On the other hand, a polynomial equation may involve several variables, in which case it is called *multivariate* and the term *polynomial equation* is usually preferred to *algebraic equation*.

For example,

$$x^5 - 3x + 1 = 0$$

is an algebraic equation with integer coefficients and

$$y^4 + \frac{xy}{2} = \frac{x^3}{3} - xy^2 + y^2 - \frac{1}{7}$$

is a multivariate polynomial equation over the rationals.

Some but not all polynomial equations with rational coefficients have a solution that is an algebraic expression that can be found using a finite number of operations that involve only those same types of coefficients (that is, can be solved algebraically). This can be done for all such equations of degree one, two, three, or four; but for degree five or more it can only be done for some equations, not for all. A large amount of research has been devoted to compute efficiently accurate approximations of the real or complex solutions of a univariate algebraic equation and of the common solutions of several multivariate polynomial equations.

History

The study of algebraic equations is probably as old as mathematics: the Babylonian mathematicians, as early as 2000 BC could solve some kinds of quadratic equations (displayed on Old Babylonian clay tablets).

Univariate algebraic equations over the rationals (i.e., with rational coefficients) have a very long history. Ancient mathematicians wanted the solutions in the form of radical expressions, like

$x = \dfrac{1+\sqrt{5}}{2}$ for the positive solution of $x^2 - x - 1 = 0$. The ancient Egyptians knew how to solve equations of degree 2 in this manner. The Indian mathematician Brahmagupta (597–668 AD) explicitly described the quadratic formula in his treatise Brāhmasphuṭasiddhānta published in 628 AD, but written in words instead of symbols. In the 9th century Muhammad ibn Musa al-Khwarizmi and other Islamic mathematicians derived the quadratic formula, the general solution of equations of degree 2, and recognized the importance of the discriminant. During the Renaissance in 1545, Gerolamo Cardano published the solution of Scipione del Ferro and Niccolò Fontana Tartaglia to equations of degree 3 and that of Lodovico Ferrari for equations of degree 4. Finally Niels Henrik Abel proved, in 1824, that equations of degree 5 and higher do not have general solutions using radicals. Galois theory, named after Évariste Galois, showed that some equations of at least degree 5 do not even have an idiosyncratic solution in radicals, and gave criteria for deciding if an equation is in fact solvable using radicals.

Areas of Study

The algebraic equations are the basis of a number of areas of modern mathematics: Algebraic number theory is the study of (univariate) algebraic equations over the rationals (that is, with rational coefficients). Galois theory has been introduced by Évariste Galois for getting criteria deciding if an algebraic equation may be solved in terms of radicals. In field theory, an algebraic extension is an extension such that every element is a root of an algebraic equation over the base field. Transcendental number theory is the study of the real numbers which are not solutions to an algebraic equation over the rationals. A Diophantine equation is a (usually multivariate) polynomial equation with integer coefficients for which one is interested in the integer solutions. Algebraic geometry is the study of the solutions in an algebraically closed field of multivariate polynomial equations.

Two equations are equivalent if they have the same set of solutions. In particular the equation $P = Q$ is equivalent with $P - Q = 0$. It follows that the study of algebraic equations is equivalent to the study of polynomials.

A polynomial equation over the rationals can always be converted to an equivalent one in which the coefficients are integers. For example, multiplying through by $42 = 2 \cdot 3 \cdot 7$ and grouping its terms in the first member, the previously mentioned polynomial equation $y^4 + \dfrac{xy}{2} = \dfrac{x^3}{3} - xy^2 + y^2 - \dfrac{1}{7}$ becomes

$$42y^4 + 21xy - 14x^3 + 42xy^2 - 42y^2 + 6 = 0.$$

Because sine, exponentiation, and $1/T$ are not polynomial functions,

$$e^T x^2 + \frac{1}{T}xy + \sin(T)z - 2 = 0$$

is *not* a polynomial equation in the four variables x, y, z, and T over the rational numbers. However, it is a polynomial equation in the three variables x, y, and z over the field of the elementary functions in the variable T.

Solutions

As for any equation, the *solutions* of an equation are the values of the variables for which the equation is true. For univariate algebraic equations these are also called roots, even if, properly speaking, one should say *the solutions of the algebraic equation P=0 are the* **roots** *of the polynomial P*. When solving an equation, it is important to specify in which set the solutions are allowed. For example, for an equation over the rationals one may look for solutions in which all the variables are integers. In this case the equation is a Diophantine equation. One may also be interested only in the real solutions. However, for univariate algebraic equations, the number of solutions is finite, and all solutions are contained in any algebraically closed field containing the coefficients—for example, the field of complex numbers in the case of equations over the rationals. It follows that without precision "root" and "solution" usually mean "solution in an algebraically closed field".

Linear Equation

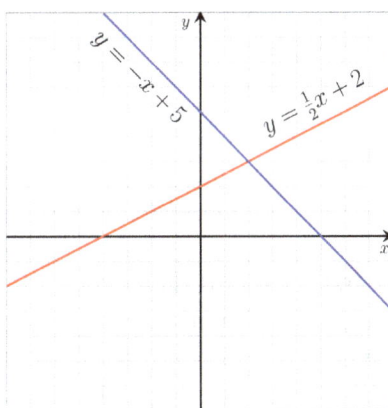

Graph sample of linear equations.

A linear equation is an algebraic equation in which each term is either a constant or the product of a constant and (the first power of) a single variable. A simple example of a linear equation with only one variable, x, may be written in the form: $ax + b = 0$, where a and b are constants and $a \neq 0$. The constants may be numbers, parameters, or even non-linear functions of parameters, and the distinction between variables and parameters may depend on the problem.

Linear equations can have one or more variables. An example of a linear equation with three variables, x, y, and z, is given by: $ax + by + cz + d = 0$, where a, b, c, and d are constants and a, b, and c are non-zero. Linear equations occur frequently in most subareas of mathematics and especially in applied mathematics. While they arise quite naturally when modeling many phenomena, they are particularly useful since many non-linear equations may be reduced to linear equations by assuming that quantities of interest vary to only a small extent from some "background" state. An equation is linear if the sum of the exponents of the variables of each term is one.

Equations with exponents greater than one are non-linear. An example of a non-linear equation of two variables is $axy + b = 0$, where a and b are constants and $a \neq 0$. It has two variables, x and y, and is non-linear because the sum of the exponents of the variables in the first term, axy, is two.

This article considers the case of a single equation for which one searches the real solutions. All its content applies for complex solutions and, more generally for linear equations with coefficients and solutions in any field.

One Variable

A linear equation in one unknown x may always be rewritten

$$ax = b.$$

If $a \neq 0$, there is a unique solution

$$x = \frac{b}{a}.$$

If $a = 0$, then, when $b = 0$ every number is a solution of the equation, but if $b \neq 0$ there are no solutions (and the equation is said to be inconsistent.)

Two Variables

A common form of a linear equation in the two variables x and y is

$$y = mx + b,$$

where m and b designate constants (parameters). The origin of the name "linear" comes from the fact that the set of solutions of such an equation forms a straight line in the plane. In this particular equation, the constant m determines the slope or gradient of that line, and the constant term b determines the point at which the line crosses the y-axis, otherwise known as the y-intercept.

Since terms of linear equations cannot contain products of distinct or equal variables, nor any power (other than 1) or other function of a variable, equations involving terms such as xy, x^2, $y^{1/3}$, and $\sin(x)$ are *nonlinear*.

Forms for Two-dimensional Linear Equations

Linear equations can be rewritten using the laws of elementary algebra into several different forms. These equations are often referred to as the "equations of the straight line." In what follows, x, y, t, and θ are variables; other letters represent constants (fixed numbers).

General (or Standard) Form

In the general (or standard) form the linear equation is written as:

$$Ax + By = C,$$

where A and B are not both equal to zero. The equation is usually written so that $A \geq 0$, by convention. The graph of the equation is a straight line, and every straight line can be represented by an equation in the above form. If A is nonzero, then the x-intercept, that is, the x-coordinate of the point where the graph crosses the x-axis (where, y is zero), is C/A. If B is nonzero, then the y-intercept, that is the y-coordinate of the point where the graph crosses the y-axis (where x is zero), is C/B, and the slope of the line is $-A/B$. The general form is sometimes written as:

$$ax + by + c = 0,$$

where a and b are not both equal to zero. The two versions can be converted from one to the other by moving the constant term to the other side of the equal sign.

Slope–intercept Form

$$y = mx + b,$$

where m is the slope of the line and b is the y intercept, which is the y coordinate of the location where the line crosses the y axis. This can be seen by letting $x = 0$, which immediately gives $y = b$. It may be helpful to think about this in terms of $y = b + mx$; where the line passes through the point $(0, b)$ and extends to the left and right at a slope of m. Vertical lines, having undefined slope, cannot be represented by this form.

Point–slope Form

$$y - y_1 = m(x - x_1),$$

where m is the slope of the line and (x_1, y_1) is any point on the line.

The point-slope form expresses the fact that the difference in the y coordinate between two points on a line (that is, $y - y_1$) is proportional to the difference in the x coordinate (that is, $x - x_1$). The proportionality constant is m (the slope of the line).

Two-point Form

$$y - y_1 = \frac{y_2 - y_1}{x_2 - x_1}(x - x_1),$$

where (x_1, y_1) and (x_2, y_2) are two points on the line with $x_2 \neq x_1$. This is equivalent to the point-slope form above, where the slope is explicitly given as $(y_2 - y_1)/(x_2 - x_1)$.

Multiplying both sides of this equation by $(x_2 - x_1)$ yields a form of the line generally referred to as the symmetric form:

$$(x_2 - x_1)(y - y_1) = (y_2 - y_1)(x - x_1).$$

Expanding the products and regrouping the terms leads to the general form:

$$x(y_2 - y_1) - y(x_2 - x_1) = x_1 y_2 - x_2 y_1$$

Using a determinant, one gets a determinant form, easy to remember:

$$\begin{vmatrix} x & y & 1 \\ x_1 & y_1 & 1 \\ x_2 & y_2 & 1 \end{vmatrix} = 0.$$

Intercept Form

$$\frac{x}{a} + \frac{y}{b} = 1,$$

where a and b must be nonzero. The graph of the equation has x-intercept a and y-intercept b. The intercept form is in standard form with $A/C = 1/a$ and $B/C = 1/b$. Lines that pass through the origin or which are horizontal or vertical violate the nonzero condition on a or b and cannot be represented in this form.

Matrix Form

Using the order of the standard form

$$Ax + By = C,$$

one can rewrite the equation in matrix form:

$$\begin{pmatrix} A & B \end{pmatrix} \begin{pmatrix} x \\ y \end{pmatrix} = \begin{pmatrix} C \end{pmatrix}.$$

Further, this representation extends to systems of linear equations.

$$A_1 x + B_1 y = C_1,$$

$$A_2 x + B_2 y = C_2,$$

becomes:

$$\begin{pmatrix} A_1 & B_1 \\ A_2 & B_2 \end{pmatrix} \begin{pmatrix} x \\ y \end{pmatrix} = \begin{pmatrix} C_1 \\ C_2 \end{pmatrix}.$$

Since this extends easily to higher dimensions, it is a common representation in linear algebra, and in computer programming. There are named methods for solving system of linear equations, like Gauss-Jordan which can be expressed as matrix elementary row operations.

Parametric Form

$$x = Tt + U$$

and

$$y = Vt + W.$$

Two simultaneous equations in terms of a variable parameter t, with slope $m = V / T$, x-intercept $(VU - WT) / V$ and y-intercept $(WT - VU) / T$. This can also be related to the two-point form, where $T = p - h$, $U = h$, $V = q - k$, and $W = k$:

$$x = (p - h)t + h$$

and

$$y = (q - k)t + k.$$

In this case t varies from 0 at point (h,k) to 1 at point (p,q), with values of t between 0 and 1 providing interpolation and other values of t providing extrapolation.

2D Vector Determinant Form

The equation of a line can also be written as the determinant of two vectors. If P_1 and P_2 are unique points on the line, then P will also be a point on the line if the following is true:

$$\det(\overrightarrow{P_1P}, \overrightarrow{P_1P_2}) = 0.$$

One way to understand this formula is to use the fact that the determinant of two vectors on the plane will give the area of the parallelogram they form. Therefore, if the determinant equals zero then the parallelogram has no area, and that will happen when two vectors are on the same line.

To expand on this we can say that $P_1 = (x_1, y_1)$, $P_2 = (x_2, y_2)$ and $P = (x, y)$. Thus $\overrightarrow{P_1P} = (x - x_1, y - y_1)$ and $\overrightarrow{P_1P_2} = (x_2 - x_1, y_2 - y_1)$, then the above equation becomes:

$$\det \begin{pmatrix} x - x_1 & y - y_1 \\ x_2 - x_1 & y_2 - y_1 \end{pmatrix} = 0.$$

Thus,

$$(x - x_1)(y_2 - y_1) - (y - y_1)(x_2 - x_1) = 0.$$

Ergo,

$$(x - x_1)(y_2 - y_1) = (y - y_1)(x_2 - x_1).$$

Then dividing both side by $(x_2 - x_1)$ would result in the "Two-point form" shown above, but leaving it here allows the equation to still be valid when $x_1 = x_2$.

Special Cases

$y = b$

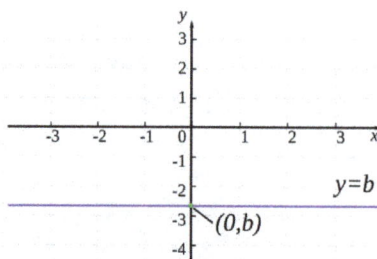

Horizontal Line $y = b$

This is a special case of the standard form where $A = 0$ and $B = 1$, or of the slope-intercept form where the slope $m = 0$. The graph is a horizontal line with y-intercept equal to b. There is no x-intercept, unless $b = 0$, in which case the graph of the line is the x-axis, and so every real number is an x-intercept.

$x = a$

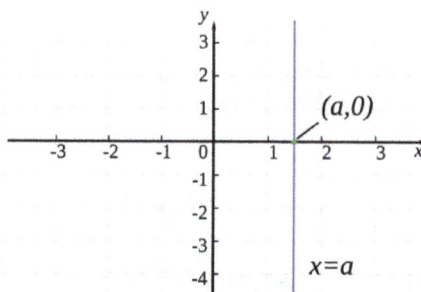

Vertical Line $x = a$

This is a special case of the standard form where $A = 1$ and $B = 0$. The graph is a vertical line with x-intercept equal to a. The slope is undefined. There is no y-intercept, unless $a = 0$, in which case the graph of the line is the y-axis, and so every real number is a y-intercept. This is the only type of straight line which is not the graph of a function (it obviously fails the vertical line test).

Connection with Linear Functions

A linear equation, written in the form $y = f(x)$ whose graph crosses the origin $(x,y) = (0,0)$, that is, whose y-intercept is 0, has the following properties:

$$f(x_1 + x_2) = f(x_1) + f(x_2)$$

and

$$f(ax) = af(x),$$

where a is any scalar. A function which satisfies these properties is called a *linear function* (or *linear operator*, or more generally a *linear map*). However, linear equations that have non-zero y-intercepts, when written in this manner, produce functions which will have neither property above and hence are not linear functions in this sense. They are known as *affine functions*.

Examples

An everyday example of the use of different forms of linear equations is computation of tax with tax brackets. This is commonly done using either point–slope form or slope–intercept form.

More than Two Variables

A linear equation can involve more than two variables. Every linear equation in n unknowns may be rewritten

$$a_1 x_1 + a_2 x_2 + \cdots + a_n x_n = b,$$

where, $a_1, a_2, ..., a_n$ represent numbers, called the *coefficients*, $x_1, x_2, ..., x_n$ are the unknowns, and b is called the *constant term*. When dealing with three or fewer variables, it is common to use x, y and z instead of x_1, x_2 and x_3.

If all the coefficients are zero, then either $b \neq 0$ and the equation does not have any solution, or $b = 0$ and every set of values for the unknowns is a solution.

If at least one coefficient is nonzero, a permutation of the subscripts allows to suppose $a_1 \neq 0$, and rewrite the equation

$$x_1 = \frac{b}{a_1} - \frac{a_2}{a_1} x_2 - \cdots - \frac{a_n}{a_1} x_n.$$

In other words, if $a_i \neq 0$, one may choose arbitrary values for all the unknowns except x_i, and express x_i in term of these values.

If $n = 3$ the set of the solutions is a plane in a three-dimensional space. More generally, the set of the solutions is an $(n - 1)$-dimensional hyperplane in a n-dimensional Euclidean space (or affine space if the coefficients are complex numbers or belong to any field).

Polynomial

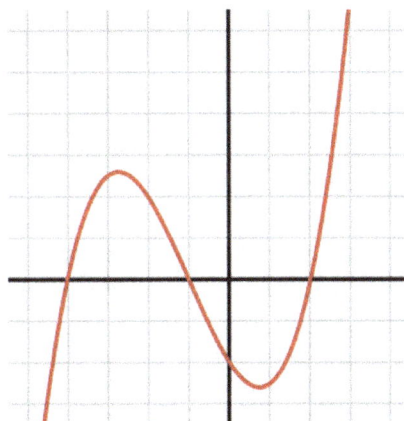

The graph of a polynomial function of degree 3

In mathematics, a polynomial is an expression consisting of variables (or indeterminates) and coefficients, that involves only the operations of addition, subtraction, multiplication, and non-negative integer exponents. An example of a polynomial of a single indeterminate x is $x^2 - 4x + 7$. An example in three variables is $x^3 + 2xyz^2 - yz + 1$.

Polynomials appear in a wide variety of areas of mathematics and science. For example, they are used to form polynomial equations, which encode a wide range of problems, from elementary word problems to complicated problems in the sciences; they are used to define polynomial functions, which appear in settings ranging from basic chemistry and physics to economics and social science; they are used in calculus and numerical analysis to approximate other functions. In advanced mathematics, polynomials are used to construct polynomial rings and algebraic varieties, central concepts in algebra and algebraic geometry.

Etymology

The word *polynomial* joins two diverse roots: the Greek *poly*, meaning "many," and the Latin *nomen*, or name. It was derived from the term *binomial* by replacing the Latin root *bi-* with the Greek *poly-*. The word *polynomial* was first used in the 17th century.

Notation and Terminology

The x occurring in a polynomial is commonly called either a *variable* or an *indeterminate*. When the polynomial is considered as an expression, x is a fixed symbol which does not have any value (its value is "indeterminate"). It is thus more correct to call it an "indeterminate". However, when one considers the function defined by the polynomial, then x represents the argument of the function, and is therefore called a "variable". Many authors use these two words interchangeably.

It is a common convention to use uppercase letters for the indeterminates and the corresponding lowercase letters for the variables (arguments) of the associated function.

It may be confusing that a polynomial P in the indeterminate x may appear in the formulas either as P or as $P(x)$.

Normally, the name of the polynomial is P, not $P(x)$. However, if a denotes a number, a variable, another polynomial, or, more generally any expression, then $P(a)$ denotes, by convention, the result of substituting x by a in P. Thus, the polynomial P defines the function

$$a \mapsto P(a),$$

which is the polynomial function associated to P.

Frequently, when using this function, one supposes that a is a number. However one may use it over any domain where addition and multiplication are defined (any ring). In particular, when a is the indeterminate x, then the image of x by this function is the polynomial P itself (substituting x to x does not change anything). In other words,

$$P(x) = P.$$

This equality allows writing "let $P(x)$ be a polynomial" as a shorthand for "let P be a polynomial in the indeterminate x". On the other hand, when it is not necessary to emphasize the name of the indeterminate, many formulas are much simpler and easier to read if the name(s) of the indeterminate(s) do not appear at each occurrence of the polynomial.

Definition

A polynomial is an expression that can be built from constants and symbols called indeterminates or variables by means of addition, multiplication and exponentiation to a non-negative power. Two such expressions that may be transformed, one to the other, by applying the usual properties of commutativity, associativity and distributivity of addition and multiplication are considered as defining the same polynomial.

A polynomial in a single indeterminate x can always be written (or rewritten) in the form

$$a_n x^n + a_{n-1} x^{n-1} + \cdots + a_2 x^2 + a_1 x + a_0,$$

where $a_0, \ldots, a_n$ are constants and x is the indeterminate. The word "indeterminate" means that x represents no particular value, although any value may be substituted for it. The mapping that associates the result of this substitution to the substituted value is a function, called a *polynomial function*.

This can be expressed more concisely by using summation notation:

$$\sum_{i=0}^{n} a_i x^i$$

That is, a polynomial can either be zero or can be written as the sum of a finite number of non-zero terms. Each term consists of the product of a number—called the coefficient of the term—and a finite number of indeterminates, raised to nonnegative integer powers. The exponent on an indeterminate in a term is called the degree of that indeterminate in that term; the degree of the term is the sum of the degrees of the indeterminates in that term, and the degree of a polynomial is the largest degree of any one term with nonzero coefficient. Because $x = x^1$, the degree of an indeterminate without a written exponent is one.

A term and a polynomial with no indeterminates are called, respectively, a constant term and a constant polynomial. The degree of a constant term and of a nonzero constant polynomial is 0. The degree of the zero polynomial, 0, (which has no terms at all) is generally treated as not defined.

For example:

$$-5x^2 y$$

is a term. The coefficient is −5, the indeterminates are x and y, the degree of x is two, while the degree of y is one. The degree of the entire term is the sum of the degrees of each indeterminate in it, so in this example the degree is $2 + 1 = 3$.

Forming a sum of several terms produces a polynomial. For example, the following is a polynomial:

$$\underbrace{-3x^2}_{\substack{\text{term} \\ 1}} \underbrace{-5x}_{\substack{\text{term} \\ 2}} \underbrace{+4}_{\substack{\text{term} \\ 3}}.$$

It consists of three terms: the first is degree two, the second is degree one, and the third is degree zero.

Polynomials of small degree have been given specific names. A polynomial of degree zero is a *constant polynomial* or simply a *constant*. Polynomials of degree one, two or three are respectively *linear polynomials, quadratic polynomials* and *cubic polynomials*. For higher degrees the specific names are not commonly used, although *quartic polynomial* (for degree four) and *quintic polynomial* (for degree five) are sometimes used. The names for the degrees may be applied to the polynomial or to its terms. For example, in $x^2 + 2x + 1$ the term $2x$ is a linear term in a quadratic polynomial.

The polynomial 0, which may be considered to have no terms at all, is called the zero polynomial. Unlike other constant polynomials, its degree is not zero. Rather the degree of the zero polynomial is either left explicitly undefined, or defined as negative (either -1 or $-\infty$). These conventions are useful when defining Euclidean division of polynomials. The zero polynomial is also unique in that it is the only polynomial having an infinite number of roots. The graph of the zero polynomial, $f(x) = 0$, is the X-axis.

In the case of polynomials in more than one indeterminate, a polynomial is called *homogeneous* of degree n if *all* its non-zero terms have degree n. The zero polynomial is homogeneous, and, as homogeneous polynomial, its degree is undefined. For example, $x^3y^2 + 7x^2y^3 - 3x^5$ is homogeneous of degree 5.

The commutative law of addition can be used to rearrange terms into any preferred order. In polynomials with one indeterminate, the terms are usually ordered according to degree, either in "descending powers of x", with the term of largest degree first, or in "ascending powers of x". The polynomial in the example above is written in descending powers of x. The first term has coefficient 3, indeterminate x, and exponent 2. In the second term, the coefficient is -5. The third term is a constant. Because the *degree* of a non-zero polynomial is the largest degree of any one term, this polynomial has degree two.

Two terms with the same indeterminates raised to the same powers are called "similar terms" or "like terms", and they can be combined, using the distributive law, into a single term whose coefficient is the sum of the coefficients of the terms that were combined. It may happen that this makes the coefficient 0. Polynomials can be classified by the number of terms with nonzero coefficients, so that a one-term polynomial is called a monomial, a two-term polynomial is called a binomial, and a three-term polynomial is called a *trinomial*. The term "quadrinomial" is occasionally used for a four-term polynomial.

A polynomial in one indeterminate is called a *univariate polynomial*, a polynomial in more than one indeterminate is called a multivariate polynomial. A polynomial with two indeterminates is called a bivariate polynomial. These notions refer more to the kind of polynomials one is generally working with than to individual polynomials; for instance when working with univariate polynomials one does not exclude constant polynomials (which may result, for instance, from the subtraction of non-constant polynomials), although strictly speaking constant polynomials do not contain any indeterminates at all. It is possible to further classify multivariate polynomials as *bivariate*, *trivariate*, and so on, according to the maximum number of indeterminates allowed. Again, so that the set of objects under consideration be closed under subtraction, a study of trivariate polynomials usually allows bivariate polynomials, and so on. It is common, also, to say simply "polynomials in x, y, and z", listing the indeterminates allowed.

The *evaluation of a polynomial* consists of substituting a numerical value to each indeterminate and carrying out the indicated multiplications and additions. For polynomials in one indeterminate, the evaluation is usually more efficient (lower number of arithmetic operations to perform) using Horner's method:

$$(((\cdots((a_n x + a_{n-1})x + a_{n-2})x + \cdots + a_3)x + a_2)x + a_1)x + a_0.$$

Arithmetic

Polynomials can be added using the associative law of addition (grouping all their terms together into a single sum), possibly followed by reordering, and combining of like terms. For example, if

$$P = 3x^2 - 2x + 5xy - 2$$
$$Q = -3x^2 + 3x + 4y^2 + 8$$

then

$$P + Q = 3x^2 - 2x + 5xy - 2 - 3x^2 + 3x + 4y^2 + 8$$

which can be simplified to

$$P + Q = x + 5xy + 4y^2 + 6$$

To work out the product of two polynomials into a sum of terms, the distributive law is repeatedly applied, which results in each term of one polynomial being multiplied by every term of the other. For example, if

$$P = 2x + 3y + 5$$
$$Q = 2x + 5y + xy + 1$$

then

$$
\begin{aligned}
PQ = \quad & (2x \cdot 2x) & + & \ (2x \cdot 5y) & + & \ (2x \cdot xy) & + & \ (2x \cdot 1) \\
+ & \ (3y \cdot 2x) & + & \ (3y \cdot 5y) & + & \ (3y \cdot xy) & + & \ (3y \cdot 1) \\
+ & \ (5 \cdot 2x) & + & \ (5 \cdot 5y) & + & \ (5 \cdot xy) & + & \ (5 \cdot 1)
\end{aligned}
$$

which can be simplified to

$$PQ = 4x^2 + 21xy + 2x^2 y + 12x + 15y^2 + 3xy^2 + 28y + 5$$

Polynomial evaluation can be used to compute the remainder of polynomial division by a polynomial of degree one, because the remainder of the division of $f(x)$ by $(x - a)$ is $f(a)$. This is more efficient than the usual algorithm of division when the quotient is not needed.

A sum of polynomials is a polynomial.

A product of polynomials is a polynomial.

A composition of two polynomials is a polynomial, which is obtained by substituting a variable of the first polynomial by the second polynomial.

- The derivative of the polynomial $a_n x^n + a_{n-1} x^{n-1} + \dots + a_2 x^2 + a_1 x + a_0$ is the polynomial $n a_n x^{n-1} + (n-1) a_{n-1} x^{n-2} + \dots + 2 a_2 x + a_1$. If the set of the coefficients does not contain the integers (for example if the coefficients are integers modulo some prime number p), then $k a_k$ should be interpreted as the sum of a_k with itself, k times. For example, over the integers modulo p, the derivative of the polynomial $x^p + 1$ is the polynomial 0.

- A primitive integral or antiderivative of the polynomial $a_n x^n + a_{n-1} x^{n-1} + \dots + a_2 x^2 + a_1 x + a_0$ is the polynomial $a_n x^{n+1}/(n+1) + a_{n-1} x^n/n + \dots + a_2 x^3/3 + a_1 x^2/2 + a_0 x + c$, where c is an arbitrary constant. For instance, the antiderivatives of $x^2 + 1$ have the form $1/3 x^3 + x + c$.

As for the integers, two kinds of divisions are considered for the polynomials. The *Euclidean division of polynomials* that generalizes the Euclidean division of the integers. It results in two polynomials, a *quotient* and a *remainder* that are characterized by the following property of the polynomials: given two polynomials a and b such that $b \neq 0$, there exists a unique pair of polynomials, q, the quotient, and r, the remainder, such that $a = b q + r$ and degree(r) < degree(b) (here the polynomial zero is supposed to have a negative degree). By hand as well as with a computer, this division can be computed by the polynomial long division algorithm.

All polynomials with coefficients in a unique factorization domain (for example, the integers or a field) also have a factored form in which the polynomial is written as a product of irreducible polynomials and a constant. This factored form is unique up to the order of the factors and their multiplication by an invertible constant. In the case of the field of complex numbers, the irreducible factors are linear. Over the real numbers, they have the degree either one or two. Over the integers and the rational numbers the irreducible factors may have any degree. For example, the factored form of

$$5x^3 - 5$$

is

$$5(x-1)\left(x^2 + x + 1\right)$$

over the integers and the reals and

$$5(x-1)\left(x + \frac{1 + i\sqrt{3}}{2}\right)\left(x + \frac{1 - i\sqrt{3}}{2}\right)$$

over the complex numbers.

The computation of the factored form, called *factorization* is, in general, too difficult to be done by hand-written computation. However, efficient polynomial factorization algorithms are available in most computer algebra systems.

A formal quotient of polynomials, that is, an algebraic fraction wherein the numerator and denominator are polynomials, is called a "rational expression" or "rational fraction" and is not, in general, a polynomial. Division of a polynomial by a number, however, yields another polynomial. For example, $x^3/12$ is considered a valid term in a polynomial (and a polynomial by itself) because it is equivalent to $(1/12)x^3$ and $1/12$ is just a constant. When this expression is used as a term, its coefficient is therefore $1/12$. For similar reasons, if complex coefficients are allowed, one may have a single term like $(2 + 3i)\,x^3$; even though it looks like it should be expanded to two terms, the complex number $2 + 3i$ is one complex number, and is the coefficient of that term. The expression $1/(x^2 + 1)$ is not a polynomial because it includes division by a non-constant polynomial. The expression $(5 + y)^x$ is not a polynomial, because it contains an indeterminate used as exponent.

Because subtraction can be replaced by addition of the opposite quantity, and because positive integer exponents can be replaced by repeated multiplication, all polynomials can be constructed from constants and indeterminates using only addition and multiplication.

Polynomial Functions

A *polynomial function* is a function that can be defined by evaluating a polynomial. A function f of one argument is thus a polynomial function if it satisfies.

$$f(x) = a_n x^n + a_{n-1} x^{n-1} + \cdots + a_2 x^2 + a_1 x + a_0$$

for all arguments x, where n is a non-negative integer and $a_0, a_1, a_2, ..., a_n$ are constant coefficients.

For example, the function f, taking real numbers to real numbers, defined by

$$f(x) = x^3 - x$$

is a polynomial function of one variable. Polynomial functions of multiple variables are similarly defined, using polynomials in multiple indeterminates, as in

$$f(x, y) = 2x^3 + 4x^2 y + xy^5 + y^2 - 7.$$

An example is also the function $f(x) = \cos(2\arccos(x))$ which, although it does not look like a polynomial, is a polynomial function on $[-1,1]$ because for every x from $f(x) = 2x^2 - 1$ it is true that $f(x) = 2x^2 - 1$.

Polynomial functions are a class of functions having many important properties. They are all continuous, smooth, entire, computable, etc.

Graphs

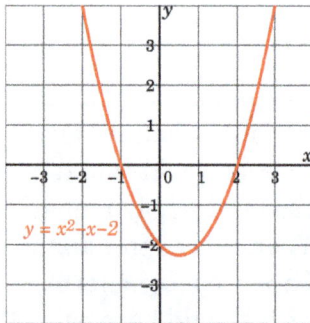

Polynomial of degree 2:
$f(x) = x^2 - x - 2$
$= (x + 1)(x - 2)$

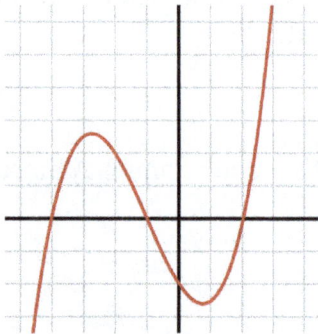

Polynomial of degree 3:
$f(x) = x^3/4 + 3x^2/4 - 3x/2 - 2$
$= 1/4\,(x + 4)(x + 1)(x - 2)$

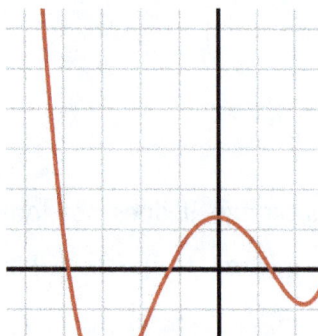

Polynomial of degree 4:
$f(x) = 1/14\,(x + 4)(x + 1)(x - 1)(x - 3)$
$+ 0.5$

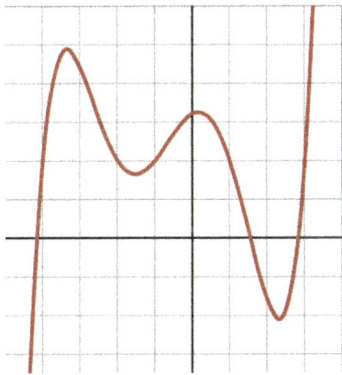

Polynomial of degree 5:
$f(x) = 1/20\ (x + 4)(x + 2)(x + 1\ )(x − 1)$
$(x − 3)+ 2$

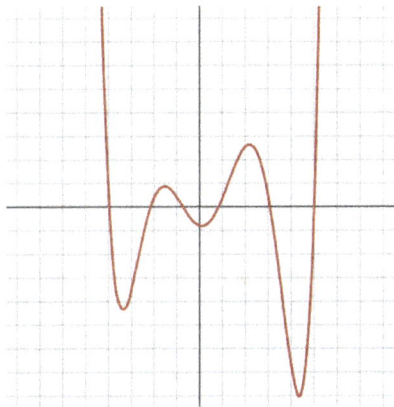

Polynomial of degree 6:
$f(x) = 1/100\ (x^6 − 2x^5 − 26x^4 + 28x^3$
$+ 145x^2 - 26x - 80)$

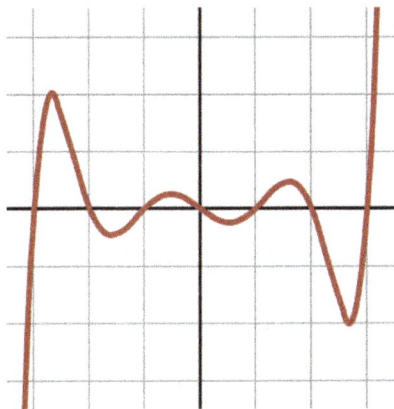

Polynomial of degree 7:
$f(x) = (x − 3)(x − 2)(x − 1)(x)(x + 1)(x + 2)$
$(x + 3)$

A polynomial function in one real variable can be represented by a graph.

- The graph of the zero polynomial

    $f(x) = 0$

    is the x-axis.

- The graph of a degree 0 polynomial

    $f(x) = a_0$, where $a_0 \neq 0$,

    is a horizontal line with y-intercept a_0

- The graph of a degree 1 polynomial (or linear function)

    $f(x) = a_0 + a_1 x$, where $a_1 \neq 0$,

    is an oblique line with y-intercept a_0 and slope a_1.

- The graph of a degree 2 polynomial

    $f(x) = a_0 + a_1 x + a_2 x^2$, where $a_2 \neq 0$

    is a parabola.

- The graph of a degree 3 polynomial

    $f(x) = a_0 + a_1 x + a_2 x^2 + a_3 x^3$, where $a_3 \neq 0$

    is a cubic curve.

- The graph of any polynomial with degree 2 or greater

    $f(x) = a_0 + a_1 x + a_2 x^2 + ... + a_n x^n$, where $a_n \neq 0$ and $n \geq 2$

    is a continuous non-linear curve.

The graph of a non-constant (univariate) polynomial always tends to infinity when the variable increases indefinitely (in absolute value).

Polynomial graphs are analyzed in calculus using intercepts, slopes, concavity, and end behavior.

Equations

A *polynomial equation*, also called *algebraic equation*, is an equation of the form

$$a_n x^n + a_{n-1} x^{n-1} + \cdots + a_2 x^2 + a_1 x + a_0 = 0.$$

For example,

$$3x^2 + 4x - 5 = 0$$

is a polynomial equation.

In case of a univariate polynomial equation, the variable is considered an unknown, and one seeks to find the possible values for which both members of the equation evaluate to the same value (in general more than one solution may exist). A polynomial equation stands in contrast to a *polynomial identity* like $(x + y)(x − y) = x^2 − y^2$, where both expressions represent the same polynomial in different forms, and as a consequence any evaluation of both members gives a valid equality.

In elementary algebra, methods such as the quadratic formula are given for solving all first degree and second degree polynomial equations in one variable. There are also formulas for the cubic and quartic equations. For higher degrees, the Abel–Ruffini theorem asserts that there can not exist a general formula in radicals. However, root-finding algorithms may be used to find numerical approximations of the roots of a polynomial expression of any degree.

The number of real solutions of a polynomial equation with real coefficients may not exceed the degree, and equals the degree when the complex solutions are counted with their multiplicity. This fact is called the fundamental theorem of algebra.

Solving Equations

Every polynomial P in x corresponds to a function, $f(x) = P$ (where the occurrences of x in P are interpreted as the argument of f), called the *polynomial function* of P; the equation in x setting $f(x) = 0$ is the *polynomial equation* corresponding to P. The solutions of this equation are called the *roots* of the polynomial; they are the *zeroes* of the function f (corresponding to the points where the graph of f meets the x-axis). A number a is a root of P if and only if the polynomial $x − a$ (of degree one in x) divides P. It may happen that $x − a$ divides P more than once: if $(x − a)^2$ divides P then a is called a *multiple root* of P, and otherwise a is called a *simple root* of P. If P is a nonzero polynomial, there is a highest power m such that $(x − a)^m$ divides P, which is called the *multiplicity* of the root a in P. When P is the zero polynomial, the corresponding polynomial equation is trivial, and this case is usually excluded when considering roots: with the above definitions every number would be a root of the zero polynomial, with undefined (or infinite) multiplicity. With this exception made, the number of roots of P, even counted with their respective multiplicities, cannot exceed the degree of P. The relation between the roots of a polynomial and its coefficients is described by Vieta's formulas.

Some polynomials, such as $x^2 + 1$, do not have any roots among the real numbers. If, however, the set of allowed candidates is expanded to the complex numbers, every non-constant polynomial has at least one root; this is the fundamental theorem of algebra. By successively dividing out factors $x − a$, one sees that any polynomial with complex coefficients can be written as a constant (its leading coefficient) times a product of such polynomial factors of degree 1; as a consequence, the number of (complex) roots counted with their multiplicities is exactly equal to the degree of the polynomial.

There is a difference between approximating roots and finding exact expressions for roots. Formulas for expressing the roots of polynomials of degree 2 in terms of square roots have been known since ancient times, and for polynomials of degree 3 or 4 similar formulas (using cube roots in addition to square roots) were found in the 16th century (cubic function and quartic function for the formulas and Niccolò Fontana Tartaglia, Lodovico Ferrari, Gerolamo Cardano, and François Viète for historical details). But formulas for degree 5 eluded researchers for several centuries. In 1824, Niels Henrik Abel proved the striking result that there can be no general (finite) formula, involving only arithmetic operations and radicals, that expresses the roots of a polynomial of degree 5 or greater in terms of its coefficients. In 1830, Évariste Galois, studying the permutations of the roots of a polynomial, extended the Abel–Ruffini theorem by showing that, given a polynomial equation, one may decide whether it is solvable by radicals, and, if it is, solve it. This result marked the start of Galois theory and group theory, two important branches of modern mathematics. Galois himself noted that the computations implied by his method were impracticable. Nevertheless, formulas for solvable equations of degrees 5 and 6 have been published.

Numerical approximation of roots of polynomials in one unknown is easily done on a computer by the Jenkins–Traub method, Laguerre's method, Durand–Kerner method, or by some other root-finding algorithm.

For polynomials in more than one indeterminate the notion of root does not exist, and there are usually infinitely many combinations of values for the variables for which the polynomial function takes the value zero. However, for certain *sets* of such polynomials it may happen that for only finitely many combinations all polynomial functions take the value zero.

For a set of polynomial equations in several unknowns, there are algorithms to decide whether they have a finite number of complex solutions. If the number of solutions is finite, there are algorithms to compute the solutions. The methods underlying these algorithms are described in the article System of polynomial equations.

The special case where all the polynomials are of degree one is called a system of linear equations, for which another range of different solution methods exist, including the classical Gaussian elimination.

A polynomial equation for which one is interested only in the solutions which are integers is called a Diophantine equation. Solving Diophantine equations is a very hard task. It has been proved that there cannot be any general algorithm for solving them, and even for deciding whether the set of solutions is empty. Some of the most famous problems that have been solved during the fifty last years are related to Diophantine equations, such as Fermat's Last Theorem.

Generalizations

There are several generalizations of the concept of polynomials.

Trigonometric Polynomials

A trigonometric polynomial is a finite linear combination of functions $\sin(nx)$ and $\cos(nx)$ with n taking on the values of one or more natural numbers. The coefficients may be taken as real numbers, for real-valued functions.

If $\sin(nx)$ and $\cos(nx)$ are expanded in terms of $\sin(x)$ and $\cos(x)$, a trigonometric polynomial becomes a polynomial in the two variables $\sin(x)$ and $\cos(x)$ (using List of trigonometric identities#-Multiple-angle formulae). Conversely, every polynomial in $\sin(x)$ and $\cos(x)$ may be converted, with Product-to-sum identities, into a linear combination of functions $\sin(nx)$ and $\cos(nx)$. This equivalence explains why linear combinations are called polynomials.

For complex coefficients, there is no difference between such a function and a finite Fourier series.

Trigonometric polynomials are widely used, for example in trigonometric interpolation applied to the interpolation of periodic functions. They are used also in the discrete Fourier transform.

Matrix Polynomials

A matrix polynomial is a polynomial with matrices as variables. Given an ordinary, scalar-valued polynomial

$$P(x) = \sum_{i=0}^{n} a_i x^i = a_0 + a_1 x + a_2 x^2 + \cdots + a_n x^n,$$

this polynomial evaluated at a matrix A is

$$P(A) = \sum_{i=0}^{n} a_i A^i = a_0 I + a_1 A + a_2 A^2 + \cdots + a_n A^n,$$

where I is the identity matrix.

A matrix polynomial equation is an equality between two matrix polynomials, which holds for the specific matrices in question. A matrix polynomial identity is a matrix polynomial equation which holds for all matrices A in a specified matrix ring $M_n(R)$.

Laurent Polynomials

Laurent polynomials are like polynomials, but allow negative powers of the variable(s) to occur.

Rational Functions

A rational fraction is the quotient (algebraic fraction) of two polynomials. Any algebraic expression that can be rewritten as a rational fraction is a rational function.

While polynomial functions are defined for all values of the variables, a rational function is defined only for the values of the variables for which the denominator is not zero.

The rational fractions include the Laurent polynomials, but do not limit denominators to powers of an indeterminate.

Power Series

Formal power series are like polynomials, but allow infinitely many non-zero terms to occur, so that they do not have finite degree. Unlike polynomials they cannot in general be explicitly and fully written down (just like irrational numbers cannot), but the rules for manipulating their terms are the same as for polynomials. Non-formal power series also generalize polynomials, but the multiplication of two power series may not converge.

Other Examples

- A bivariate polynomial where the second variable is substituted by an exponential function applied to the first variable, for example $P(x, e^x)$, may be called an exponential polynomial.

Applications

Calculus

The simple structure of polynomial functions makes them quite useful in analyzing general functions using polynomial approximations. An important example in calculus is Taylor's theorem, which roughly states that every differentiable function locally looks like a polynomial function, and the Stone–Weierstrass theorem, which states that every continuous function defined on a compact interval of the real axis can be approximated on the whole interval as closely as desired by a polynomial function.

Calculating derivatives and integrals of polynomial functions is particularly simple. For the polynomial function

$$\sum_{i=0}^{n} a_i x^i$$

the derivative with respect to x is

$$\sum_{i=1}^{n} a_i i x^{i-1}$$

and the indefinite integral is

$$\sum_{i=0}^{n} \frac{a_i}{i+1} x^{i+1} + c.$$

Abstract Algebra

In abstract algebra, one distinguishes between *polynomials* and *polynomial functions*. A *polynomial f* in one indeterminate x over a ring R is defined as a formal expression of the form

$$f = a_n x^n + a_{n-1} x^{n-1} + \cdots + a_1 x^1 + a_0 x^0$$

where n is a natural number, the coefficients $a_0, \ldots, a_n$ are elements of R, and x is a formal symbol, whose powers x^i are just placeholders for the corresponding coefficients a_i, so that the given formal expression is just a way to encode the sequence $(a_0, a_1, \ldots)$, where there is an n such that $a_i = 0$ for all $i > n$. Two polynomials sharing the same value of n are considered equal if and only if the sequences of their coefficients are equal; furthermore any polynomial is equal to any polynomial with greater value of n obtained from it by adding terms in front whose coefficient is zero. These polynomials can be added by simply adding corresponding coefficients (the rule for extending by terms with zero coefficients can be used to make sure such coefficients exist). Thus each polynomial is actually equal to the sum of the terms used in its formal expression, if such a term $a_i x^i$ is interpreted as a polynomial that has zero coefficients at all powers of x other than x^i. Then to define multiplication, it suffices by the distributive law to describe the product of any two such terms, which is given by the rule

$ax^k bx^l = abx^{k+l}$ for all elements a, b of the ring R and all natural numbers k and l.

Thus the set of all polynomials with coefficients in the ring R forms itself a ring, the *ring of polynomials* over R, which is denoted by $R[x]$. The map from R to $R[x]$ sending r to rx^o is an injective homomorphism of rings, by which R is viewed as a subring of $R[x]$. If R is commutative, then $R[x]$ is an algebra over R.

One can think of the ring $R[x]$ as arising from R by adding one new element x to R, and extending in a minimal way to a ring in which x satisfies no other relations than the obligatory ones, plus commutation with all elements of R (that is $xr = rx$). To do this, one must add all powers of x and their linear combinations as well.

Formation of the polynomial ring, together with forming factor rings by factoring out ideals, are important tools for constructing new rings out of known ones. For instance, the ring (in fact field) of complex numbers, which can be constructed from the polynomial ring $R[x]$ over the real numbers by factoring out the ideal of multiples of the polynomial $x^2 + 1$. Another example is the construction of finite fields, which proceeds similarly, starting out with the field of integers modulo some prime number as the coefficient ring R.

If R is commutative, then one can associate to every polynomial P in $R[x]$, a *polynomial function f* with domain and range equal to R (more generally one can take domain and range to be the same unital associative algebra over R). One obtains the value $f(r)$ by substitution of the value R for the symbol x in P. One reason to distinguish between polynomials and polynomial functions is that over some rings different polynomials may give rise to the same polynomial function. This is not the case when R is the real or complex numbers, whence the two concepts are not always distinguished in analysis. An even more important reason to distinguish between polynomials and polynomial functions is that many operations on polynomials (like Euclidean division) require looking at what a polynomial is composed of as an expression rather than evaluating it at some constant value for x.

Divisibility

In commutative algebra, one major focus of study is *divisibility* among polynomials. If R is an integral domain and f and g are polynomials in $R[x]$, it is said that f *divides* g or f is a divisor of g if there exists a polynomial q in $R[x]$ such that $fq = g$. One can show that every zero gives rise to a linear divisor, or more formally, if f is a polynomial in $R[x]$ and r is an element of R such that $f(r) = 0$, then the polynomial $(x - r)$ divides f. The converse is also true. The quotient can be computed using the polynomial long division.

If F is a field and f and g are polynomials in $F[x]$ with $g \neq 0$, then there exist unique polynomials q and r in $F[x]$ with

$$f = qg + r$$

and such that the degree of r is smaller than the degree of g (using the convention that the polynomial o has a negative degree). The polynomials q and r are uniquely determined by f and g. This is called *Euclidean division, division with remainder* or *polynomial long division* and shows that the ring $F[x]$ is a Euclidean domain.

Analogously, *prime polynomials* (more correctly, *irreducible polynomials*) can be defined as *non-zero polynomials which cannot be factorized into the product of two non constant polynomials*. In the case of coefficients in a ring, *"non constant"* must be replaced by *"non constant or non unit"* (both definitions agree in the case of coefficients in a field). Any polynomial may be decomposed into the product of an invertible constant by a product of irreducible polynomials. If the coefficients belong to a field or a unique factorization domain this decomposition is unique up to the order of the factors and the multiplication of any non unit factor by a unit (and division of the unit factor by the same unit). When the coefficients belong to integers, rational numbers or a finite field, there are algorithms to test irreducibility and to compute the factorization into irreducible polynomials. These algorithms are not practicable for hand written computation, but are available in any computer algebra system. Eisenstein's criterion can also be used in some cases to determine irreducibility.

Other Applications

Polynomials serve to approximate other functions, such as the use of splines.

Polynomials are frequently used to encode information about some other object. The characteristic polynomial of a matrix or linear operator contains information about the operator's eigenvalues. The minimal polynomial of an algebraic element records the simplest algebraic relation satisfied by that element. The chromatic polynomial of a graph counts the number of proper colourings of that graph.

The term "polynomial", as an adjective, can also be used for quantities or functions that can be written in polynomial form. For example, in computational complexity theory the phrase *polynomial time* means that the time it takes to complete an algorithm is bounded by a polynomial function of some variable, such as the size of the input.

History

Determining the roots of polynomials, or "solving algebraic equations", is among the oldest problems in mathematics. However, the elegant and practical notation we use today only developed beginning in the 15th century. Before that, equations were written out in words. For example, an algebra problem from the Chinese Arithmetic in Nine Sections, circa 200 BCE, begins "Three sheafs of good crop, two sheafs of mediocre crop, and one sheaf of bad crop are sold for 29 dou." We would write $3x + 2y + z = 29$.

History of the Notation

The earliest known use of the equal sign is in Robert Recorde's *The Whetstone of Witte*, 1557. The signs + for addition, − for subtraction, and the use of a letter for an unknown appear in Michael Stifel's *Arithemetica integra*, 1544. René Descartes, in *La géometrie*, 1637, introduced the concept of the graph of a polynomial equation. He popularized the use of letters from the beginning of the alphabet to denote constants and letters from the end of the alphabet to denote variables, as can be seen above, in the general formula for a polynomial in one variable, where the a's denote constants and x denotes a variable. Descartes introduced the use of superscripts to denote exponents as well.

Differential Equation

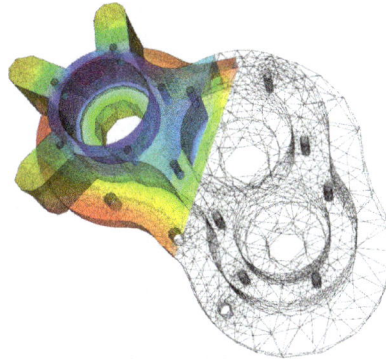

Visualization of heat transfer in a pump casing, created by solving the heat equation. Heat is being generated internally in the casing and being cooled at the boundary, providing a steady state temperature distribution.

A differential equation is a mathematical equation that relates some function with its derivatives. In applications, the functions usually represent physical quantities, the derivatives represent their rates of change, and the equation defines a relationship between the two. Because such relations are extremely common, differential equations play a prominent role in many disciplines including engineering, physics, economics, and biology.

In pure mathematics, differential equations are studied from several different perspectives, mostly concerned with their solutions—the set of functions that satisfy the equation. Only the simplest differential equations are solvable by explicit formulas; however, some properties of solutions of a given differential equation may be determined without finding their exact form.

If a self-contained formula for the solution is not available, the solution may be numerically approximated using computers. The theory of dynamical systems puts emphasis on qualitative analysis of systems described by differential equations, while many numerical methods have been developed to determine solutions with a given degree of accuracy.

History

Differential equations first came into existence with the invention of calculus by Newton and Leibniz. In Chapter 2 of his 1671 work "Methodus fluxionum et Serierum Infinitarum", Isaac Newton listed three kinds of differential equations:

$$\frac{dy}{dx} = f(x)$$

$$\frac{dy}{dx} = f(x, y)$$

$$x_1 \frac{\partial y}{\partial x_1} + x_2 \frac{\partial y}{\partial x_2} = y$$

He solves these examples and others using infinite series and discusses the non-uniqueness of solutions.

Jacob Bernoulli proposed the Bernoulli differential equation in 1695. This is an ordinary differential equation of the form

$$y' + P(x)y = Q(x)y^n$$

for which the following year Leibniz obtained solutions by simplifying it.

Historically, the problem of a vibrating string such as that of a musical instrument was studied by Jean le Rond d'Alembert, Leonhard Euler, Daniel Bernoulli, and Joseph-Louis Lagrange. In 1746, d'Alembert discovered the one-dimensional wave equation, and within ten years Euler discovered the three-dimensional wave equation.

The Euler–Lagrange equation was developed in the 1750s by Euler and Lagrange in connection with their studies of the tautochrone problem. This is the problem of determining a curve on which a weighted particle will fall to a fixed point in a fixed amount of time, independent of the starting point.

Lagrange solved this problem in 1755 and sent the solution to Euler. Both further developed Lagrange's method and applied it to mechanics, which led to the formulation of Lagrangian mechanics.

Fourier published his work on heat flow in *Théorie analytique de la chaleur* (The Analytic Theory of Heat), in which he based his reasoning on Newton's law of cooling, namely, that the flow of heat between two adjacent molecules is proportional to the extremely small difference of their temperatures. Contained in this book was Fourier's proposal of his heat equation for conductive diffusion of heat. This partial differential equation is now taught to every student of mathematical physics.

Example

For example, in classical mechanics, the motion of a body is described by its position and velocity as the time value varies. Newton's laws allow (given the position, velocity, acceleration and various forces acting on the body) one to express these variables dynamically as a differential equation for the unknown position of the body as a function of time.

In some cases, this differential equation (called an equation of motion) may be solved explicitly.

An example of modelling a real world problem using differential equations is the determination of the velocity of a ball falling through the air, considering only gravity and air resistance. The ball's acceleration towards the ground is the acceleration due to gravity minus the acceleration due to air resistance.

Gravity is considered constant, and air resistance may be modeled as proportional to the ball's velocity. This means that the ball's acceleration, which is a derivative of its velocity, depends on the velocity (and the velocity depends on time). Finding the velocity as a function of time involves solving a differential equation and verifying its validity.

Types

Differential equations can be divided into several types. Apart from describing the properties of the equation itself, these classes of differential equations can help inform the choice of approach to a solution. Commonly used distinctions include whether the equation is: Ordinary/Partial, Linear/Non-linear, and Homogeneous/Inhomogeneous. This list is far from exhaustive; there are many other properties and subclasses of differential equations which can be very useful in specific contexts.

Ordinary Differential Equations

An ordinary differential equation (*ODE*) is an equation containing a function of one independent variable and its derivatives. The term *"ordinary"* is used in contrast with the term partial differential equation which may be with respect to *more than* one independent variable.

Linear differential equations, which have solutions that can be added and multiplied by coefficients, are well-defined and understood, and exact closed-form solutions are obtained. By contrast, ODEs that lack additive solutions are nonlinear, and solving them is far more intricate, as one can rarely represent them by elementary functions in closed form: Instead, exact and analytic solutions of ODEs are in series or integral form. Graphical and numerical methods, applied by hand or by computer, may approximate solutions of ODEs and perhaps yield useful information, often sufficing in the absence of exact, analytic solutions.

Partial Differential Equations

A partial differential equation (*PDE*) is a differential equation that contains unknown multivariable functions and their partial derivatives. (This is in contrast to ordinary differential equations, which deal with functions of a single variable and their derivatives.) PDEs are used to formulate problems involving functions of several variables, and are either solved in closed form, or used to create a relevant computer model.

PDEs can be used to describe a wide variety of phenomena such as sound, heat, electrostatics, electrodynamics, fluid flow, elasticity, or quantum mechanics. These seemingly distinct physical phenomena can be formalised similarly in terms of PDEs. Just as ordinary differential equations often model one-dimensional dynamical systems, partial differential equations often model multidimensional systems. PDEs find their generalisation in stochastic partial differential equations.

Linear Differential Equations

A differential equation is *linear* if the unknown function and its derivatives have *degree* 1 (products of the unknown function and its derivatives are not allowed) and *nonlinear* otherwise. The characteristic property of linear equations is that their solutions form an affine subspace of an appropriate function space, which results in much more developed theory of linear differential equations.

Homogeneous linear differential equations are a subclass of linear differential equations for which the space of solutions is a linear subspace i.e. the sum of any set of solutions or multiples of solutions is also a solution. The coefficients of the unknown function and its derivatives in a linear differential equation are allowed to be (known) functions of the independent variable or variables; if these coefficients are constants then one speaks of a *constant coefficient linear differential equation*.

Non-linear Differential Equations

Non-linear differential equations are formed by the *products of the unknown function and its derivatives* are allowed and its degree is > 1.There are very few methods of solving nonlinear differential equations exactly; those that are known typically depend on the equation having particular symmetries. Nonlinear differential equations can exhibit very complicated behavior over extended time intervals, characteristic of chaos. Even the fundamental questions of existence, uniqueness, and extendability of solutions for nonlinear differential equations, and well-posedness of initial and boundary value problems for nonlinear PDEs are hard problems and their resolution in special cases is considered to be a significant advance in the mathematical theory (cf. Navier–Stokes existence and smoothness). However, if the differential equation is a correctly formulated representation of a meaningful physical process, then one expects it to have a solution.

Linear differential equations frequently appear as approximations to nonlinear equations. These approximations are only valid under restricted conditions. For example, the harmonic oscillator equation is an approximation to the nonlinear pendulum equation that is valid for small amplitude oscillations.

Equation Order

Differential equations are described by their order, determined by the term with the highest derivatives. An equation containing only first derivatives is a *first-order differential equation*, an equation containing the second derivative is a *second-order differential equation*, and so on.

Examples

In the first group of examples, let u be an unknown function of x, and c and ω are known constants. Note both ordinary and partial differential equations are broadly classified as *linear* and *nonlinear*.

- Inhomogeneous first-order linear constant coefficient ordinary differential equation:

$$\frac{du}{dx} = cu + x^2.$$

- Homogeneous second-order linear ordinary differential equation:

$$\frac{d^2u}{dx^2} - x\frac{du}{dx} + u = 0.$$

- Homogeneous second-order linear constant coefficient ordinary differential equation describing the harmonic oscillator:

$$\frac{d^2u}{dx^2} + \omega^2 u = 0.$$

- Inhomogeneous first-order nonlinear ordinary differential equation:

$$\frac{du}{dx} = u^2 + 4.$$

- Second-order nonlinear (due to sine function) ordinary differential equation describing the motion of a pendulum of length L:

$$L\frac{d^2u}{dx^2} + g\sin u = 0.$$

In the next group of examples, the unknown function u depends on two variables x and t or x and y.

- Homogeneous first-order linear partial differential equation:

$$\frac{\partial u}{\partial t} + t\frac{\partial u}{\partial x} = 0.$$

- Homogeneous second-order linear constant coefficient partial differential equation of elliptic type, the Laplace equation:

$$\frac{\partial^2 u}{\partial x^2} + \frac{\partial^2 u}{\partial y^2} = 0.$$

$$\frac{\partial u}{\partial t} = 6u\frac{\partial u}{\partial x} - \frac{\partial^3 u}{\partial x^3}.$$

Existence of Solutions

Solving differential equations is not like solving algebraic equations. Not only are their solutions oftentimes unclear, but whether solutions are unique or exist at all are also notable subjects of interest.

For first order initial value problems, the Peano existence theorem gives one set of circumstances in which a solution exists. Given any point (a,b) in the xy-plane, define some rectangular region Z, such that $Z = [l,m] \times [n,p]$ and (a,b) is in the interior of Z. If we are given a differential equation $\frac{dy}{dx} = g(x,y)$ and the condition that $y = b$ when $x = a$, then there is locally a solution to this problem if $g(x,y)$ and $\frac{\partial g}{\partial x}$ are both continuous on Z. This solution exists on some interval with its center at a. The solution may not be unique.

However, this only helps us with first order initial value problems. Suppose we had a linear initial value problem of the nth order:

$$f_n(x)\frac{d^n y}{dx^n} + \cdots + f_1(x)\frac{dy}{dx} + f_0(x)y = g(x)$$

such that

$$y(x_0) = y_0, y'(x_0) = y_0', y''(x_0) = y_0'', \cdots$$

For any nonzero $f_n(x)$, if $\{f_0, f_1, \cdots\}$ and g are continuous on some interval containing x_0, y is unique and exists.

Related Concepts

- A delay differential equation (DDE) is an equation for a function of a single variable, usually called time, in which the derivative of the function at a certain time is given in terms of the values of the function at earlier times.

- A stochastic differential equation (SDE) is an equation in which the unknown quantity is a stochastic process and the equation involves some known stochastic processes, for example, the Wiener process in the case of diffusion equations.

- A differential algebraic equation (DAE) is a differential equation comprising differential and algebraic terms, given in implicit form.

Connection to Difference Equations

The theory of differential equations is closely related to the theory of difference equations, in which the coordinates assume only discrete values, and the relationship involves values of the unknown function or functions and values at nearby coordinates. Many methods to compute numerical solutions of differential equations or study the properties of differential equations involve approximation of the solution of a differential equation by the solution of a corresponding difference equation.

Applications

The study of differential equations is a wide field in pure and applied mathematics, physics, and engineering. All of these disciplines are concerned with the properties of differential equations of various types. Pure mathematics focuses on the existence and uniqueness of solutions, while applied mathematics emphasizes the rigorous justification of the methods for approximating solutions. Differential equations play an important role in modelling virtually every physical, technical, or biological process, from celestial motion, to bridge design, to interactions between neurons. Differential equations such as those used to solve real-life problems may not necessarily be directly solvable, i.e. do not have closed form solutions. Instead, solutions can be approximated using numerical methods.

Many fundamental laws of physics and chemistry can be formulated as differential equations. In biology and economics, differential equations are used to model the behavior of complex systems. The mathematical theory of differential equations first developed together with the sciences where the equations had originated and where the results found application. However, diverse problems, sometimes originating in quite distinct scientific fields, may give rise to identical differential equations. Whenever this happens, mathematical theory behind the equations can be viewed as a unifying principle behind diverse phenomena. As an example, consider propagation of light and sound in the atmosphere, and of waves on the surface of a pond. All of them may be described by the same second-order partial differential equation, the wave equation, which allows us to think of light and sound as forms of waves, much like familiar waves in the water. Conduction of heat, the theory of which was developed by Joseph Fourier, is governed by another second-order partial differential equation, the heat equation. It turns out that many diffusion processes, while seemingly different, are described by the same equation; the Black–Scholes equation in finance is, for instance, related to the heat equation.

Physics

- Euler–Lagrange equation in classical mechanics
- Hamilton's equations in classical mechanics
- Radioactive decay in nuclear physics
- Newton's law of cooling in thermodynamics
- The wave equation
- The heat equation in thermodynamics
- Laplace's equation, which defines harmonic functions
- Poisson's equation
- The geodesic equation
- The Navier–Stokes equations in fluid dynamics
- The Diffusion equation in stochastic processes
- The Convection–diffusion equation in fluid dynamics
- The Cauchy–Riemann equations in complex analysis
- The Poisson–Boltzmann equation in molecular dynamics
- The shallow water equations
- Universal differential equation
- The Lorenz equations whose solutions exhibit chaotic flow.

Classical Mechanics

So long as the force acting on a particle is known, Newton's second law is sufficient to describe the motion of a particle. Once independent relations for each force acting on a particle are available, they can be substituted into Newton's second law to obtain an ordinary differential equation, which is called the *equation of motion*.

Electrodynamics

Maxwell's equations are a set of partial differential equations that, together with the Lorentz force law, form the foundation of classical electrodynamics, classical optics, and electric circuits. These fields in turn underlie modern electrical and communications technologies. Maxwell's equations describe how electric and magnetic fields are generated and altered by each other and by charges and currents. They are named after the Scottish physicist and mathematician James Clerk Maxwell, who published an early form of those equations between 1861 and 1862.

General Relativity

The Einstein field equations (EFE; also known as "Einstein's equations") are a set of ten partial differential equations in Albert Einstein's general theory of relativity which describe the fundamental interaction of gravitation as a result of spacetime being curved by matter and energy. First published by Einstein in 1915 as a tensor equation, the EFE equate local spacetime curvature (expressed by the Einstein tensor) with the local energy and momentum within that spacetime (expressed by the stress–energy tensor).

Quantum Mechanics

In quantum mechanics, the analogue of Newton's law is Schrödinger's equation (a partial differential equation) for a quantum system (usually atoms, molecules, and subatomic particles whether free, bound, or localized). It is not a simple algebraic equation, but in general a linear partial differential equation, describing the time-evolution of the system's wave function (also called a "state function").

Biology

- Verhulst equation – biological population growth
- von Bertalanffy model – biological individual growth
- Replicator dynamics – found in theoretical biology
- Hodgkin–Huxley model – neural action potentials

Predator-prey Equations

The Lotka–Volterra equations, also known as the predator–prey equations, are a pair of first-order, non-linear, differential equations frequently used to describe the dynamics of biological systems in which two species interact, one as a predator and the other as prey.

Chemistry

The *rate law* or rate equation for a chemical reaction is a differential equation that links the reaction rate with concentrations or pressures of reactants and constant parameters (normally rate coefficients and partial reaction orders). To determine the rate equation for a particular system one combines the reaction rate with a mass balance for the system.

Economics

- The key equation of the Solow–Swan model is $\dfrac{\partial k(t)}{\partial t} = s[k(t)]^{\alpha} - \delta k(t)$

- The Black–Scholes PDE
- Malthusian growth model
- The Vidale–Wolfe advertising model

Integral Equation

In mathematics, an integral equation is an equation in which an unknown function appears under an integral sign.

There is a close connection between differential and integral equations, and some problems may be formulated either way.

Overview

The most basic type of integral equation is called a *Fredholm equation of the first type,*

$$f(x) \quad \int K(x,t) \ (t)dt.$$

The notation follows Arfken. Here φ is an unknown function, f is a known function, and K is another known function of two variables, often called the kernel function. Note that the limits of integration are constant: this is what characterizes a Fredholm equation.

If the unknown function occurs both inside and outside of the integral, the equation is known as a *Fredholm equation of the second type,*

$$\varphi(x) = f(x) + \lambda \int_a^b K(x,t)\varphi(t)dt.$$

The parameter λ is an unknown factor, which plays the same role as the eigenvalue in linear algebra.

If one limit of integration is a variable, the equation is called a Volterra equation. The following are called *Volterra equations of the first and second types,* respectively,

$$f(x) = \int_a^x K(x,t)\varphi(t)dt$$

$$\varphi(x) = f(x) + \lambda \int_a^x K(x,t)\varphi(t)dt.$$

In all of the above, if the known function f is identically zero, the equation is called a *homogeneous integral equation.* If f is nonzero, it is called an *inhomogeneous integral equation.*

Numerical Solution

It is worth noting that integral equations often do not have an analytical solution, and must be solved numerically. An example of this is evaluating the Electric-Field Integral Equation (EFIE) or Magnetic-Field Integral Equation (MFIE) over an arbitrarily shaped object in an electromagnetic scattering problem.

One method to solve numerically requires discretizing variables and replacing integral by a quadrature rule

$$\sum_{j=1}^{n} w_j K\left(s_i, t_j\right) u(t_j) = f(s_i), \qquad i = 0, 1, \cdots, n.$$

Then we have a system with n equations and n variables. By solving it we get the value of the n variables

$$u(t_0), u(t_1), \cdots, u(t_n).$$

Classification

Integral equations are classified according to three different dichotomies, creating eight different kinds:

Limits of integration

- both fixed: Fredholm equation

- one variable: Volterra equation

Placement of unknown function

- only inside integral: first kind

- both inside and outside integral: second kind

Nature of known function f

- identically zero: homogeneous

- not identically zero: inhomogeneous

Integral equations are important in many applications. Problems in which integral equations are encountered include radiative transfer, and the oscillation of a string, membrane, or axle. Oscillation problems may also be solved as differential equations.

Both Fredholm and Volterra equations are linear integral equations, due to the linear behaviour of $\varphi(x)$ under the integral. A nonlinear Volterra integral equation has the general form:

$$\varphi(x) = f(x) + \lambda \int_a^x K(x,t) F(x,t,\varphi(t)) dt,$$

where F is a known function.

Wiener–Hopf Integral Equations

$$y(t) = \lambda x(t) + \int_0^{\infty} k(t-s) x(s) ds, \qquad 0 \le t < \infty.$$

Originally, such equations were studied in connection with problems in radiative transfer, and more recently, they have been related to the solution of boundary integral equations for planar problems in which the boundary is only piecewise smooth.

Power Series Solution for Integral Equations

In many cases, if the Kernel of the integral equation is of the form $K(xt)$ and the Mellin transform of $K(t)$ exists, we can find the solution of the integral equation

$$g(s) = s \int_0^\infty dt K(st) f(t)$$

in the form of a power series

$$f(t) = \sum_{n=0}^\infty \frac{a_n}{M(n+1)} t^n$$

where

$$g(s) = \sum_{n=0}^\infty a_n s^{-n}, \qquad M(n+1) = \int_0^\infty dt K(t) t^n$$

are the Z-transform of the function $g(s)$, and $M(n+1)$ is the Mellin transform of the Kernel.

Integral Equations as a Generalization of Eigenvalue Equations

Certain homogeneous linear integral equations can be viewed as the continuum limit of eigenvalue equations. Using index notation, an eigenvalue equation can be written as

$$\sum_j M_{i,j} v_j = \lambda v_i$$

where M = $[M_{i,j}]$ is a matrix, v is one of its eigenvectors, and λ is the associated eigenvalue.

Taking the continuum limit, i.e., replacing the discrete indices i and j with continuous variables x and y, yields

$$\int K(x, y) \varphi(y) \mathrm{d}y = \lambda \varphi(x),$$

where the sum over j has been replaced by an integral over y and the matrix M and the vector v have been replaced by the *kernel* $K(x, y)$ and the eigenfunction $\varphi(y)$. (The limits on the integral are fixed, analogously to the limits on the sum over j.) This gives a linear homogeneous Fredholm equation of the second type.

In general, $K(x, y)$ can be a distribution, rather than a function in the strict sense. If the distribution K has support only at the point $x = y$, then the integral equation reduces to a differential eigenfunction equation.

In general, Volterra and Fredholm integral equations can arise from a single differential equation, depending on which sort of conditions are applied at the boundary of the domain of its solution.

Diophantine Equation

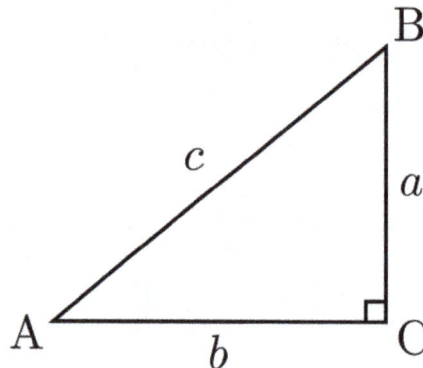

Finding all right triangles with integer side-lengths is equivalent to solving the Diophantine equation $a^2 + b^2 = c^2$.

In mathematics, a Diophantine equation is a polynomial equation, usually in two or more unknowns, such that only the integer solutions are sought or studied (an integer solution is a solution such that all the unknowns take integer values). A linear Diophantine equation is an equation between two sums of monomials of degree zero or one. An exponential Diophantine equation is one in which exponents on terms can be unknowns.

Diophantine problems have fewer equations than unknown variables and involve finding integers that work correctly for all equations. In more technical language, they define an algebraic curve, algebraic surface, or more general object, and ask about the lattice points on it.

The word *Diophantine* refers to the Hellenistic mathematician of the 3rd century, Diophantus of Alexandria, who made a study of such equations and was one of the first mathematicians to introduce symbolism into algebra. The mathematical study of Diophantine problems that Diophantus initiated is now called Diophantine analysis.

While individual equations present a kind of puzzle and have been considered throughout history, the formulation of general theories of Diophantine equations (beyond the theory of quadratic forms) was an achievement of the twentieth century.

Examples

In the following Diophantine equations, w, x, y, and z are the unknowns and the other letters are given constants:

$ax + by = 1$	This is a linear Diophantine equation.

$w^3 + x^3 = y^3 + z^3$	The smallest nontrivial solution in positive integers is $12^3 + 1^3 = 9^3 + 10^3 = 1729$. It was famously given as an evident property of 1729, a taxicab number (also named Hardy–Ramanujan number) by Ramanujan to Hardy while meeting in 1917. There are infinitely many nontrivial solutions.
$x^n + y^n = z^n$	For $n = 2$ there are infinitely many solutions (x,y,z): the Pythagorean triples. For larger integer values of n, Fermat's Last Theorem (initially claimed in 1637 by Fermat and proved by Wiles in 1995) states there are no positive integer solutions (x,y,z).
$x^2 - ny^2 = \pm 1$	This is Pell's equation, which is named after the English mathematician John Pell. It was studied by Brahmagupta in the 7th century, as well as by Fermat in the 17th century.
$4/n = 1/x + 1/y + 1/z$	The Erdős–Straus conjecture states that, for every positive integer $n \geq 2$, there exists a solution in x, y, and z, all as positive integers. Although not usually stated in polynomial form, this example is equivalent to the polynomial equation $4xyz = yzn + xzn + xyn = n(yz + xz + xy)$.
$x^4 + y^4 + z^4 = w^4$	Conjectured incorrectly by Euler to have no nontrivial solutions. Proved by Elkies to have infinitely many nontrivial solutions, with a computer search by Frye determining the smallest nontrivial solution.

Linear Diophantine Equations

One Equation

The simplest linear Diophantine equation takes the form $ax + by = c$, where a, b and c are given integers. The solutions are described by the following theorem:

> *This Diophantine equation has a solution* (where x and y are integers) *if and only if c is a multiple of the greatest common divisor of a and b. Moreover, if (x, y) is a solution, then the other solutions have the form $(x + kv, y - ku)$, where k is an arbitrary integer, and u and v are the quotients of a and b (respectively) by the greatest common divisor of a and b.*

Proof: If d is this greatest common divisor, Bézout's identity asserts the existence of integers e and f such that $ae + bf = d$. If c is a multiple of d, then $c = dh$ for some integer h, and (eh, fh) is a solution. On the other hand, for every pair of integers x and y, the greatest common divisor d of a and b divides $ax + by$. Thus, if the equation has a solution, then c must be a multiple of d. If $a = ud$ and $b = vd$, then for every solution (x, y), we have

$$a(x + kv) + b(y - ku) = ax + by + k(av - bu) = ax + by + k(udv - vdu) = ax + by,$$

showing that $(x + kv, y - ku)$ is another solution. Finally, given two solutions such that $ax_1 + by_1 = ax_2 + by_2 = c$, one deduces that $u(x_2 - x_1) + v(y_2 - y_1) = 0$. As u and v are coprime, Euclid's lemma shows that there exists an integer k such that $x_2 - x_1 = kv$ and $y_2 - y_1 = -ku$. Therefore, $x_2 = x_1 + kv$ and $y_2 = y_1 - ku$, which completes the proof.

Chinese Remainder Theorem

The Chinese remainder theorem describes an important class of linear Diophantine systems of equations: let $n_1, ..., n_k$ be k pairwise coprime integers greater than one, $a_1, ..., a_k$ be k arbitrary integers, and N be the product $n_1 \cdots n_k$. The Chinese remainder theorem asserts that the following linear Diophantine system has exactly one solution $(x, x_1, ..., x_k)$ such that $0 \leq x < N$, and that the other solutions are obtained by adding to x a multiple of N:

$$x = a_1 + n_1 x_1$$
$$\vdots$$
$$x = a_k + n_k x_k$$

System of Linear Diophantine Equations

More generally, every system of linear Diophantine equations may be solved by computing the Smith normal form of its matrix, in a way that is similar to the use of the reduced row echelon form to solve a system of linear equations over a field. Using matrix notation every system of linear Diophantine equations may be written

$$AX = C,$$

where A is an $m \times n$ matrix of integers, X is an $n \times 1$ column matrix of unknowns and C is an $m \times 1$ column matrix of integers.

The computation of the Smith normal form of A provides two unimodular matrices (that is matrices that are invertible over the integers and have ± 1 as determinant) U and V of respective dimensions $m \times m$ and $n \times n$, such that the matrix

$$B = [b_{i,j}] = UAV$$

is such that $b_{i,i}$ is not zero for i not greater than some integer k, and all the other entries are zero. The system to be solved may thus be rewritten as

$$B(V^{-1}X) = UC.$$

Calling y_i the entries of $V^{-1}X$ and d_i those of $D = UC$, this leads to the system

$$b_{i,i} y_i = d_i \text{ for } 1 \leq i \leq k,$$

$$0 y_i = d_i \text{ for } k < i \leq n.$$

This system is equivalent to the given one in the following sense: A column matrix of integers x is a solution of the given system if and only if $x = Vy$ for some column matrix of integers y such that $By = D$.

It follows that the system has a solution if and only if $b_{i,i}$ divides d_i for $i \leq k$ and $d_i = 0$ for $i > k$. If this condition is fulfilled, the solutions of the given system are

$$V \begin{bmatrix} \dfrac{d_1}{b_{1,1}} \\ \vdots \\ \dfrac{d_k}{b_{k,k}} \\ h_{k+1} \\ \vdots \\ h_n \end{bmatrix},$$

where $h_{k+1}, ..., h_n$ are arbitrary integers.

Hermite normal form may also be used for solving systems of linear Diophantine equations. However, Hermite normal form does not directly provide the solutions; to get the solutions from the Hermite normal form, one has to successively solve several linear equations. Nevertheless, Richard Zippel wrote that the Smith normal form "is somewhat more than is actually needed to solve linear diophantine equations. Instead of reducing the equation to diagonal form, we only need to make it triangular, which is called the Hermite normal form. The Hermite normal form is substantially easier to compute than the Smith normal form."

Integer linear programming amounts to finding some integer solutions (optimal in some sense) of linear systems that include also inequations. Thus systems of linear Diophantine equations are basic in this context, and textbooks on integer programming usually have a treatment of systems of linear Diophantine equations.

Diophantine Analysis

Typical Questions

The questions asked in Diophantine analysis include:

1. Are there any solutions?
2. Are there any solutions beyond some that are easily found by inspection?
3. Are there finitely or infinitely many solutions?
4. Can all solutions be found in theory?
5. Can one in practice compute a full list of solutions?

These traditional problems often lay unsolved for centuries, and mathematicians gradually came to understand their depth (in some cases), rather than treat them as puzzles.

Typical Problem

The given information is that a father's age is 1 less than twice that of his son, and that the digits AB making up the father's age are reversed in the son's age (i.e. BA). This leads to the equation $10A + B = 2(10B + A) - 1$, thus $19B - 8A = 1$. Inspection gives the result $A = 7$, $B = 3$, and thus AB equals 73 years and BA equals 37 years. One may easily show that there is not any other solution with A and B positive integers less than 10.

17th and 18th Centuries

In 1637, Pierre de Fermat scribbled on the margin of his copy of *Arithmetica*: "It is impossible to separate a cube into two cubes, or a fourth power into two fourth powers, or in general, any power higher than the second into two like powers." Stated in more modern language, "The equation $a^n + b^n = c^n$ has no solutions for any n higher than 2." And then he wrote, intriguingly: "I have discovered a truly marvelous proof of this proposition, which this margin is too narrow to contain." Such a proof eluded mathematicians for centuries, however, and as such his statement became famous as Fermat's Last Theorem. It wasn't until 1995 that it was proven by the British mathematician Andrew Wiles.

In 1657, Fermat attempted to solve the Diophantine equation $61x^2 + 1 = y^2$ (solved by Brahmagupta over 1000 years earlier). The equation was eventually solved by Euler in the early 18th century, who also solved a number of other Diophantine equations. The smallest solution of this equation in positive integers is $x = 226153980$, $y = 1766319049$.

Hilbert's Tenth Problem

In 1900, David Hilbert proposed the solvability of all Diophantine equations as the tenth of his fundamental problems. In 1970, Yuri Matiyasevich solved it negatively, by proving that a general algorithm for solving all Diophantine equations cannot exist.

Diophantine Geometry

Diophantine geometry, which is the application of techniques from algebraic geometry in this field, has continued to grow as a result; since treating arbitrary equations is a dead end, attention turns to equations that also have a geometric meaning. The central idea of Diophantine geometry is that of a rational point, namely a solution to a polynomial equation or a system of polynomial equations, which is a vector in a prescribed field K, when K is *not* algebraically closed.

Modern Research

One of the few general approaches is through the Hasse principle. Infinite descent is the traditional method, and has been pushed a long way.

The depth of the study of general Diophantine equations is shown by the characterisation of Diophantine sets as equivalently described as recursively enumerable. In other words, the general problem of Diophantine analysis is blessed or cursed with universality, and in any case is not something that will be solved except by re-expressing it in other terms.

The field of Diophantine approximation deals with the cases of *Diophantine inequalities*. Here variables are still supposed to be integral, but some coefficients may be irrational numbers, and the equality sign is replaced by upper and lower bounds.

The most celebrated single question in the field, the conjecture known as Fermat's Last Theorem, was solved by Andrew Wiles but using tools from algebraic geometry developed during the last century rather than within number theory where the conjecture was originally formulated. Other major results, such as Faltings' theorem, have disposed of old conjectures.

Infinite Diophantine Equations

An example of an infinite diophantine equation is:

$$n = a^2 + 2b^2 + 3c^2 + 4d^2 + 5e^2 + ...,$$

which can be expressed as "How many ways can a given integer n be written as the sum of a square plus twice a square plus thrice a square and so on?" The number of ways this can be done for each n forms an integer sequence. Infinite Diophantine equations are related to theta functions and infinite dimensional lattices. This equation always has a solution for any positive n. Compare this to:

$$n = a^2 + 4b^2 + 9c^2 + 16d^2 + 25e^2 + ...,$$

which does not always have a solution for positive n.

Exponential Diophantine Equations

If a Diophantine equation has as an additional variable or variables occurring as exponents, it is an exponential Diophantine equation. Examples include the Ramanujan–Nagell equation, $2^n - 7 = x^2$, and the equation of the Fermat-Catalan conjecture and Beal's conjecture, $a^m + b^n = c^k$ with inequality restrictions on the exponents. A general theory for such equations is not available; particular cases such as Catalan's conjecture have been tackled. However, the majority are solved via ad hoc methods such as Størmer's theorem or even trial and error.

Quadratic Formula

In elementary algebra, the quadratic formula is the solution of the quadratic equation. There are other ways to solve the quadratic equation instead of using the quadratic formula, such as factoring, completing the square, or graphing. Using the quadratic formula is often the most convenient way.

$$x = \frac{-b \pm \sqrt{b^2 - 4ac}}{2a}$$

The quadratic formula

The general quadratic equation is

$$ax^2 + bx + c = 0.$$

Here x represents an unknown, while a, b, and c are constants with a not equal to 0. One can verify that the quadratic formula satisfies the quadratic equation, by inserting the former into the latter. With the above parameterization, the quadratic formula is:

$$x = \frac{-b \pm \sqrt{b^2 - 4ac}}{2a}.$$

Each of the solutions given by the quadratic formula is called a root of the quadratic equation. Geometrically, these roots represent the x values at which *any* parabola, explicitly given as $y = ax^2 + bx + c$, crosses the x-axis. As well as being a formula that will yield the zeros of any parabola, the quadratic equation will give the axis of symmetry of the parabola, and it can be used to immediately determine how many zeros it has.

Derivation of the Formula

The quadratic formula can be derived with a simple application of technique of completing the square. For this reason, the derivation is sometimes left as an exercise for students, who can thus experience rediscovery of this important formula. The explicit derivation is as follows.

Divide the quadratic equation by a, which is allowed because a is non-zero:

$$x^2 + \frac{b}{a}x + \frac{c}{a} = 0.$$

Subtract c/a from both sides of the equation, yielding:

$$x^2 + \frac{b}{a}x = -\frac{c}{a}.$$

The quadratic equation is now in a form to which the method of completing the square can be applied. Thus, add a constant to both sides of the equation such that the left hand side becomes a complete square:

$$x^2 + \frac{b}{a}x + \left(\frac{b}{2a}\right)^2 = -\frac{c}{a} + \left(\frac{b}{2a}\right)^2,$$

which produces:

$$\left(x + \frac{b}{2a}\right)^2 = -\frac{c}{a} + \frac{b^2}{4a^2}.$$

Accordingly, after rearranging the terms on the right hand side to have a common denominator, we obtain this:

$$\left(x + \frac{b}{2a}\right)^2 = \frac{b^2 - 4ac}{4a^2}.$$

The square has thus been completed. Taking the square root of both sides yields the following equation:

$$x + \frac{b}{2a} = \pm\frac{\sqrt{b^2 - 4ac}}{2a}.$$

Isolating x gives the quadratic formula:

$$x = \frac{-b \pm \sqrt{b^2 - 4ac}}{2a}.$$

The plus-minus symbol "$\pm$" indicates that both

$$x = \frac{-b + \sqrt{b^2 - 4ac}}{2a} \quad \text{and} \quad x = \frac{-b - \sqrt{b^2 - 4ac}}{2a}$$

are solutions of the quadratic equation. There are many alternatives of this derivation with minor differences, mostly concerning the manipulation of a.

Some sources, particularly older ones, use alternative parameterizations of the quadratic equation such as $ax^2 - 2bx + c = 0$ or $ax^2 + 2bx + c = 0$, where b has a magnitude one half of the more common one. These result in slightly different forms for the solution, but are otherwise equivalent.

A lesser known quadratic formula, as used in Muller's method, and which can be found from Vieta's formulas, provides the same roots via the equation:

$$x = \frac{-2c}{b \pm \sqrt{b^2 - 4ac}}.$$

Geometrical Significance

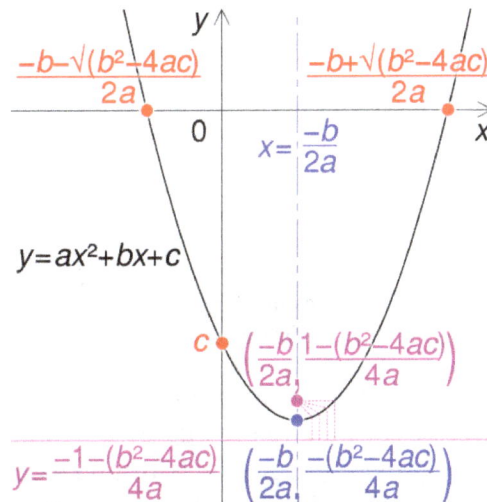

Graph of $y = ax^2 + bx + c$, where a and the discriminant (b^2 - $4ac$) are positive, with

- Roots and y-intercept in red
- Vertex and axis of symmetry in blue
- Focus and directrix in pink

Without going into parabolas as geometrical objects on a cone, a parabola is any curve described by a second-degree polynomial, i.e. any equation of the form:

$$p_2(x) = a_2 x^2 + a_1 x + a_0$$

where p_2 represents the polynomial of degree 2 and a_0, a_1, and a_2 are constant coefficients whose subscripts correspond to their respective term's degree. The first and foremost geometrical application of the quadratic formula is that it will define the points along the x-axis where the parabola will cross it. Additionally, if the quadratic formula were broken into two terms,

$$x = \frac{-b \pm \sqrt{b^2 - 4ac}}{2a} = -\frac{b}{2a} \pm \frac{\sqrt{b^2 - 4ac}}{2a}$$

the definition of the axis of symmetry appears as the $-b/2a$ term. The other term, $\sqrt{b^2 - 4ac}/2a$, must then be the distance the zeros are away from the axis of symmetry, where the plus sign represents the distance away to the right, and the minus sign represents the distance away to the left.

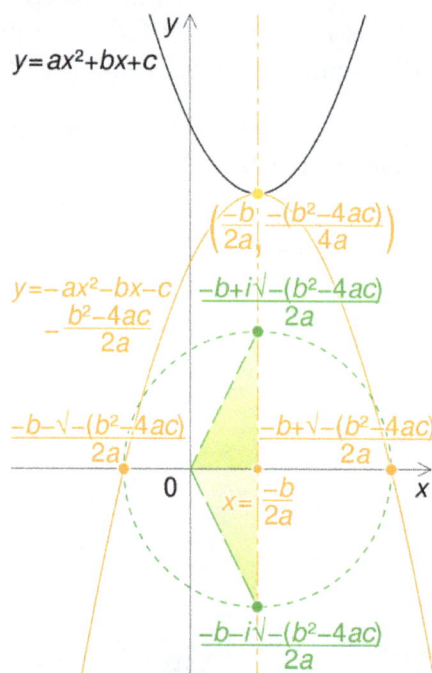

Visualisation of the complex roots of $y = ax^2 + bx + c$: the parabola is rotated 180° about its vertex (yellow). Its x-intercepts are rotated 90° around their mid-point, and the Cartesian plane is interpreted as the complex plane (green).

If this distance term were to decrease to zero, the axis of symmetry would be the x value of the zero, indicating there is only one possible solution to the quadratic equation. Algebraically, this means that $\sqrt{b^2 - 4ac} = 0$, or simply $b^2 - 4ac = 0$ (where the left-hand side is referred to as the *discriminant*), for its term to be reduced to zero. This is simply one of three cases, where the discriminant can indicate how many zeros the parabola will have. If the discriminant were positive, the distance would be non-zero, and there will be two solutions, as expected. However, there is the case where the discriminant is less than zero, and this indicates the distance will be *imaginary* — or some multiple of the unit i, such that $i = \sqrt{-1}$ — and the parabola's zeros will be a complex number. The complex roots will be complex conjugates, and by definition cannot be entirely real, where the real part of the complex root will be the axis of symmetry, therefore geometrical interpretation is that there are no real values of x such that the parabola will be observed to cross the x-axis.

Historical Development

The earliest methods for solving quadratic equations were geometric. Babylonian cuneiform tablets contain problems reducible to solving quadratic equations. The Egyptian Berlin Papyrus, dating back to the Middle Kingdom (2050 BC to 1650 BC), contains the solution to a two-term quadratic equation.

Euclid in Raphael's *School of Athens*

The Greek mathematician Euclid (circa 300 BC) used geometric methods to solve quadratic equations in Book 2 of his *Elements*, an influential mathematical treatise. Rules for quadratic equations appear in the Chinese *The Nine Chapters on the Mathematical Art* circa 200 BC. In his work *Arithmetica*, the Greek mathematician Diophantus (circa 250 BC) solved quadratic equations with a method more recognizably algebraic than the geometric algebra of Euclid. His solution gives only one root, even when both roots are positive.

The Indian mathematician Brahmagupta (597–668 AD) explicitly described the quadratic formula in his treatise *Brāhmasphu asiddhānta* published in 628 AD, but written in words instead of symbols. His solution of the quadratic equation $ax^2 + bx = c$ was as follows: "To the absolute number multiplied by four times the [coefficient of the] square, add the square of the [coefficient of the] middle term; the square root of the same, less the [coefficient of the] middle term, being divided by twice the [coefficient of the] square is the value." This is equivalent to:

$$x = \frac{\sqrt{4ac + b^2} - b}{2a}.$$

The 9th-century Persian mathematician al-Khwārizmī, influenced by earlier Greek and Indian mathematicians, solved quadratic equations algebraically. The quadratic formula covering all cases was first obtained by Simon Stevin in 1594. In 1637 René Descartes published *La Géométrie* containing the quadratic formula in the form we know today. The first appearance of the general solution in the modern mathematical literature appeared in an 1896 paper by Henry Heaton.

Other Derivations

Many alternative derivations of the quadratic formula are in the literature. These derivations may be simpler than the standard completing the square method, may represent interesting applications of other alegraic techniques, or may offer insight into other areas of mathematics.

Alternate Method of Completing the Square

The great majority of algebra texts published over the last several decades teach completing the square using the sequence presented earlier: (1) divide each side by a to make the equation monic, (2) rearrange, (3) then add $(b/2a)^2$ to both sides to complete the square.

As pointed out by Larry Hoehn in 1975, completing the square can be accomplished by a different sequence that leads to a simpler sequence of intermediate terms: (1) multiply each side by $4a$, (2) rearrange, (3) then add b^2.

In other words, the quadratic formula can be derived as follows:

$$ax^2 + bx + c = 0$$
$$4a^2x^2 + 4abx + 4ac = 0$$
$$4a^2x^2 + 4abx = -4ac$$
$$4a^2x^2 + 4abx + b^2 = b^2 - 4ac$$
$$(2ax + b)^2 = b^2 - 4ac$$
$$2ax + b = \pm\sqrt{b^2 - 4ac}$$
$$2ax = -b \pm \sqrt{b^2 - 4ac}$$
$$x = \frac{-b \pm \sqrt{b^2 - 4ac}}{2a}.$$

This actually represents an ancient derivation of the quadratic formula, and was known to the Hindus at least as far back as 1025. Compared with the derivation in standard usage, this alternate derivation is shorter, involves fewer computations with literal coefficients, avoids fractions until the last step, has simpler expressions, and uses simpler mathematics. As Hoehn states, "it is easier 'to add the square of b' than it is 'to add the square of half the coefficient of the x term'".

By Substitution

Another technique is solution by substitution. In this technique, we substitute $x = y + m$ into the quadratic to get:

$$a(y + m)^2 + b(y + m) + c = 0.$$

Expanding the result and then collecting the powers of y produces:

$$ay^2 + y(2am + b) + (am^2 + bm + c) = 0.$$

We have not yet imposed a second condition on y and m, so we now choose m so that the middle term vanishes. That is, $2am + b = 0$ or $m = -b/2a$. Subtracting the constant term from both sides of the equation (to move it to the right hand side) and then dividing by a gives:

$$y^2 = \frac{-(am^2 + bm + c)}{a}.$$

Substituting for m gives:

$$y^2 = \frac{-(\frac{b^2}{4a} + \frac{-b^2}{2a} + c)}{a} = \frac{b^2 - 4ac}{4a^2}.$$

Therefore,

$$y = \pm \frac{\sqrt{b^2 - 4ac}}{2a};$$

substituting $x = y + m = y - b/2a$ provides the quadratic formula.

By using Algebraic Identities

The following method was used by many historical mathematicians:

Let the roots of the standard quadratic equation be r_1 and r_2. The derivation starts by recalling the identity:

$$(r_1 - r_2)^2 = (r_1 + r_2)^2 - 4r_1 r_2.$$

Taking the square root on both sides, we get:

$$r_1 - r_2 = \pm\sqrt{(r_1 + r_2)^2 - 4r_1 r_2}.$$

Since the coefficient $a \neq 0$, we can divide the standard equation by a to obtain a quadratic polynomial having the same roots. Namely,

$$x^2 + \frac{b}{a}x + \frac{c}{a} = (x - r_1)(x - r_2) = x^2 - (r_1 + r_2)x + r_1 r_2.$$

From this we can see that the sum of the roots of the standard quadratic equation is given by $-b/a$, and the product of those roots is given by c/a. Hence the identity can be rewritten as:

$$r_1 - r_2 = \pm\sqrt{\left(-\frac{b}{a}\right)^2 - 4\frac{c}{a}} = \pm\sqrt{\frac{b^2}{a^2} - \frac{4ac}{a^2}} = \pm\frac{\sqrt{b^2 - 4ac}}{a}.$$

Now,

$$r_1 = \frac{(r_1 + r_2) + (r_1 - r_2)}{2} = \frac{-\frac{b}{a} \pm \frac{\sqrt{b^2 - 4ac}}{a}}{2} = \frac{-b \pm \sqrt{b^2 - 4ac}}{2a}.$$

Since $r_2 = -r_1 - b/a$, if we take

$$r_1 = \frac{-b + \sqrt{b^2 - 4ac}}{2a}$$

then we obtain

$$r_2 = \frac{-b - \sqrt{b^2 - 4ac}}{2a};$$

and if we instead take

$$r_1 = \frac{-b - \sqrt{b^2 - 4ac}}{2a}$$

then we calculate that

$$r_2 = \frac{-b + \sqrt{b^2 - 4ac}}{2a}.$$

Combining these results by using the standard shorthand ±, we have that the solutions of the quadratic equation are given by:

$$x = \frac{-b \pm \sqrt{b^2 - 4ac}}{2a}.$$

By Lagrange Resolvents.

An alternative way of deriving the quadratic formula is via the method of Lagrange resolvents, which is an early part of Galois theory. This method can be generalized to give the roots of cubic polynomials and quartic polynomials, and leads to Galois theory, which allows one to understand the solution of algebraic equations of any degree in terms of the symmetry group of their roots, the Galois group.

This approach focuses on the *roots* more than on rearranging the original equation. Given a monic quadratic polynomial

$$x^2 + px + q,$$

assume that it factors as

$$x^2 + px + q = (x - \alpha)(x - \beta),$$

Expanding yields

$$x^2 + px + q = x^2 - (\alpha + \beta)x + \alpha\beta,$$

where $p = -(\alpha + \beta)$ and $q = \alpha\beta$.

Since the order of multiplication does not matter, one can switch α and β and the values of p and q will not change: one can say that p and q are symmetric polynomials in α and β. In fact, they are the elementary symmetric polynomials – any symmetric polynomial in α and β can be expressed in terms of $\alpha + \beta$ and $\alpha\beta$ The Galois theory approach to analyzing and solving polynomials is: given the coefficients of a polynomial, which are symmetric functions in the roots, can one "break the symmetry" and recover the roots? Thus solving a polynomial of degree n is related to the ways of rearranging ("permuting") n terms, which is called the symmetric group on n letters, and denoted S_n. For the quadratic polynomial, the only way to rearrange two terms is to swap them ("transpose" them), and thus solving a quadratic polynomial is simple.

To find the roots α and β, consider their sum and difference:

$$
\begin{aligned}
r_1 &= \alpha + \beta \\
r_2 &= \alpha - \beta.
\end{aligned}
$$

These are called the Lagrange resolvents of the polynomial; notice that one of these depends on the order of the roots, which is the key point. One can recover the roots from the resolvents by inverting the above equations:

$$
\begin{aligned}
\alpha &= \frac{1}{2}(r_1 + r_2) \\
\beta &= \frac{1}{2}(r_1 - r_2).
\end{aligned}
$$

Thus, solving for the resolvents gives the original roots.

Now $r_1 = \alpha + \beta$ is a symmetric function in α and β, so it can be expressed in terms of p and q, and in fact $r_1 = -p$ as noted above. But $r_2 = \alpha - \beta$ is not symmetric, since switching α and β yields $-r_2 = \beta - \alpha$ (formally, this is termed a group action of the symmetric group of the roots). Since r_2 is not symmetric, it cannot be expressed in terms of the coefficients p and q, as these are symmetric in the roots and thus so is any polynomial expression involving them. Changing the order of the roots only changes r_2 by a factor of -1, and thus the square $r_2^2 = (\alpha - \beta)^2$ is symmetric in the roots, and thus expressible in terms of p and q. Using the equation

$$(\alpha - \beta)^2 = (\alpha + \beta)^2 - 4\alpha\beta$$

yields

$$r_2^2 = p^2 - 4q$$

and thus

$$r_2 = \pm\sqrt{p^2 - 4q}.$$

If one takes the positive root, breaking symmetry, one obtains:

$$
\begin{aligned}
r_1 &= -p \\
r_2 &= \sqrt{p^2 - 4q}
\end{aligned}
$$

and thus

$$
\begin{aligned}
\alpha &= \frac{1}{2}\left(-p + \sqrt{p^2 - 4q}\right) \\
\beta &= \frac{1}{2}\left(-p - \sqrt{p^2 - 4q}\right)
\end{aligned}
$$

Thus the roots are

$$\frac{1}{2}\left(-p \pm \sqrt{p^2 - 4q}\right)$$

which is the quadratic formula. Substituting $p = b/a$, $q = c/a$ yields the usual form for when a quadratic is not monic. The resolvents can be recognized as $r_1/2 = -p/2 = -b/2a$ being the vertex, and $r_2^2 = p^2 - 4q$ is the discriminant (of a monic polynomial).

A similar but more complicated method works for cubic equations, where one has three resolvents and a quadratic equation (the "resolving polynomial") relating r_2 and r_3, which one can solve by the quadratic equation, and similarly for a quartic equation (degree 4), whose resolving polynomial is a cubic, which can in turn be solved. The same method for a quintic equation yields a polynomial of degree 24, which does not simplify the problem, and in fact solutions to quintic equations in general cannot be expressed using only roots.

By Extrema

Knowing the value of x in the functional extreme point makes it possible to solve only for the increase (or decrease) needed in x to solve the quadratic equation. This method first uses differentiation to find the x value at the extremum, called x_{ext}. We then solve for the value, q, that ensures that $f(x_{ext} + q) = 0$. While this may not be the most intuitive method, it ensures that the mathematics is straightforward.

$$f(x) = ax^2 + bx + c$$

$$\frac{\partial f}{\partial x} = 2ax_{ext} + b$$

Setting the above differential to zero will give us the extrema of the quadratic function

$$x_{ext} = \frac{-b}{2a}$$

We define q as follows:

$$q = x_0 - x_{ext}$$

Here x_0 is the value of x that solves the quadratic equation. The sum of x_{ext} and the variable of interest, q, is plugged in to the quadratic equation

$$a\left(\frac{-b}{2a} + q\right)^2 + b\left(\frac{-b}{2a} + q\right) + c = 0$$

$$\Leftrightarrow \left(\frac{-b}{2a} + q\right)^2 + \frac{b}{a}\left(\frac{-b}{2a} + q\right) + \frac{c}{a} = 0$$

$$\Leftrightarrow \frac{b^2}{4a^2} + q^2 - \frac{bq}{a} - \frac{b^2}{2a^2} + \frac{bq}{a} + \frac{c}{a} = 0$$

$$\Leftrightarrow \frac{-b^2}{4a^2} + q^2 + \frac{c}{a} = 0$$

$$\Leftrightarrow q^2 = \frac{b^2 - 4ac}{4a^2}$$

$$\Leftrightarrow q = \frac{\pm\sqrt{b^2 - 4ac}}{2a}$$

The value of x in the extremepoint is then added to both sides of the equation

$$\Leftrightarrow x_0 = \frac{-b \pm \sqrt{b^2 - 4ac}}{2a}.$$

This gives the quadratic formula. This way one avoids the technique of completing the square, and much more complicated math is not needed. Note this solution is very similar to solving deriving the formula by substitution.

Dimensional Analysis

If the constants a, b, and/or c are not unitless, then the units of x must be equal to the units of b/a, due to the requirement that ax^2 and bx agree on their units. Furthermore, by the same logic, the units of c must be equal to the units of $b^{2/a}$, which can be verified without solving for x. This can be a powerful tool for verifying that a quadratic expression of physical quantities has been set up correctly, prior to solving it.

References

- Barnett, R.A.; Ziegler, M.R.; Byleen, K.E. (2008), College Mathematics for Business, Economics, Life Sciences and the Social Sciences (11th ed.), Upper Saddle River, N.J.: Pearson, ISBN 0-13-157225-3

- Bronstein, Manuel; et al., eds. (2006). Solving Polynomial Equations: Foundations, Algorithms, and Applications. Springer. ISBN 978-3-540-27357-8.

- Cahen, Paul-Jean; Chabert, Jean-Luc (1997). Integer-Valued Polynomials. American Mathematical Society. ISBN 978-0-8218-0388-2.

- Lang, Serge (2002), Algebra, Graduate Texts in Mathematics, 211 (Revised third ed.), New York: Springer-Verlag, ISBN 978-0-387-95385-4, MR 1878556. This classical book covers most of the content of this article.

- Sethuraman, B.A. (1997). "Polynomials". Rings, Fields, and Vector Spaces: An Introduction to Abstract Algebra Via Geometric Constructibility. Springer. ISBN 978-0-387-94848-5.

- Hairer, Ernst; Nørsett, Syvert Paul; Wanner, Gerhard (1993), Solving ordinary differential equations I: Nonstiff problems, Berlin, New York: Springer-Verlag, ISBN 978-3-540-56670-0

- Misner, Charles W.; Thorne, Kip S.; Wheeler, John Archibald (1973). Gravitation. San Francisco: W. H. Freeman. ISBN 978-0-7167-0344-0 Chapter 34, p. 916.

- Andrei D. Polyanin and Alexander V. Manzhirov Handbook of Integral Equations. CRC Press, Boca Raton, 1998. ISBN 0-8493-2876-4.

- Mordell, L. J. (1969). Diophantine equations. Pure and Applied Mathematics. 30. Academic Press. ISBN 0-12-506250-8. Zbl 0188.34503.

- Schmidt, Wolfgang M. (1991). Diophantine approximations and Diophantine equations. Lecture Notes in Mathematics. 1467. Berlin: Springer-Verlag. ISBN 3-540-54058-X. Zbl 0754.11020.

- Shorey, T. N.; Tijdeman, R. (1986). Exponential Diophantine equations. Cambridge Tracts in Mathematics. 87. Cambridge University Press. ISBN 0-521-26826-5. Zbl 0606.10011.

- Smart, Nigel P. (1998). The algorithmic resolution of Diophantine equations. London Mathematical Society Student Texts. 41. Cambridge University Press. ISBN 0-521-64156-X. Zbl 0907.11001.

- Stillwell, John (2004). Mathematics and its History (Second ed.). Springer Science + Business Media Inc. ISBN 0-387-95336-1.

- Rich, Barnett; Schmidt, Philip (2004), Schaum's Outline of Theory and Problems of Elementary Algebra, The McGraw–Hill Companies, ISBN 0-07-141083-X, Chapter 13 §4.4, p. 291

- The Cambridge Ancient History Part 2 Early History of the Middle East. Cambridge University Press. 1971. p. 530. ISBN 978-0-521-07791-0.

Applications of Algebra

Algebraic geometry is the branch of mathematics, mainly studying zeros whereas the algebraic construction is used in geometry to study areas and volumes is known as exterior algebra. Some other applications of algebra that have been discussed in the following chapter are symbolic computation, permutation, algebraic number theory, algebraic K-theory etc. The diverse applications of algebra in the current scenario have been thoroughly discussed in this text.

Algebraic Geometry

Algebraic geometry is a branch of mathematics, classically studying zeros of multivariate polynomials. Modern algebraic geometry is based on the use of abstract algebraic techniques, mainly from commutative algebra, for solving geometrical problems about these sets of zeros.

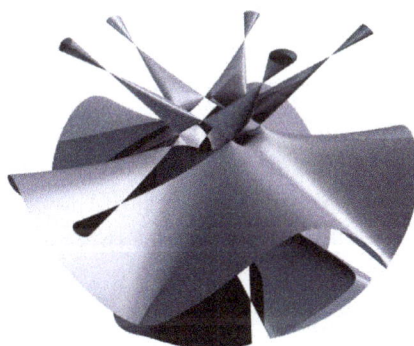

This Togliatti surface is an algebraic surface of degree five. The picture represents a portion of its real locus.

The fundamental objects of study in algebraic geometry are algebraic varieties, which are geometric manifestations of solutions of systems of polynomial equations. Examples of the most studied classes of algebraic varieties are: plane algebraic curves, which include lines, circles, parabolas, ellipses, hyperbolas, cubic curves like elliptic curves and quartic curves like lemniscates, and Cassini ovals. A point of the plane belongs to an algebraic curve if its coordinates satisfy a given polynomial equation. Basic questions involve the study of the points of special interest like the singular points, the inflection points and the points at infinity. More advanced questions involve the topology of the curve and relations between the curves given by different equations.

Algebraic geometry occupies a central place in modern mathematics and has multiple conceptual connections with such diverse fields as complex analysis, topology and number theory. Initially a study of systems of polynomial equations in several variables, the subject of algebraic geometry starts where equation solving leaves off, and it becomes even more important to understand the intrinsic properties of the totality of solutions of a system of equations, than to find a specific solution; this leads into some of the deepest areas in all of mathematics, both conceptually and in terms of technique.

In the 20th century, algebraic geometry split into several subareas.

- The mainstream of algebraic geometry is devoted to the study of the complex points of the algebraic varieties and more generally to the points with coordinates in an algebraically closed field.

- The study of the points of an algebraic variety with coordinates in the field of the rational numbers or in a number field became arithmetic geometry (or more classically Diophantine geometry), a subfield of algebraic number theory.

- The study of the real points of an algebraic variety is the subject of real algebraic geometry.

- A large part of singularity theory is devoted to the singularities of algebraic varieties.

- With the rise of the computers, a computational algebraic geometry area has emerged, which lies at the intersection of algebraic geometry and computer algebra. It consists essentially in developing algorithms and software for studying and finding the properties of explicitly given algebraic varieties.

Much of the development of the mainstream of algebraic geometry in the 20th century occurred within an abstract algebraic framework, with increasing emphasis being placed on "intrinsic" properties of algebraic varieties not dependent on any particular way of embedding the variety in an ambient coordinate space; this parallels developments in topology, differential and complex geometry. One key achievement of this abstract algebraic geometry is Grothendieck's scheme theory which allows one to use sheaf theory to study algebraic varieties in a way which is very similar to its use in the study of differential and analytic manifolds. This is obtained by extending the notion of point: In classical algebraic geometry, a point of an affine variety may be identified, through Hilbert's Nullstellensatz, with a maximal ideal of the coordinate ring, while the points of the corresponding affine scheme are all prime ideals of this ring. This means that a point of such a scheme may be either a usual point or a subvariety. This approach also enables a unification of the language and the tools of classical algebraic geometry, mainly concerned with complex points, and of algebraic number theory. Wiles's proof of the longstanding conjecture called Fermat's last theorem is an example of the power of this approach.

Basic Notions

Zeros of Simultaneous Polynomials

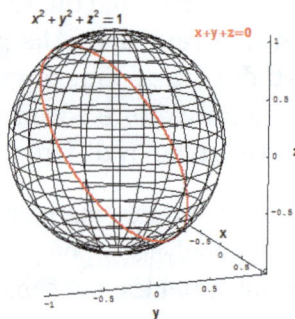

Sphere and slanted circle

In classical algebraic geometry, the main objects of interest are the vanishing sets of collections of polynomials, meaning the set of all points that simultaneously satisfy one or more polynomial equations. For instance, the two-dimensional sphere in three-dimensional Euclidean space R³ could be defined as the set of all points (x,y,z) with

$$x^2 + y^2 + z^2 - 1 = 0.$$

A "slanted" circle in R³ can be defined as the set of all points (x,y,z) which satisfy the two polynomial equations

$$x^2 + y^2 + z^2 - 1 = 0,$$

$$x + y + z = 0.$$

Affine Varieties

First we start with a field k. In classical algebraic geometry, this field was always the complex numbers C, but many of the same results are true if we assume only that k is algebraically closed. We consider the affine space of dimension n over k, denoted $A^n(k)$ (or more simply A^n, when k is clear from the context). When one fixes a coordinates system, one may identify $A^n(k)$ with k^n. The purpose of not working with k^n is to emphasize that one "forgets" the vector space structure that k^n carries.

A function $f: A^n \to A^1$ is said to be *polynomial* (or *regular*) if it can be written as a polynomial, that is, if there is a polynomial p in $k[x_1,...,x_n]$ such that $f(M) = p(t_1,...,t_n)$ for every point M with coordinates $(t_1,...,t_n)$ in A^n. The property of a function to be polynomial (or regular) does not depend on the choice of a coordinate system in A^n.

When a coordinate system is chosen, the regular functions on the affine n-space may be identified with the ring of polynomial functions in n variables over k. Therefore, the set of the regular functions on A^n is a ring, which is denoted $k[A^n]$.

We say that a polynomial *vanishes* at a point if evaluating it at that point gives zero. Let S be a set of polynomials in $k[A^n]$. The *vanishing set of S* (or *vanishing locus* or *zero set*) is the set $V(S)$ of all points in A^n where every polynomial in S vanishes. In other words,

$$V(S) = \{(t_1,...,t_n) \mid \forall p \in S, p(t_1,...,t_n) = 0\}.$$

A subset of A^n which is $V(S)$, for some S, is called an *algebraic set*. The V stands for *variety* (a specific type of algebraic set to be defined below).

Given a subset U of A^n, can one recover the set of polynomials which generate it? If U is *any* subset of A^n, define $I(U)$ to be the set of all polynomials whose vanishing set contains U. The I stands for ideal: if two polynomials f and g both vanish on U, then $f+g$ vanishes on U, and if h is any polynomial, then hf vanishes on U, so $I(U)$ is always an ideal of the polynomial ring $k[A^n]$.

Two natural questions to ask are:

- Given a subset U of A^n, when is $U = V(I(U))$?

- Given a set S of polynomials, when is $S = I(V(S))$?

The answer to the first question is provided by introducing the Zariski topology, a topology on A^n whose closed sets are the algebraic sets, and which directly reflects the algebraic structure of $k[A^n]$. Then $U = V(I(U))$ if and only if U is an algebraic set or equivalently a Zariski-closed set. The answer to the second question is given by Hilbert's Nullstellensatz. In one of its forms, it says that $I(V(S))$ is the radical of the ideal generated by S. In more abstract language, there is a Galois connection, giving rise to two closure operators; they can be identified, and naturally play a basic role in the theory; the example is elaborated at Galois connection.

For various reasons we may not always want to work with the entire ideal corresponding to an algebraic set U. Hilbert's basis theorem implies that ideals in $k[A^n]$ are always finitely generated.

An algebraic set is called *irreducible* if it cannot be written as the union of two smaller algebraic sets. Any algebraic set is a finite union of irreducible algebraic sets and this decomposition is unique. Thus its elements are called the *irreducible components* of the algebraic set. An irreducible algebraic set is also called a *variety*. It turns out that an algebraic set is a variety if and only if it may be defined as the vanishing set of a prime ideal of the polynomial ring.

Some authors do not make a clear distinction between algebraic sets and varieties and use *irreducible variety* to make the distinction when needed.

Regular Functions

Just as continuous functions are the natural maps on topological spaces and smooth functions are the natural maps on differentiable manifolds, there is a natural class of functions on an algebraic set, called *regular functions* or *polynomial functions*. A regular function on an algebraic set V contained in A^n is the restriction to V of a regular function on A^n. For an algebraic set defined on the field of the complex numbers, the regular functions are smooth and even analytic.

It may seem unnaturally restrictive to require that a regular function always extend to the ambient space, but it is very similar to the situation in a normal topological space, where the Tietze extension theorem guarantees that a continuous function on a closed subset always extends to the ambient topological space.

Just as with the regular functions on affine space, the regular functions on V form a ring, which we denote by $k[V]$. This ring is called the *coordinate ring of V*.

Since regular functions on V come from regular functions on A^n, there is a relationship between the coordinate rings. Specifically, if a regular function on V is the restriction of two functions f and g in $k[A^n]$, then $f - g$ is a polynomial function which is null on V and thus belongs to $I(V)$. Thus $k[V]$ may be identified with $k[A^n]/I(V)$.

Morphism of Affine Varieties

Using regular functions from an affine variety to A^1, we can define regular maps from one affine variety to another. First we will define a regular map from a variety into affine space: Let V be a variety contained in A^n. Choose m regular functions on V, and call them $f_1, ..., f_m$. We define a *regular map f* from V to A^m by letting $f = (f_1, ..., f_m)$. In other words, each f_i determines one coordinate of the range of f.

If V' is a variety contained in $\mathbb{A}^m$, we say that f is a *regular map* from V to V' if the range of f is contained in V'.

The definition of the regular maps apply also to algebraic sets. The regular maps are also called *morphisms*, as they make the collection of all affine algebraic sets into a category, where the objects are the affine algebraic sets and the morphisms are the regular maps. The affine varieties is a subcategory of the category of the algebraic sets.

Given a regular map g from V to V' and a regular function f of $k[V']$, then $f \circ g \in k[V]$. The map $f \rightarrow f \circ g$ is a ring homomorphism from $k[V']$ to $k[V]$. Conversely, every ring homomorphism from $k[V']$ to $k[V]$ defines a regular map from V to V'. This defines an equivalence of categories between the category of algebraic sets and the opposite category of the finitely generated reduced k-algebras. This equivalence is one of the starting points of scheme theory.

Rational Function and Birational Equivalence

Contrarily to the preceding ones, this section concerns only varieties and not algebraic sets. On the other hand, the definitions extend naturally to projective varieties (next section), as an affine variety and its projective completion have the same field of functions.

If V is an affine variety, its coordinate ring is an integral domain and has thus a field of fractions which is denoted $k(V)$ and called the *field of the rational functions* on V or, shortly, the *function field* of V. Its elements are the restrictions to V of the rational functions over the affine space containing V. The domain of a rational function f is not V but the complement of the subvariety (a hypersurface) where the denominator of f vanishes.

Like for regular maps, one may define a *rational map* from a variety V to a variety V'. Like for the regular maps, the rational maps from V to V' may be identified to the field homomorphisms from $k(V')$ to $k(V)$.

Two affine varieties are *birationally equivalent* if there are two rational functions between them which are inverse one to the other in the regions where both are defined. Equivalently, they are birationally equivalent if their function fields are isomorphic.

An affine variety is a *rational variety* if it is birationally equivalent to an affine space. This means that the variety admits a rational parameterization. For example, the circle of equation $x^2 + y^2 - 1 = 0$ is a rational curve, as it has the parameterization

$$x = \frac{2t}{1+t^2}$$

$$y = \frac{1-t^2}{1+t^2},$$

which may also be viewed as a rational map from the line to the circle.

The problem of resolution of singularities is to know if every algebraic variety is birationally equivalent to a variety whose projective completion is nonsingular. It has been positively solved in characteristic 0 by Heisuke Hironaka in 1964 and is yet unsolved in finite characteristic.

Projective Variety

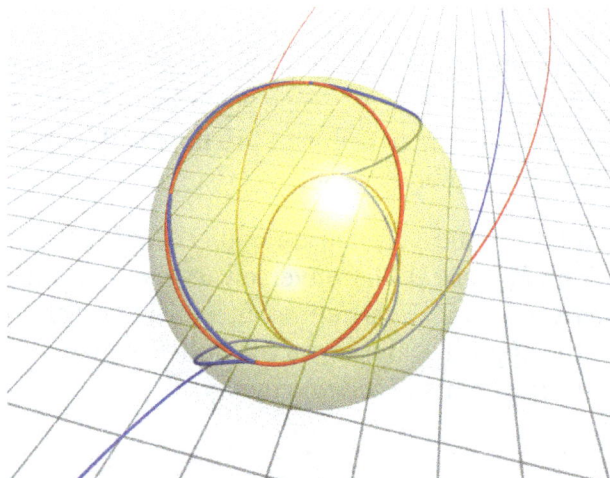

parabola ($y = x^2$, red) and cubic ($y = x^3$, blue) in projective space

Just as the formulas for the roots of 2nd, 3rd and 4th degree polynomials suggest extending real numbers to the more algebraically complete setting of the complex numbers, many properties of algebraic varieties suggest extending affine space to a more geometrically complete projective space. Whereas the complex numbers are obtained by adding the number i, a root of the polynomial x^2 + 1, projective space is obtained by adding in appropriate points "at infinity", points where parallel lines may meet.

To see how this might come about, consider the variety $V(y - x^2)$. If we draw it, we get a parabola. As x goes to positive infinity, the slope of the line from the origin to the point (x, x^2) also goes to positive infinity. As x goes to negative infinity, the slope of the same line goes to negative infinity.

Compare this to the variety $V(y - x^3)$. This is a cubic curve. As x goes to positive infinity, the slope of the line from the origin to the point (x, x^3) goes to positive infinity just as before. But unlike before, as x goes to negative infinity, the slope of the same line goes to positive infinity as well; the exact opposite of the parabola. So the behavior "at infinity" of $V(y - x^3)$ is different from the behavior "at infinity" of $V(y - x^2)$.

The consideration of the *projective completion* of the two curves, which is their prolongation "at infinity" in the projective plane, allows to quantify this difference: the point at infinity of the parabola is a regular point, whose tangent is the line at infinity, while the point at infinity of the cubic curve is a cusp. Also, both curves are rational, as they are parameterized by x, and Riemann-Roch theorem implies that the cubic curve must have a singularity, which must be at infinity, as all its points in the affine space are regular.

Thus many of the properties of algebraic varieties, including birational equivalence and all the topological properties, depend on the behavior "at infinity" and so it is natural to study the varieties in projective space. Furthermore, the introduction of projective techniques made many theorems in algebraic geometry simpler and sharper: For example, Bézout's theorem on the number of intersection points between two varieties can be stated in its sharpest form only in projective space. For these reasons, projective space plays a fundamental role in algebraic geometry.

Nowadays, the *projective space* P^n of dimension n is usually defined as the set of the lines passing through a point, considered as the origin, in the affine space of dimension $n+1$, or equivalently to the set of the vector lines in a vector space of dimension $n+1$. When a coordinate system has been chosen in the space of dimension $n+1$, all the points of a line have the same set of coordinates, up to the multiplication by an element of k. This defines the homogeneous coordinates of a point of P^n as a sequence of $n+1$ elements of the base field k, defined up to the multiplication by a nonzero element of k (the same for the whole sequence).

Given a polynomial in $n+1$ variables, it vanishes at all the point of a line passing through the origin if and only if it is homogeneous. In this case, one says that the polynomial *vanishes* at the corresponding point of P^n. This allows to define a *projective algebraic set* in P^n as the set $V(f_1, ..., f_k)$ where a finite set of homogeneous polynomials $\{f_1, ..., f_k\}$ vanishes. Like for affine algebraic sets, there is a bijection between the projective algebraic sets and the reduced homogeneous ideals which define them. The *projective varieties* are the projective algebraic sets whose defining ideal is prime. In other words, a projective variety is a projective algebraic set, whose homogeneous coordinate ring is an integral domain, the *projective coordinates ring* being defined as the quotient of the graded ring or the polynomials in $n+1$ variables by the homogeneous (reduced) ideal defining the variety. Every projective algebraic set may be uniquely decomposed into a finite union of projective varieties.

The only regular functions which may be defined properly on a projective variety are the constant functions. Thus this notion is not used in projective situations. On the other hand, the *field of the rational functions* or *function field* is a useful notion, which, similarly as in the affine case, is defined as the set of the quotients of two homogeneous elements of the same degree in the homogeneous coordinate ring.

Real Algebraic Geometry

The real algebraic geometry is the study of the real points of the algebraic geometry.

The fact that the field of the reals number is an ordered field should not be ignored in such a study. For example, the curve of equation $x^2 + y^2 - a = 0$ is a circle if $a > 0$, but does not have any real point if $a < 0$. It follows that real algebraic geometry is not only the study of the real algebraic varieties, but has been generalized to the study of the *semi-algebraic sets*, which are the solutions of systems of polynomial equations and polynomial inequalities. For example, a branch of the hyperbola of equation $xy - 1 = 0$ is not an algebraic variety, but is a semi-algebraic set defined by $xy - 1 = 0$ and $x > 0$ or by $xy - 1 = 0$ and $x + y > 0$.

One of the challenging problems of real algebraic geometry is the unsolved Hilbert's sixteenth problem: Decide which respective positions are possible for the ovals of a nonsingular plane curve of degree 8.

Computational Algebraic Geometry

One may date the origin of computational algebraic geometry to meeting EUROSAM'79 (International Symposium on Symbolic and Algebraic Manipulation) held at Marseille, France in June 1979. At this meeting,

- Dennis S. Arnon showed that George E. Collins's Cylindrical algebraic decomposition (CAD) allows the computation of the topology of semi-algebraic sets,

- Bruno Buchberger presented the Gröbner bases and his algorithm to compute them,

- Daniel Lazard presented a new algorithm for solving systems of homogeneous polynomial equations with a computational complexity which is essentially polynomial in the expected number of solutions and thus simply exponential in the number of the unknowns. This algorithm is strongly related with Macaulay's multivariate resultant.

Since then, most results in this area are related to one or several of these items either by using or improving one of these algorithms, or by finding algorithms whose complexity is simply exponential in the number of the variables.

Gröbner Basis

A Gröbner basis is a system of generators of a polynomial ideal whose computation allows the deduction of many properties of the affine algebraic variety defined by the ideal.

Given an ideal I defining an algebraic set V:

- V is empty (over an algebraically closed extension of the basis field), if and only if the Gröbner basis for any monomial ordering is reduced to {1}.

- By means of the Hilbert series one may compute the dimension and the degree of V from any Gröbner basis of I for a monomial ordering refining the total degree.

- If the dimension of V is 0, one may compute the points (finite in number) of V from any Gröbner basis of I.

- A Gröbner basis computation allows to remove from V all irreducible components which are contained in a given hyper surface.

- A Gröbner basis computation allows to compute the Zariski closure of the image of V by the projection on the k first coordinates, and the subset of the image where the projection is not proper.

- More generally Gröbner basis computations allows to compute the Zariski closure of the image and the critical points of a rational function of V into another affine variety.

Gröbner basis computations do not allow to compute directly the primary decomposition of I nor the prime ideals defining the irreducible components of V, but most algorithms for this involve Gröbner basis computation. The algorithms which are not based on Gröbner bases use regular chains but may need Gröbner bases in some exceptional situations.

Gröbner base are deemed to be difficult to compute. In fact they may contain, in the worst case, polynomials whose degree is doubly exponential in the number of variables and a number of polynomials which is also doubly exponential. However, this is only a worst case complexity, and the complexity bound of Lazard's algorithm of 1979 may frequently apply. Faugère's F4 and F5 algorithms realize this complexity, as F5 algorithm may be viewed as an improvement of Lazard's 1979 algorithm. It follows that the best implementations allow to compute almost routinely with algebraic sets of degree more than 100. This means that, presently, the difficulty of computing a Gröbner basis is strongly related to the intrinsic difficulty of the problem.

Cylindrical Algebraic Decomposition (CAD)

CAD is an algorithm which had been introduced in 1973 by G. Collins to implement with an acceptable complexity the Tarski–Seidenberg theorem on quantifier elimination over the real numbers.

This theorem concerns the formulas of the first-order logic whose atomic formulas are polynomial equalities or inequalities between polynomials with real coefficients. These formulas are thus the formulas which may be constructed from the atomic formulas by the logical operators *and* ($\wedge$), *or* ($\vee$), *not* ($\neg$), *for all* ($\forall$) and *exists* ($\exists$). Tarski's theorem asserts that, from such a formula, one may compute an equivalent formula without quantifier ($\forall$, $\exists$).

The complexity of CAD is doubly exponential in the number of variables. This means that CAD allow, in theory, to solve every problem of real algebraic geometry which may be expressed by such a formula, that is almost every problem concerning explicitly given varieties and semi-algebraic sets.

While Gröbner basis computation has doubly exponential complexity only in rare cases, CAD has almost always this high complexity. This implies that, unless if most polynomials appearing in the input are linear, it may not solve problems with more than four variables.

Since 1973, most of the research on this subject is devoted either to improve CAD or to find alternate algorithms in special cases of general interest.

As an example of the state of art, there are efficient algorithms to find at least a point in every connected component of a semi-algebraic set, and thus to test if a semi-algebraic set is empty. On the other hand, CAD is yet, in practice, the best algorithm to count the number of connected components.

Asymptotic Complexity vs. Practical Efficiency

The basic general algorithms of computational geometry have a double exponential worst case complexity. More precisely, if d is the maximal degree of the input polynomials and n the number of variables, their complexity is at most $d^{2^{cn}}$ for some constant c, and, for some inputs, the complexity is at least $d^{2^{c'n}}$ for another constant c'.

During the last 20 years of 20th century, various algorithms have been introduced to solve specific subproblems with a better complexity. Most of these algorithms have a complexity $d^{O(n^2)}$.

Among these algorithms which solve a sub problem of the problems solved by Gröbner bases, one may cite *testing if an affine variety is empty* and *solving nonhomogeneous polynomial systems which have a finite number of solutions*. Such algorithms are rarely implemented because, on most entries Faugère's F4 and F5 algorithms have a better practical efficiency and probably a similar or better complexity (*probably* because the evaluation of the complexity of Gröbner basis algorithms on a particular class of entries is a difficult task which has been done only in a few special cases).

The main algorithms of real algebraic geometry which solve a problem solved by CAD are related to the topology of semi-algebraic sets. One may cite *counting the number of connected compo-*

nents, testing if two points are in the same components or *computing a Whitney stratification of a real algebraic set.* They have a complexity of $d^{O(n^2)}$, but the constant involved by O notation is so high that using them to solve any nontrivial problem effectively solved by CAD, is impossible even if one could use all the existing computing power in the world. Therefore, these algorithms have never been implemented and this is an active research area to search for algorithms with have together a good asymptotic complexity and a good practical efficiency.

Abstract Modern Viewpoint

The modern approaches to algebraic geometry redefine and effectively extend the range of basic objects in various levels of generality to schemes, formal schemes, ind-schemes, algebraic spaces, algebraic stacks and so on. The need for this arises already from the useful ideas within theory of varieties, e.g. the formal functions of Zariski can be accommodated by introducing nilpotent elements in structure rings; considering spaces of loops and arcs, constructing quotients by group actions and developing formal grounds for natural intersection theory and deformation theory lead to some of the further extensions.

Most remarkably, in late 1950s, algebraic varieties were subsumed into Alexander Grothendieck's concept of a scheme. Their local objects are affine schemes or prime spectra which are locally ringed spaces which form a category which is antiequivalent to the category of commutative unital rings, extending the duality between the category of affine algebraic varieties over a field k, and the category of finitely generated reduced k-algebras. The gluing is along Zariski topology; one can glue within the category of locally ringed spaces, but also, using the Yoneda embedding, within the more abstract category of presheaves of sets over the category of affine schemes. The Zariski topology in the set theoretic sense is then replaced by a Grothendieck topology. Grothendieck introduced Grothendieck topologies having in mind more exotic but geometrically finer and more sensitive examples than the crude Zariski topology, namely the étale topology, and the two flat Grothendieck topologies: fppf and fpqc; nowadays some other examples became prominent including Nisnevich topology. Sheaves can be furthermore generalized to stacks in the sense of Grothendieck, usually with some additional representability conditions leading to Artin stacks and, even finer, Deligne-Mumford stacks, both often called algebraic stacks.

Sometimes other algebraic sites replace the category of affine schemes. For example, Nikolai Durov has introduced commutative algebraic monads as a generalization of local objects in a generalized algebraic geometry. Versions of a tropical geometry, of an absolute geometry over a field of one element and an algebraic analogue of Arakelov's geometry were realized in this setup.

Another formal generalization is possible to Universal algebraic geometry in which every variety of algebras has its own algebraic geometry. The term *variety of algebras* should not be confused with *algebraic variety*.

The language of schemes, stacks and generalizations has proved to be a valuable way of dealing with geometric concepts and became cornerstones of modern algebraic geometry.

Algebraic stacks can be further generalized and for many practical questions like deformation theory and intersection theory, this is often the most natural approach. One can extend the Grothendieck site of affine schemes to a higher categorical site of derived affine schemes, by replacing the commutative rings with an infinity category of differential graded commutative algebras, or of simplicial commutative rings or a similar category with an appropriate variant of a Grothendieck topology. One can also replace presheaves of sets by presheaves of simplicial sets (or of infinity groupoids). Then, in presence of an appropriate homotopic machinery one can develop a notion of derived stack as such a presheaf on the infinity category of derived affine schemes, which is satisfying certain infinite categorical version of a sheaf axiom (and to be algebraic, inductively a sequence of representability conditions). Quillen model categories, Segal categories and quasicategories are some of the most often used tools to formalize this yielding the *derived algebraic geometry*, introduced by the school of Carlos Simpson, including Andre Hirschowitz, Bertrand Toën, Gabrielle Vezzosi, Michel Vaquié and others; and developed further by Jacob Lurie, Bertrand Toën, and Gabrielle Vezzosi. Another (noncommutative) version of derived algebraic geometry, using A-infinity categories has been developed from early 1990s by Maxim Kontsevich and followers.

History

Prehistory: Before the 16th Century

Some of the roots of algebraic geometry date back to the work of the Hellenistic Greeks from the 5th century BC. The Delian problem, for instance, was to construct a length x so that the cube of side x contained the same volume as the rectangular box a^2b for given sides a and b. Menaechmus (circa 350 BC) considered the problem geometrically by intersecting the pair of plane conics $ay = x^2$ and $xy = ab$. The later work, in the 3rd century BC, of Archimedes and Apollonius studied more systematically problems on conic sections, and also involved the use of coordinates. The Arab mathematicians were able to solve by purely algebraic means certain cubic equations, and then to interpret the results geometrically. This was done, for instance, by Ibn al-Haytham in the 10th century AD. Subsequently, Persian mathematician Omar Khayyám (born 1048 A.D.) discovered the general method of solving cubic equations by intersecting a parabola with a circle. Each of these early developments in algebraic geometry dealt with questions of finding and describing the intersections of algebraic curves.

Renaissance

Such techniques of applying geometrical constructions to algebraic problems were also adopted by a number of Renaissance mathematicians such as Gerolamo Cardano and Niccolò Fontana "Tartaglia" on their studies of the cubic equation. The geometrical approach to construction problems, rather than the algebraic one, was favored by most 16th and 17th century mathematicians, notably Blaise Pascal who argued against the use of algebraic and analytical methods in geometry. The French mathematicians Franciscus Vieta and later René Descartes and Pierre de Fermat revolutionized the conventional way of thinking about construction problems through the introduction of coordinate geometry. They were interested primarily in the properties of *algebraic curves*, such as those defined by Diophantine equations (in the case of Fermat), and the algebraic reformulation of the classical Greek works on conics and cubics (in the case of Descartes).

During the same period, Blaise Pascal and Gérard Desargues approached geometry from a different perspective, developing the synthetic notions of projective geometry. Pascal and Desargues also studied curves, but from the purely geometrical point of view: the analog of the Greek *ruler and compass construction*. Ultimately, the analytic geometry of Descartes and Fermat won out, for it supplied the 18th century mathematicians with concrete quantitative tools needed to study physical problems using the new calculus of Newton and Leibniz. However, by the end of the 18th century, most of the algebraic character of coordinate geometry was subsumed by the *calculus of infinitesimals* of Lagrange and Euler.

19th and Early 20th Century

It took the simultaneous 19th century developments of non-Euclidean geometry and Abelian integrals in order to bring the old algebraic ideas back into the geometrical fold. The first of these new developments was seized up by Edmond Laguerre and Arthur Cayley, who attempted to ascertain the generalized metric properties of projective space. Cayley introduced the idea of *homogeneous polynomial forms*, and more specifically quadratic forms, on projective space. Subsequently, Felix Klein studied projective geometry (along with other types of geometry) from the viewpoint that the geometry on a space is encoded in a certain class of transformations on the space. By the end of the 19th century, projective geometers were studying more general kinds of transformations on figures in projective space. Rather than the projective linear transformations which were normally regarded as giving the fundamental Kleinian geometry on projective space, they concerned themselves also with the higher degree birational transformations. This weaker notion of congruence would later lead members of the 20th century Italian school of algebraic geometry to classify algebraic surfaces up to birational isomorphism.

The second early 19th century development, that of Abelian integrals, would lead Bernhard Riemann to the development of Riemann surfaces.

In the same period began the algebraization of the algebraic geometry through commutative algebra. The prominent results in this direction are Hilbert's basis theorem and Hilbert's Nullstellensatz, which are the basis of the connexion between algebraic geometry and commutative algebra, and Macaulay's multivariate resultant, which is the basis of elimination theory. Probably because of the size of the computation which is implied by multivariate resultants, elimination theory was forgotten during the middle of the 20th century until it was renewed by singularity theory and computational algebraic geometry.

20th Century

B. L. van der Waerden, Oscar Zariski and André Weil developed a foundation for algebraic geometry based on contemporary commutative algebra, including valuation theory and the theory of ideals. One of the goals was to give a rigorous framework for proving the results of Italian school of algebraic geometry. In particular, this school used systematically the notion of generic point without any precise definition, which was first given by these authors during the 1930s.

In the 1950s and 1960s Jean-Pierre Serre and Alexander Grothendieck recast the foundations making use of sheaf theory. Later, from about 1960, and largely led by Grothendieck, the idea of schemes was worked out, in conjunction with a very refined apparatus of homological techniques. After a decade of rapid development the field stabilized in the 1970s, and new applications were made, both to number theory and to more classical geometric questions on algebraic varieties, singularities and moduli.

An important class of varieties, not easily understood directly from their defining equations, are the abelian varieties, which are the projective varieties whose points form an abelian group. The prototypical examples are the elliptic curves, which have a rich theory. They were instrumental in the proof of Fermat's last theorem and are also used in elliptic curve cryptography.

In parallel with the abstract trend of the algebraic geometry, which is concerned with general statements about varieties, methods for effective computation with concretely-given varieties have also been developed, which lead to the new area of computational algebraic geometry. One of the founding methods of this area is the theory of Gröbner bases, introduced by Bruno Buchberger in 1965. Another founding method, more specially devoted to real algebraic geometry, is the cylindrical algebraic decomposition, introduced by George E. Collins in 1973.

Analytic Geometry

An analytic variety is defined locally as the set of common solutions of several equations involving analytic functions. It is analogous to the included concept of real or complex algebraic variety. Any complex manifold is an analytic variety. Since analytic varieties may have singular points, not all analytic varieties are manifolds.

Modern analytic geometry is essentially equivalent to real and complex algebraic geometry, as has been shown by Jean-Pierre Serre in his paper *GAGA*, the name of which is French for *Algebraic geometry and analytic geometry*. Nevertheless, the two fields remain distinct, as the methods of proof are quite different and algebraic geometry includes also geometry in finite characteristic.

Applications

Algebraic geometry now finds applications in statistics, control theory, robotics, error-correcting codes, phylogenetics and geometric modelling. There are also connections to string theory, game theory, graph matchings, solitons and integer programming.

Exterior Algebra

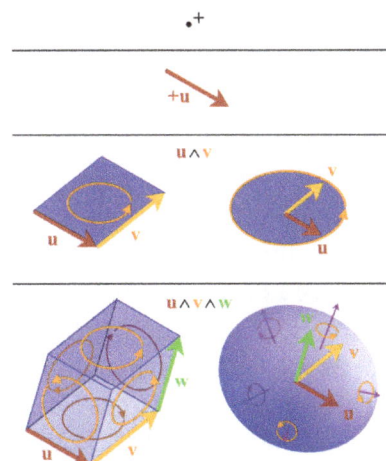

Orientation defined by an ordered set of vectors.

Geometric interpretation of grade n elements in a real exterior algebra for $n = 0$ (signed point), 1 (directed line segment, or vector), 2 (oriented plane element), 3 (oriented volume). The exterior product of n vectors can be visualized as any n-dimensional shape (e.g. n-parallelotope, n-ellipsoid); with magnitude (hypervolume), and orientation defined by that on its $(n-1)$-dimensional boundary and on which side the interior is.

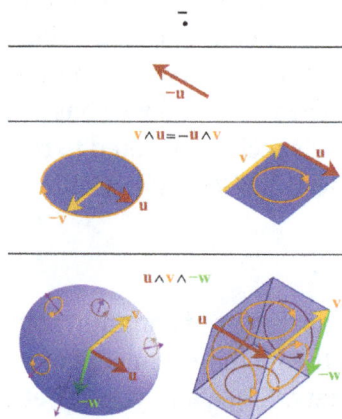

Reversed orientation corresponds to negating the exterior product.

In mathematics, the exterior product or wedge product of vectors is an algebraic construction used in geometry to study areas, volumes, and their higher-dimensional analogs. The exterior product of two vectors u and v, denoted by $u \wedge v$, is called a bivector and lives in a space called the *exterior square*, a vector space that is distinct from the original space of vectors. The magnitude of $u \wedge v$ can be interpreted as the area of the parallelogram with sides u and v, which in three dimensions can also be computed using the cross product of the two vectors. Like the cross product, the exterior product is anticommutative, meaning that $u \wedge v = -(v \wedge u)$ for all vectors u and v. One way to visualize a bivector is as a family of parallelograms all lying in the same plane, having the same area, and with the same orientation—a choice of clockwise or counterclockwise.

When regarded in this manner, the exterior product of two vectors is called a 2-blade. More generally, the exterior product of any number k of vectors can be defined and is sometimes called a k-blade. It lives in a space known as the kth exterior power. The magnitude of the resulting k-blade is the volume of the k-dimensional parallelotope whose edges are the given vectors, just as the magnitude of the scalar triple product of vectors in three dimensions gives the volume of the parallelepiped generated by those vectors.

The exterior algebra, or Grassmann algebra after Hermann Grassmann, is the algebraic system whose product is the exterior product. The exterior algebra provides an algebraic setting in which to answer geometric questions. For instance, blades have a concrete geometric interpretation, and objects in the exterior algebra can be manipulated according to a set of unambiguous rules. The exterior algebra contains objects that are not only k-blades, but sums of k-blades; such a sum is called a k-vector. The k-blades, because they are simple products of vectors, are called the simple elements of the algebra. The *rank* of any k-vector is defined to be the smallest number of simple elements of which it is a sum. The exterior product extends to the full exterior algebra, so that it makes sense to multiply any two elements of the algebra. Equipped with this product, the exterior algebra is an associative algebra, which means that $\alpha \wedge (\beta \wedge \gamma) = (\alpha \wedge \beta) \wedge \gamma$ for any elements α, β, γ. The k-vectors have degree k, meaning that they are sums of products of k vectors. When elements of different degrees are multiplied, the degrees add like multiplication of polynomials. This means that the exterior algebra is a graded algebra.

The definition of the exterior algebra makes sense for spaces not just of geometric vectors, but of other vector-like objects such as vector fields or functions. In full generality, the exterior algebra can be defined for modules over a commutative ring, and for other structures of interest in abstract algebra. It is one of these more general constructions where the exterior algebra finds one of its most important applications, where it appears as the algebra of differential forms that is fundamental in areas that use differential geometry. Differential forms are mathematical objects that represent infinitesimal areas of infinitesimal parallelograms (and higher-dimensional bodies), and so can be integrated over surfaces and higher dimensional manifolds in a way that generalizes the line integrals from calculus. The exterior algebra also has many algebraic properties that make it a convenient tool in algebra itself. The association of the exterior algebra to a vector space is a type of functor on vector spaces, which means that it is compatible in a certain way with linear transformations of vector spaces. The exterior algebra is one example of a bialgebra, meaning that its dual space also possesses a product, and this dual product is compatible with the exterior product. This dual algebra is precisely the algebra of alternating multilinear forms, and the pairing between the exterior algebra and its dual is given by the interior product.

Motivating Examples

Areas in the Plane

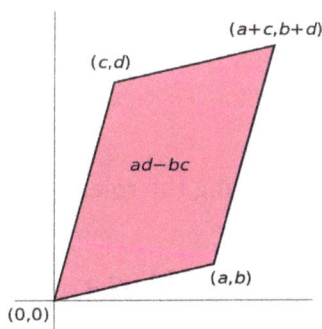

The area of a parallelogram in terms of the determinant of the matrix of coordinates of two of its vertices.

The Cartesian plane $\mathbf{R}^2$ is a vector space equipped with a basis consisting of a pair of unit vectors

$$\mathbf{e}_1 = \begin{bmatrix} 1 \\ 0 \end{bmatrix}, \quad \mathbf{e}_2 = \begin{bmatrix} 0 \\ 1 \end{bmatrix}.$$

Suppose that

$$\mathbf{v} = \begin{bmatrix} a \\ b \end{bmatrix} = a\mathbf{e}_1 + b\mathbf{e}_2, \quad \mathbf{w} = \begin{bmatrix} c \\ d \end{bmatrix} = c\mathbf{e}_1 + d\mathbf{e}_2$$

are a pair of given vectors in $\mathbf{R}^2$, written in components. There is a unique parallelogram having v and w as two of its sides. The *area* of this parallelogram is given by the standard determinant formula:

$$\text{Area} = \left| \det \begin{bmatrix} \mathbf{v} & \mathbf{w} \end{bmatrix} \right| = \left| \det \begin{bmatrix} a & c \\ b & d \end{bmatrix} \right| = |ad - bc|.$$

Consider now the exterior product of v and w:

$$\begin{aligned}
\mathbf{v} \wedge \mathbf{w} &= (a\mathbf{e}_1 + b\mathbf{e}_2) \wedge (c\mathbf{e}_1 + d\mathbf{e}_2) \\
&= ac\mathbf{e}_1 \wedge \mathbf{e}_1 + ad\mathbf{e}_1 \wedge \mathbf{e}_2 + bc\mathbf{e}_2 \wedge \mathbf{e}_1 + bd\mathbf{e}_2 \wedge \mathbf{e}_2 \\
&= (ad - bc)\mathbf{e}_1 \wedge \mathbf{e}_2
\end{aligned}$$

where the first step uses the distributive law for the exterior product, and the last uses the fact that the exterior product is alternating, and in particular $\mathbf{e}_2 \wedge \mathbf{e}_1 = -(\mathbf{e}_1 \wedge \mathbf{e}_2)$. Note that the coefficient in this last expression is precisely the determinant of the matrix [v w]. The fact that this may be positive or negative has the intuitive meaning that v and w may be oriented in a counterclockwise or clockwise sense as the vertices of the parallelogram they define. Such an area is called the *signed area* of the parallelogram: the absolute value of the signed area is the ordinary area, and the sign determines its orientation.

The fact that this coefficient is the signed area is not an accident. In fact, it is relatively easy to see that the exterior product should be related to the signed area if one tries to axiomatize this area as an algebraic construct. In detail, if A(v, w) denotes the signed area of the parallelogram determined by the pair of vectors v and w, then A must satisfy the following properties:

1. $A(j\mathbf{v}, k\mathbf{w}) = jkA(\mathbf{v}, \mathbf{w})$ for any real numbers j and k, since rescaling either of the sides rescales the area by the same amount (and reversing the direction of one of the sides reverses the orientation of the parallelogram).

2. $A(\mathbf{v}, \mathbf{v}) = 0$, since the area of the degenerate parallelogram determined by v (i.e., a line segment) is zero.

3. $A(\mathbf{w}, \mathbf{v}) = -A(\mathbf{v}, \mathbf{w})$, since interchanging the roles of v and w reverses the orientation of the parallelogram.

4. $A(\mathbf{v} + j\mathbf{w}, \mathbf{w}) = A(\mathbf{v}, \mathbf{w})$, for real j, since adding a multiple of w to v affects neither the base nor the height of the parallelogram and consequently preserves its area.

5. $A(\mathbf{e}_1, \mathbf{e}_2) = 1$, since the area of the unit square is one.

With the exception of the last property, the exterior product satisfies the same formal properties as the area. In a certain sense, the exterior product generalizes the final property by allowing the area of a parallelogram to be compared to that of any "standard" chosen parallelogram (here, the one with sides $\mathbf{e}_1$ and $\mathbf{e}_2$). In other words, the exterior product in two dimensions provides a *basis-independent* formulation of area.

Cross and Triple Products

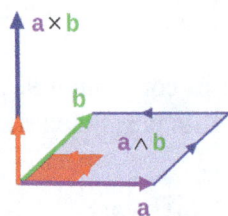

The cross product (**blue** vector) in relation to the exterior product (**light blue** parallelogram). The length of the cross product is to the length of the parallel unit vector (**red**) as the size of the exterior product is to the size of the reference parallelogram (**light red**).

For vectors in R³, the exterior algebra is closely related to the cross product and triple product. Using the standard basis $\{e_1, e_2, e_3\}$, the exterior product of a pair of vectors

$$\mathbf{u} = u_1\mathbf{e}_1 + u_2\mathbf{e}_2 + u_3\mathbf{e}_3$$

and

$$\mathbf{v} = v_1\mathbf{e}_1 + v_2\mathbf{e}_2 + v_3\mathbf{e}_3$$

is

$$\mathbf{u} \wedge \mathbf{v} = (u_1v_2 - u_2v_1)(\mathbf{e}_1 \wedge \mathbf{e}_2) + (u_3v_1 - u_1v_3)(\mathbf{e}_3 \wedge \mathbf{e}_1) + (u_2v_3 - u_3v_2)(\mathbf{e}_2 \wedge \mathbf{e}_3)$$

where $\{e_1 \wedge e_2, e_3 \wedge e_1, e_2 \wedge e_3\}$ is the basis for the three-dimensional space $\Lambda^2(R^3)$. The coefficients above are the same as those in the usual definition of the cross product of vectors in three dimensions, the only difference being that the exterior product is not an ordinary vector, but instead is a 2-vector.

Bringing in a third vector

$$\mathbf{w} = w_1\mathbf{e}_1 + w_2\mathbf{e}_2 + w_3\mathbf{e}_3,$$

the exterior product of three vectors is

$$\mathbf{u} \wedge \mathbf{v} \wedge \mathbf{w} = (u_1v_2w_3 + u_2v_3w_1 + u_3v_1w_2 - u_1v_3w_2 - u_2v_1w_3 - u_3v_2w_1)(\mathbf{e}_1 \wedge \mathbf{e}_2 \wedge \mathbf{e}_3)$$

where $e_1 \wedge e_2 \wedge e_3$ is the basis vector for the one-dimensional space $\Lambda^3(R^3)$. The scalar coefficient is the triple product of the three vectors.

The cross product and triple product in three dimensions each admit both geometric and algebraic interpretations. The cross product u × v can be interpreted as a vector which is perpendicular to both u and v and whose magnitude is equal to the area of the parallelogram determined by the two vectors. It can also be interpreted as the vector consisting of the minors of the matrix with columns u and v. The triple product of u, v, and w is geometrically a (signed) volume. Algebraically, it is the determinant of the matrix with columns u, v, and w. The exterior product in three dimensions allows for similar interpretations. In fact, in the presence of a positively oriented orthonormal basis, the exterior product generalizes these notions to higher dimensions.

Formal Definitions and Algebraic Properties

The exterior algebra $\Lambda(V)$ over a vector space V over a field K is defined as the quotient algebra of the tensor algebra $T(V)$ by the two-sided ideal I generated by all elements of the form $x \otimes x$ for $x \in V$ (i.e. all tensors which can be expressed as the tensor product of any vector in V by itself). Symbolically,

$$\Lambda(V) := T(V)\,/\,I.$$

The exterior product $\wedge$ of two elements of $\Lambda(V)$ is defined by

$$\alpha \wedge \beta = \alpha \otimes \beta + I,$$

where the mod I means that we do the tensor product in the usual way and then declare every element of the tensor that is in the ideal to be zero. Equivalently, any two tensors which differ only by a member of the ideal are handled as the same element in the exterior algebra.

As $T^0 = K$, $T^1 = V$, and $\left(T^0(V) \oplus T^1(V)\right) \cap I = \{0\}$, the inclusions of K and V in $T(V)$ induce injections of K and V into $\Lambda(V)$. These injections are commonly considered as inclusions, and called *natural embeddings*, *natural injections* or *natural inclusions*.

Anticommutativity of the Exterior Product

The exterior product is *alternating* on elements of V, which means that $x \wedge x = 0$ for all $x \in V$, by the above construction. It follows that the product is also anticommutative on elements of V, for supposing that $x, y \in V$,

$$0 = (x + y) \wedge (x + y) = x \wedge x + x \wedge y + y \wedge x + y \wedge y = x \wedge y + y \wedge x$$

hence

$$x \wedge y = -(y \wedge x).$$

More generally, if σ is a permutation of the integers [1, ..., k], and x1, x2, ..., xk are elements of V, it follows that

$$x_{\sigma(1)} \wedge x_{\sigma(2)} \wedge \cdots \wedge x_{\sigma(k)} = \operatorname{sgn}(\sigma) x_1 \wedge x_2 \wedge \cdots \wedge x_k,$$

where $\operatorname{sgn}(\sigma)$ is the signature of the permutation σ.

In particular, if $x_i = x_j$ for some i ≠ j, then the following generalization of the alternating property also holds:

$$x_1 \wedge x_2 \wedge \cdots \wedge x_k = 0.$$

Exterior Power

The kth exterior power of V, denoted $\Lambda^k(V)$, is the vector subspace of $\Lambda(V)$ spanned by elements of the form

$$x_1 \wedge x_2 \wedge \cdots \wedge x_k, \quad x_i \in V, i = 1, 2, \ldots, k.$$

If $\alpha \in \Lambda^k(V)$, then α is said to be a **k-vector**. If, furthermore, α can be expressed as an exterior product of k elements of V, then α is said to be decomposable. Although decomposable k-vectors span $\Lambda^k(V)$, not every element of $\Lambda^k(V)$ is decomposable. For example, in R⁴, the following 2-vector is not decomposable:

$$\alpha = e_1 \wedge e_2 + e_3 \wedge e_4.$$

(This is a symplectic form, since $\alpha \wedge \alpha \neq 0$.)

Basis and Dimension

If the dimension of V is n and $\{e_1, ..., e_n\}$ is a basis of V, then the set

$$\{e_{i_1} \wedge e_{i_2} \wedge \cdots \wedge e_{i_k} \mid 1 \leq i_1 < i_2 < \cdots < i_k \leq n\}$$

is a basis for $\Lambda^k(V)$. The reason is the following: given any exterior product of the form

$$v_1 \wedge \cdots \wedge v_k,$$

every vector v_j can be written as a linear combination of the basis vectors e_i; using the bilinearity of the exterior product, this can be expanded to a linear combination of exterior products of those basis vectors. Any exterior product in which the same basis vector appears more than once is zero; any exterior product in which the basis vectors do not appear in the proper order can be reordered, changing the sign whenever two basis vectors change places. In general, the resulting coefficients of the basis k-vectors can be computed as the minors of the matrix that describes the vectors v_j in terms of the basis e_i.

By counting the basis elements, the dimension of $\Lambda^k(V)$ is equal to a binomial coefficient:

$$\dim(\Lambda^k(V)) = \binom{n}{k}$$

In particular, $\Lambda^k(V) = \{0\}$ for $k > n$.

Any element of the exterior algebra can be written as a sum of k-vectors. Hence, as a vector space the exterior algebra is a direct sum

$$\Lambda(V) = \Lambda^0(V) \oplus \Lambda^1(V) \oplus \Lambda^2(V) \oplus \cdots \oplus \Lambda^n(V)$$

(where by convention $\Lambda^0(V) = K$ and $\Lambda^1(V) = V$), and therefore its dimension is equal to the sum of the binomial coefficients, which is 2^n.

Rank of a k-vector

If $\alpha \in \Lambda^k(V)$, then it is possible to express α as a linear combination of decomposable k-vectors:

$$\alpha = \alpha^{(1)} + \alpha^{(2)} + \cdots + \alpha^{(s)}$$

where each $\alpha^{(i)}$ is decomposable, say

$$\alpha^{(i)} = \alpha_1^{(i)} \wedge \cdots \wedge \alpha_k^{(i)}, \quad i = 1, 2, \ldots, s.$$

The rank of the k-vector α is the minimal number of decomposable k-vectors in such an expansion of α. This is similar to the notion of tensor rank.

Rank is particularly important in the study of 2-vectors (Sternberg 1974, §III.6) (Bryant et al. 1991). The rank of a 2-vector α can be identified with half the rank of the matrix of coefficients of α in a basis. Thus if e_i is a basis for V, then α can be expressed uniquely as

$$\alpha = \sum_{i,j} a_{ij} e_i \wedge e_j$$

where $a_{ij} = -a_{ji}$ (the matrix of coefficients is skew-symmetric). The rank of the matrix a_{ij} is therefore even, and is twice the rank of the form α.

In characteristic 0, the 2-vector α has rank p if and only if

$$\underbrace{\alpha \wedge \cdots \wedge \alpha}_{p} \neq 0$$

and

$$\underbrace{\alpha \wedge \cdots \wedge \alpha}_{p+1} = 0.$$

Graded Structure

The exterior product of a k-vector with a p-vector is a $(k + p)$-vector, once again invoking bilinearity. As a consequence, the direct sum decomposition of the preceding section

$$\Lambda(V) = \Lambda^0(V) \oplus \Lambda^1(V) \oplus \Lambda^2(V) \oplus \cdots \oplus \Lambda^n(V)$$

gives the exterior algebra the additional structure of a graded algebra, that is

$$\Lambda^k(V) \wedge \Lambda^p(V) \subset \Lambda^{k+p}(V).$$

Moreover, if K is the basis field, we have

$$\Lambda^0(V) = K$$

and

$$\Lambda^1(V) = V.$$

The exterior product is graded anticommutative, meaning that if $\alpha \in \Lambda^k(V)$ and $\beta \in \Lambda^p(V)$, then

$$\alpha \wedge \beta = (-1)^{kp} \beta \wedge \alpha.$$

In addition to studying the graded structure on the exterior algebra, Bourbaki (1989) studies additional graded structures on exterior algebras, such as those on the exterior algebra of a graded module (a module that already carries its own gradation).

Universal Property

Let V be a vector space over the field K. Informally, multiplication in $\Lambda(V)$ is performed by manipulating symbols and imposing a distributive law, an associative law, and using the identity $v \wedge v = 0$ for $v \in V$. Formally, $\Lambda(V)$ is the "most general" algebra in which these rules hold for the multiplication, in the sense that any unital associative K-algebra containing V with alternating multiplication on V must contain a homomorphic image of $\Lambda(V)$. In other words, the exterior algebra has the following universal property:

Given any unital associative K-algebra A and any K-linear map $j : V \to A$ such that $j(v)j(v) = 0$ for every v in V, then there exists *precisely one* unital algebra homomorphism $f : \Lambda(V) \to A$ such that $j(v) = f(i(v))$ for all v in V (here i is the natural inclusion of V in $\Lambda(V)$.

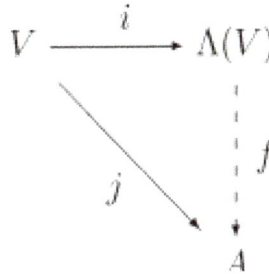

To construct the most general algebra that contains V and whose multiplication is alternating on V, it is natural to start with the most general associative algebra that contains V, the tensor algebra $T(V)$, and then enforce the alternating property by taking a suitable quotient. We thus take the two-sided ideal I in $T(V)$ generated by all elements of the form $v \otimes v$ for v in V, and define $\Lambda(V)$ as the quotient

$$\Lambda(V) = T(V) / I$$

(and use $\wedge$ as the symbol for multiplication in $\Lambda(V)$). It is then straightforward to show that $\Lambda(V)$ contains V and satisfies the above universal property.

As a consequence of this construction, the operation of assigning to a vector space V its exterior algebra $\Lambda(V)$ is a functor from the category of vector spaces to the category of algebras.

Rather than defining $\Lambda(V)$ first and then identifying the exterior powers $\Lambda^k(V)$ as certain subspaces, one may alternatively define the spaces $\Lambda^k(V)$ first and then combine them to form the algebra $\Lambda(V)$. This approach is often used in differential geometry and is described in the next section.

Generalizations

Given a commutative ring R and an R-module M, we can define the exterior algebra $\Lambda(M)$ just as above, as a suitable quotient of the tensor algebra $T(M)$. It will satisfy the analogous universal property. Many of the properties of $\Lambda(M)$ also require that M be a projective module. Where finite dimensionality is used, the properties further require that M be finitely generated and projective. Generalizations to the most common situations can be found in (Bourbaki 1989).

Exterior algebras of vector bundles are frequently considered in geometry and topology. There are no essential differences between the algebraic properties of the exterior algebra of finite-dimensional vector bundles and those of the exterior algebra of finitely generated projective modules, by the Serre–Swan theorem. More general exterior algebras can be defined for sheaves of modules.

Duality

Alternating Operators

Given two vector spaces V and X, an alternating operator from V^k to X is a multilinear map

$$f : V^k \to X$$

such that whenever $v_1, \ldots, v_k$ are linearly dependent vectors in V, then

$$f(v_1, \ldots, v_k) = 0$$

The map

$$w : V^k \to \Lambda^k(V)$$

which associates to k vectors from V their exterior product, i.e. their corresponding k-vector, is also alternating. In fact, this map is the "most general" alternating operator defined on V^k: given any other alternating operator $f : V^k \to X$, there exists a unique linear map $\varphi : \Lambda^k(V) \to X$ with $f = \varphi \circ w$. This universal property characterizes the space $\Lambda^k(V)$ and can serve as its definition.

Alternating Multilinear Forms

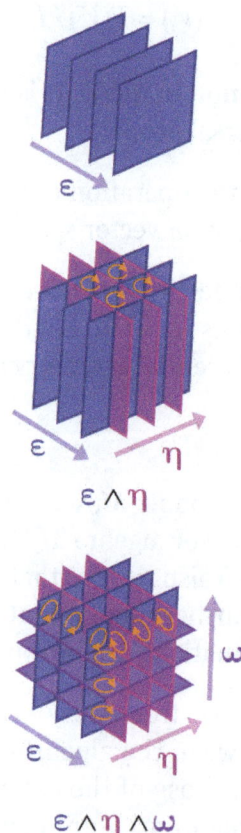

Geometric interpretation for the **exterior product** of n 1-forms (ε, η, ω) to obtain an n-form ("mesh" of coordinate surfaces, here planes), for $n = 1, 2, 3$. The "circulations" show orientation.

The above discussion specializes to the case when $X = K$, the base field. In this case an alternating multilinear function

$$f : V^k \to K$$

is called an alternating multilinear form. The set of all alternating multilinear forms is a vector space, as the sum of two such maps, or the product of such a map with a scalar, is again alternating. By the universal property of the exterior power, the space of alternating forms of degree k on V is naturally isomorphic with the dual vector space $(\Lambda^k V)^*$. If V is finite-dimensional, then the latter is naturally isomorphic to $\Lambda^k(V^*)$. In particular, the dimension of the space of anti-symmetric maps from V^k to K is the binomial coefficient n choose k.

Under this identification, the exterior product takes a concrete form: it produces a new anti-symmetric map from two given ones. Suppose $\omega : V^k \to K$ and $\eta : V^m \to K$ are two anti-symmetric maps. As in the case of tensor products of multilinear maps, the number of variables of their exterior product is the sum of the numbers of their variables. It is defined as follows:

$$\omega \wedge \eta = \frac{(k+m)!}{k!m!} \mathrm{Alt}(\omega \otimes \eta)$$

where the alternation Alt of a multilinear map is defined to be the signed average of the values over all the permutations of its variables:

$$\mathrm{Alt}(\omega)(x_1,\ldots,x_k) = \frac{1}{k!} \sum_{\sigma \in S_k} \mathrm{sgn}(\sigma)\omega(x_{\sigma(1)},\ldots,x_{\sigma(k)}).$$

This definition of the exterior product is well-defined even if the field K has finite characteristic, if one considers an equivalent version of the above that does not use factorials or any constants:

$$\omega \wedge \eta(x_1,\ldots,x_{k+m}) = \sum_{\sigma \in Sh_{k,m}} \mathrm{sgn}(\sigma)\omega(x_{\sigma(1)},\ldots,x_{\sigma(k)})\eta(x_{\sigma(k+1)},\ldots,x_{\sigma(k+m)}),$$

where here $\mathrm{Sh}_{k,m} \subset S_{k+m}$ is the subset of (k,m) shuffles: permutations σ of the set $\{1, 2, ..., k + m\}$ such that $\sigma(1) < \sigma(2) < ... < \sigma(k)$, and $\sigma(k + 1) < \sigma(k + 2) < ... < \sigma(k + m)$.

Bialgebra Structure

In formal terms, there is a correspondence between the graded dual of the graded algebra $\Lambda(V)$ and alternating multilinear forms on V. The exterior algebra (as well as the symmetric algebra) inherits a bialgebra structure, and, indeed, a Hopf algebra structure, from the tensor algebra.

The exterior product of multilinear forms defined above is dual to a coproduct defined on $\Lambda(V)$, giving the structure of a coalgebra. The coproduct is a linear function $\Delta : \Lambda(V) \to \Lambda(V) \otimes \Lambda(V)$ which is given by

$$\Delta(v) = 1 \otimes v + v \otimes 1$$

on elements $v \in V$. The symbol 1 stands for the unit element of the field K. Recall that $K \subset \Lambda(V)$, so that the above really does lie in $\Lambda(V) \otimes \Lambda(V)$. This definition of the coproduct is extended to the full space $\Lambda(V)$ by (linear) homomorphism. That is, for $v, w \in V$, one has, by definition, the homomorphism

$$\Delta(v \wedge w) = \Delta(v) \wedge \Delta(w)$$

The correct form of this homomorphism is not what one might naively write, but has to be the one carefully defined in the coalgebra article. In this case, one obtains

$$\Delta(v \wedge w) = 1 \otimes (v \wedge w) + v \otimes w - w \otimes v + (v \wedge w) \otimes 1.$$

Extending to the full space $\Lambda(V)$, one has, in general,

$$\Delta(x_1 \wedge \cdots \wedge x_k) = \Delta(x_1) \wedge \cdots \wedge \Delta(x_k)$$

Expanding this out in detail, one obtains the following expression on decomposable elements:

$$\Delta(x_1 \wedge \cdots \wedge x_k) = \sum_{p=0}^{k} \sum_{\sigma \in Sh(p+1, k-p)} \mathrm{sgn}(\sigma)(x_{\sigma(0)} \wedge \cdots \wedge x_{\sigma(p)}) \otimes (x_{\sigma(p+1)} \wedge \cdots \wedge x_{\sigma(k)}).$$

where the second summation is taken over all (p+1,k-p)-shuffles. The above is written with a notational trick, to keep track of the field element 1: the trick is to write $x_0 = 1$, and this is shuffled into various locations during the expansion of the sum over shuffles. The shuffle follows directly from the first axiom of a co-algebra: the relative order of the elements x_k is *preserved* in the riffle shuffle: the riffle shuffle merely splits the ordered sequence into two ordered sequences, one on the left, and one on the right.

Observe that the coproduct preserves the grading of the algebra. That is, one has that

$$\Delta : \Lambda^k(V) \to \bigoplus_{p=0}^{k} \Lambda^p(V) \otimes \Lambda^{k-p}(V)$$

The tensor symbol $\otimes$ used in this section should be understood with some caution: it is *not* the same tensor symbol as the one being used in the definition of the alternating product. Intuitively, it is perhaps easiest to think it as just another, but different, tensor product: it is still (bi-)linear, as tensor products should be, but it is the product that is appropriate for the definition of a bialgebra, that is, for creating the object $\Lambda(V) \otimes \Lambda(V)$. Any lingering doubt can be shaken by pondering the equalities $(1 \otimes v) \wedge (1 \otimes w) = 1 \otimes (v \wedge w)$ and $(v \otimes 1) \wedge (1 \otimes w) = v \otimes w$, which follow from the definition of the coalgebra, as opposed to naive manipulations involving the tensor and wedge symbols. This distinction is developed in greater detail in the article on tensor algebras. Here, there is much less of a problem, in that the alternating product $\wedge$ clearly corresponds to multiplication in the bialgebra, leaving the symbol $\otimes$ free for use in the definition of the bialgebra. In practice, this presents no particular problem, as long as one avoids the fatal trap of replacing alternating sums of $\otimes$ by the wedge symbol, with one exception. One can construct an alternating product from $\otimes$, with the understanding that it works in a different space. Immediately below, an example is given: the alternating product for the *dual space* can be given in terms of the coproduct. The construction of the bialgebra here parallels the construction in the tensor algebra article almost exactly, except for the need to correctly track the alternating signs for the exterior algebra.

In terms of the coproduct, the exterior product on the dual space is just the graded dual of the coproduct:

$$(\alpha \wedge \beta)(x_1 \wedge \cdots \wedge x_k) = (\alpha \otimes \beta)\big(\Delta(x_1 \wedge \cdots \wedge x_k)\big)$$

where the tensor product on the right-hand side is of multilinear linear maps (extended by zero on elements of incompatible homogeneous degree: more precisely, $\alpha \wedge \beta = \varepsilon \circ (\alpha \otimes \beta) \circ \Delta$, where ε is the counit, as defined presently).

The counit is the homomorphism $\varepsilon : \Lambda(V) \to K$ that returns the 0-graded component of its argument. The coproduct and counit, along with the exterior product, define the structure of a bialgebra on the exterior algebra.

With an antipode defined on homogeneous elements by $S(x) = (-1)^{\deg x}x$, the exterior algebra is furthermore a Hopf algebra.

Interior Product

Suppose that V is finite-dimensional. If V^* denotes the dual space to the vector space V, then for each $\alpha \in V^*$, it is possible to define an antiderivation on the algebra $\Lambda(V)$,

$$i_\alpha : \Lambda^k V \to \Lambda^{k-1} V.$$

This derivation is called the interior product with α, or sometimes the insertion operator, or contraction by α.

Suppose that $w \in \Lambda^k V$. Then w is a multilinear mapping of V^* to K, so it is defined by its values on the k-fold Cartesian product $V^* \times V^* \times \ldots \times V^*$. If $u_1, u_2, \ldots, u_{k-1}$ are $k - 1$ elements of V^*, then define

$$(i_\alpha \mathbf{w})(u_1, u_2, \ldots, u_{k-1}) = \mathbf{w}(\alpha, u_1, u_2, \ldots, u_{k-1}).$$

Additionally, let $i_\alpha f = 0$ whenever f is a pure scalar (i.e., belonging to $\Lambda^0 V$).

Axiomatic Characterization and Properties

The interior product satisfies the following properties:

- For each k and each $\alpha \in V^*$,

  $$i_\alpha : \Lambda^k V \to \Lambda^{k-1} V.$$

  (By convention, $\Lambda^{-1} = \{0\}$.)

- If v is an element of $V (= \Lambda^1 V)$, then $i_\alpha v = \alpha(v)$ is the dual pairing between elements of V and elements of V^*.

- For each $\alpha \in V^*$, i_α is a graded derivation of degree -1:
  $$i_\alpha(a \wedge b) = (i_\alpha a) \wedge b + (-1)^{\deg a} a \wedge (i_\alpha b).$$

These three properties are sufficient to characterize the interior product as well as define it in the general infinite-dimensional case.

Further properties of the interior product include:

- $i_\alpha{}^\circ i_\alpha = 0$.

- $i_\alpha{}^\circ i_\beta = -i_\beta{}^\circ i_\alpha$.

Hodge Duality

Suppose that V has finite dimension n. Then the interior product induces a canonical isomorphism of vector spaces

$$\Lambda^k(V^*) \otimes \Lambda^n(V) \to \Lambda^{n-k}(V)$$

by the recursive definition

$$i_{\alpha \wedge \beta} = i_\beta{}^\circ i_\alpha.$$

In the geometrical setting, a non-zero element of the top exterior power $\Lambda^n(V)$ (which is a one-dimensional vector space) is sometimes called a volume form (or orientation form, although this term may sometimes lead to ambiguity). Relative to a given volume form σ, the isomorphism is given explicitly by

$$\alpha \in \Lambda^k(V^*) \mapsto i_\alpha \sigma \in \Lambda^{n-k}(V).$$

If, in addition to a volume form, the vector space V is equipped with an inner product identifying V with V^*, then the resulting isomorphism is called the Hodge dual (or more commonly the Hodge star operator)

$$*: \Lambda^k(V) \to \Lambda^{n-k}(V).$$

The composite of $*$ with itself maps $\Lambda^k(V) \to \Lambda^k(V)$ and is always a scalar multiple of the identity map. In most applications, the volume form is compatible with the inner product in the sense that it is an exterior product of an orthonormal basis of V. In this case,

$$*\circ *: \Lambda^k(V) \to \Lambda^k(V) = (-1)^{k(n-k)+q} I$$

where I is the identity, and the inner product has metric signature (p, q) — p plusses and q minuses.

Inner Product

For V a finite-dimensional space, an inner product on V defines an isomorphism of V with V^*, and so also an isomorphism of $\Lambda^k V$ with $(\Lambda^k V)^*$. The pairing between these two spaces also takes the form of an inner product. On decomposable k-vectors,

$$\langle v_1 \wedge \cdots \wedge v_k, w_1 \wedge \cdots \wedge w_k \rangle = \det(\langle v_i, w_j \rangle),$$

the determinant of the matrix of inner products. In the special case $v_i = w_i$, the inner product is the square norm of the k-vector, given by the determinant of the Gramian matrix $(\langle v_i, v_j \rangle)$. This is then extended bilinearly (or sesquilinearly in the complex case) to a non-degenerate inner product on $\Lambda^k V$. If e_i, $i = 1, 2, \ldots, n$, form an orthonormal basis of V, then the vectors of the form

$$e_{i_1} \wedge \cdots \wedge e_{i_k}, \quad i_1 < \cdots < i_k,$$

constitute an orthonormal basis for $\Lambda^k(V)$.

With respect to the inner product, exterior multiplication and the interior product are mutually adjoint. Specifically, for $v \in \Lambda^{k-1}(V)$, $w \in \Lambda^k(V)$, and $x \in V$,

$$\langle x \wedge \mathbf{v}, \mathbf{w} \rangle = \langle \mathbf{v}, i_{x^\flat} \mathbf{w} \rangle$$

where $\mathbf{x}^\flat \in V^*$ is the linear functional defined by

$$x^\flat(y) = \langle x, y \rangle$$

for all $y \in V$. This property completely characterizes the inner product on the exterior algebra.

Indeed, more generally for $v \in \Lambda^{k-l}(V)$, $w \in \Lambda^k(V)$, and $x \in \Lambda^l(V)$, iteration of the above adjoint properties gives

$$\langle \mathbf{x} \wedge \mathbf{v}, \mathbf{w} \rangle = \langle \mathbf{v}, i_{x^\flat} \mathbf{w} \rangle$$

where now $\mathbf{x}^\flat \in \Lambda^l(V^*) \simeq (\Lambda^l(V))^*$ is the dual l-vector defined by

$$\mathbf{x}^\flat(\mathbf{y}) = \langle \mathbf{x}, \mathbf{y} \rangle$$

for all $y \in \Lambda^l(V)$.

Functoriality

Suppose that V and W are a pair of vector spaces and $f : V \to W$ is a linear transformation. Then, by the universal construction, there exists a unique homomorphism of graded algebras

$$\Lambda(f) : \Lambda(V) \to \Lambda(W)$$

such that

$$\Lambda(f)\big|_{\Lambda^1(V)} = f : V = \Lambda^1(V) \to W = \Lambda^1(W).$$

In particular, $\Lambda(f)$ preserves homogeneous degree. The k-graded components of $\Lambda(f)$ are given on decomposable elements by

$$\Lambda(f)(x_1 \wedge \cdots \wedge x_k) = f(x_1) \wedge \cdots \wedge f(x_k).$$

Let

$$\Lambda^k(f) = \Lambda(f)_{\Lambda^k(V)} : \Lambda^k(V) \to \Lambda^k(W).$$

The components of the transformation $\Lambda(k)$ relative to a basis of V and W is the matrix of $k \times k$ minors of f. In particular, if $V = W$ and V is of finite dimension n, then $\Lambda^n(f)$ is a mapping of a one-dimensional vector space Λ^n to itself, and is therefore given by a scalar: the determinant of f.

Exactness

If

$$0 \to U \to V \to W \to 0$$

is a short exact sequence of vector spaces, then

$$0 \to \Lambda^1(U) \wedge \Lambda(V) \to \Lambda(V) \to \Lambda(W) \to 0$$

is an exact sequence of graded vector spaces as is

$$0 \to \Lambda(U) \to \Lambda(V).$$

Direct Sums

In particular, the exterior algebra of a direct sum is isomorphic to the tensor product of the exterior algebras:

$$\Lambda(V \oplus W) \cong \Lambda(V) \otimes \Lambda(W).$$

This is a graded isomorphism; i.e.,

$$\Lambda^k(V \oplus W) \cong \bigoplus_{p+q=k} \Lambda^p(V) \otimes \Lambda^q(W).$$

Slightly more generally, if

$$0 \to U \to V \to W \to 0$$

is a short exact sequence of vector spaces then $\Lambda^k(V)$ has a filtration

$$0 = F^0 \subseteq F^1 \subseteq \cdots \subseteq F^k \subseteq F^{k+1} = \Lambda^k(V)$$

with quotients : $F^{p+1} / F^p = \Lambda^{k-p}(U) \otimes \Lambda^p(W)$. In particular, if U is 1-dimensional then

$$0 \to U \otimes \Lambda^{k-1}(W) \to \Lambda^k(V) \to \Lambda^k(W) \to 0$$

is exact, and if W is 1-dimensional then

$$0 \to \Lambda^k(U) \to \Lambda^k(V) \to \Lambda^{k-1}(U) \otimes W \to 0$$

is exact.

The Alternating Tensor Algebra

If K is a field of characteristic 0, then the exterior algebra of a vector space V can be canonically identified with the vector subspace of $T(V)$ consisting of antisymmetric tensors. Recall that the exterior algebra is the quotient of $T(V)$ by the ideal I generated by $x \otimes x$.

Let $T^r(V)$ be the space of homogeneous tensors of degree r. This is spanned by decomposable tensors

$$v_1 \otimes \cdots \otimes v_r, \quad v_i \in V.$$

The antisymmetrization (or sometimes the skew-symmetrization) of a decomposable tensor is defined by

$$\mathrm{Alt}(v_1 \otimes \cdots \otimes v_r) = \frac{1}{r!} \sum_{\sigma \in \mathfrak{S}_r} \mathrm{sgn}(\sigma) v_{\sigma(1)} \otimes \cdots \otimes v_{\sigma(r)}$$

where the sum is taken over the symmetric group of permutations on the symbols $\{1, ..., r\}$. This extends by linearity and homogeneity to an operation, also denoted by Alt, on the full tensor algebra $T(V)$. The image $\mathrm{Alt}(T(V))$ is the alternating tensor algebra, denoted $A(V)$. This is a vector subspace of $T(V)$, and it inherits the structure of a graded vector space from that on $T(V)$. It carries an associative graded product $\hat{\otimes}$ defined by

$$t \,\hat{\otimes}\, s = \mathrm{Alt}(t \otimes s).$$

Although this product differs from the tensor product, the kernel of *Alt* is precisely the ideal I (again, assuming that K has characteristic 0), and there is a canonical isomorphism

$$A(V) \cong \Lambda(V).$$

Index Notation

Suppose that V has finite dimension n, and that a basis $e_1, ..., e_n$ of V is given. then any alternating tensor $t \in A^r(V) \subset T^r(V)$ can be written in index notation as

$$t = t^{i_1 i_2 \cdots i_r} \mathbf{e}_{i_1} \otimes \mathbf{e}_{i_2} \otimes \cdots \otimes \mathbf{e}_{i_r}$$

where $t^{i_1 \cdots i_r}$ is completely antisymmetric in its indices.

The exterior product of two alternating tensors t and s of ranks r and p is given by

$$t \hat{\otimes} s = \frac{1}{(r+p)!} \sum_{\sigma \in \mathfrak{S}_{r+p}} \text{sgn}(\sigma) t^{i_{\sigma(1)} \cdots i_{\sigma(r)}} s^{i_{\sigma(r+1)} \cdots i_{\sigma(r+p)}} \mathbf{e}_{i_1} \otimes \mathbf{e}_{i_2} \otimes \cdots \otimes \mathbf{e}_{i_{r+p}}, t \otimes s = \frac{}{(r \quad p !} \sum_{r \ p} \text{sgn}(\ ) t^{i_{\sigma(1)} \quad i_{\sigma(r)}} s^{i_{\sigma(r+1)} \quad i_{\sigma(r)}}$$

The components of this tensor are precisely the skew part of the components of the tensor product $s \otimes t$, denoted by square brackets on the indices:

$$(t \hat{\otimes} s)^{i_1 \cdots i_{r+p}} = t^{[i_1 \cdots i_r} s^{i_{r+1} \cdots i_{r+p}]}.$$

The interior product may also be described in index notation as follows. Let $t = t^{i_0 i_1 \cdots i_{r-1}}$ be an anti-symmetric tensor of rank r. Then, for $\alpha \in V^*$, $i_\alpha t$ is an alternating tensor of rank $r - 1$, given by

$$(i_\alpha t)^{i_1 \cdots i_{r-1}} = r \sum_{j=0}^{n} \alpha_j t^{j i_1 \cdots i_{r-1}}.$$

where n is the dimension of V.

Applications

Linear Algebra

In applications to linear algebra, the exterior product provides an abstract algebraic manner for describing the determinant and the minors of a matrix. For instance, it is well known that the magnitude of the determinant of a square matrix is equal to the volume of the parallelotope whose sides are the columns of the matrix. This suggests that the determinant can be *defined* in terms of the exterior product of the column vectors. Likewise, the $k \times k$ minors of a matrix can be defined by looking at the exterior products of column vectors chosen k at a time. These ideas can be extended not just to matrices but to linear transformations as well: the magnitude of the determinant of a linear transformation is the factor by which it scales the volume of any given reference parallelotope. So the determinant of a linear transformation can be defined in terms of what the transformation does to the top exterior power. The action of a transformation on the lesser exterior powers gives a basis-independent way to talk about the minors of the transformation.

Physics

In physics, many quantities are naturally represented by alternating operators. For example, if the motion of a charged particle is described by velocity and acceleration vectors in four-dimensional spacetime, then normalization of the velocity vector requires that the electromagnetic force must be an alternating operator on the velocity. Its six degrees of freedom are identified with the electric and magnetic fields.

Linear Geometry

The decomposable k-vectors have geometric interpretations: the bivector $u \wedge v$ represents the plane spanned by the vectors, "weighted" with a number, given by the area of the oriented parallelogram with sides u and v. Analogously, the 3-vector $u \wedge v \wedge w$ represents the spanned 3-space weighted by the volume of the oriented parallelepiped with edges u, v, and w.

Projective Geometry

Decomposable k-vectors in $\Lambda^k V$ correspond to weighted k-dimensional linear subspaces of V. In particular, the Grassmannian of k-dimensional subspaces of V, denoted $Gr_k(V)$, can be naturally identified with an algebraic subvariety of the projective space $P(\Lambda^k V)$. This is called the Plücker embedding.

Differential Geometry

The exterior algebra has notable applications in differential geometry, where it is used to define differential forms. A differential form at a point of a differentiable manifold is an alternating multilinear form on the tangent space at the point. Equivalently, a differential form of degree k is a linear functional on the k-th exterior power of the tangent space. As a consequence, the exterior product of multilinear forms defines a natural exterior product for differential forms. Differential forms play a major role in diverse areas of differential geometry.

In particular, the exterior derivative gives the exterior algebra of differential forms on a manifold the structure of a differential algebra. The exterior derivative commutes with pullback along smooth mappings between manifolds, and it is therefore a natural differential operator. The exterior algebra of differential forms, equipped with the exterior derivative, is a cochain complex whose cohomology is called the de Rham cohomology of the underlying manifold and plays a vital role in the algebraic topology of differentiable manifolds.

Representation Theory

In representation theory, the exterior algebra is one of the two fundamental Schur functors on the category of vector spaces, the other being the symmetric algebra. Together, these constructions are used to generate the irreducible representations of the general linear group.

Superspace

The exterior algebra over the complex numbers is the archetypal example of a superalgebra, which plays a fundamental role in physical theories pertaining to fermions and supersymmetry. A single element of the exterior algebra is called a supernumber or Grassmann number. The exterior algebra itself is then just a one-dimensional superspace: it is just the set of all of the points in the exterior algebra. The topology on this space is essentially the weak topology, the open sets being the cylinder sets. An n-dimensional superspace is just the n-fold product of exterior algebras.

Lie Algebra Homology

Let L be a Lie algebra over a field K, then it is possible to define the structure of a chain complex on the exterior algebra of L. This is a K-linear mapping

$$\partial : \Lambda^{p+1} L \to \Lambda^p L$$

defined on decomposable elements by

$$\partial(x_1 \wedge \cdots \wedge x_{p+1}) = \frac{1}{p+1} \sum_{j < \ell} (-1)^{j+\ell+1} [x_j, x_\ell] \wedge x_1 \wedge \cdots \wedge \hat{x}_j \wedge \cdots \wedge \hat{x}_\ell \wedge \cdots \wedge x_{p+1}.$$

The Jacobi identity holds if and only if $\partial\partial = 0$, and so this is a necessary and sufficient condition for an anticommutative nonassociative algebra L to be a Lie algebra. Moreover, in that case ΛL is a chain complex with boundary operator ∂. The homology associated to this complex is the Lie algebra homology.

Homological Algebra

The exterior algebra is the main ingredient in the construction of the Koszul complex, a fundamental object in homological algebra.

History

The exterior algebra was first introduced by Hermann Grassmann in 1844 under the blanket term of *Ausdehnungslehre*, or *Theory of Extension*. This referred more generally to an algebraic (or axiomatic) theory of extended quantities and was one of the early precursors to the modern notion of a vector space. Saint-Venant also published similar ideas of exterior calculus for which he claimed priority over Grassmann.

The algebra itself was built from a set of rules, or axioms, capturing the formal aspects of Cayley and Sylvester's theory of multivectors. It was thus a *calculus*, much like the propositional calculus, except focused exclusively on the task of formal reasoning in geometrical terms. In particular, this new development allowed for an *axiomatic* characterization of dimension, a property that had previously only been examined from the coordinate point of view.

The import of this new theory of vectors and multivectors was lost to mid 19th century mathematicians, until being thoroughly vetted by Giuseppe Peano in 1888. Peano's work also remained somewhat obscure until the turn of the century, when the subject was unified by members of the French geometry school (notably Henri Poincaré, Élie Cartan, and Gaston Darboux) who applied Grassmann's ideas to the calculus of differential forms.

A short while later, Alfred North Whitehead, borrowing from the ideas of Peano and Grassmann, introduced his universal algebra. This then paved the way for the 20th century developments of abstract algebra by placing the axiomatic notion of an algebraic system on a firm logical footing.

Symbolic Computation

In computational mathematics, computer algebra, also called symbolic computation or algebraic computation, is a scientific area that refers to the study and development of algorithms and software for manipulating mathematical expressions and other mathematical objects. Although, properly speaking, computer algebra should be a subfield of scientific computing, they are generally considered as distinct fields because scientific computing is usually based on numerical computation with approximate floating point numbers, while symbolic computation emphasizes *exact* computation with expressions containing variables that have no given value and are manipulated as symbols, hence the name *symbolic computation*.

Software applications that perform symbolic calculations are called *computer algebra systems*, with the term *system* alluding to the complexity of the main applications that include, at least, a method to represent mathematical data in a computer, a user programming language (usually different from the language used for the implementation), a dedicated memory manager, a user interface for the input/output of mathematical expressions, a large set of routines to perform usual operations, like simplification of expressions, differentiation using chain rule, polynomial factorization, indefinite integration, etc.

At the beginning of computer algebra, circa 1970, when the long-known algorithms were first put on computers, they turned out to be highly inefficient. Therefore, a large part of the work of the researchers in the field consisted in revisiting classical algebra in order to make it effective and to discover efficient algorithms to implement this effectiveness. A typical example of this kind of work is the computation of polynomial greatest common divisors, which is required to simplify fractions. Surprisingly, the classical Euclid's algorithm turned out to be inefficient for polynomials over infinite fields, and thus new algorithms needed to be developed. The same was also true for the classical algorithms from linear algebra.

Computer algebra is widely used to experiment in mathematics and to design the formulas that are used in numerical programs. It is also used for complete scientific computations, when purely numerical methods fail, like in public key cryptography or for some non-linear problems.

Terminology

Some authors distinguish *computer algebra* from *symbolic computation* using the latter name to refer to kinds of symbolic computation other than the computation with mathematical formulas. Some authors use *symbolic computation* for the computer science aspect of the subject and "computer algebra" for the mathematical aspect. In some languages the name of the field is not a direct translation of its English name. Typically, it is called *calcul formel* in French, which means "formal computation".

Symbolic computation has also been referred to, in the past, as *symbolic manipulation, algebraic manipulation, symbolic processing, symbolic mathematics,* or *symbolic algebra,* but these terms, which also refer to non-computational manipulation, are no more in use for referring to computer algebra.

Scientific Community

There is no learned society that is specific to computer algebra, but this function is assumed by the special interest group of the Association for Computing Machinery named SIGSAM (Special Interest Group on Symbolic and Algebraic Manipulation).

There are several annual conferences on computer algebra, the premier being ISSAC (International Symposium on Symbolic and Algebraic Computation), which is regularly sponsored by SIGSAM.

There are several journals specializing in computer algebra, the top one being Journal of Symbolic Computation founded in 1985 by Bruno Buchberger. There are also several other journals that regularly publish articles in computer algebra.

Computer Science Aspects

Data Representation

As numerical software are highly efficient for approximate numerical computation, it is common, in computer algebra, to emphasize on *exact* computation with exactly represented data. Such an exact representation implies that, even when the size of the output is small, the intermediate data generated during a computation may grow in an unpredictable way. This behavior is called *expression swell*. To obviate this problem, various methods are used in the representation of the data, as well as in the algorithms that manipulate them.

Numbers

The usual numbers systems used in numerical computation are either the floating point numbers and the integers of a fixed bounded size, that are improperly called *integers* by most programming languages. None is convenient for computer algebra, because of the expression swell.

Therefore, the basic numbers used in computer algebra are the integers of the mathematicians, commonly represented by an unbounded signed sequence of digits in some base of numeration, usually the largest base allowed by the machine word. These integers allow to define the rational numbers, which are irreducible fractions of two integers.

Programming an efficient implementation of the arithmetic operations is a hard task. Therefore, most free computer algebra systems and some commercial ones, like Maple (software), use the GMP library, which is thus a *de facto* standard.

Expressions

Except for numbers and variables, every mathematical expression may be viewed as the symbol of an operator followed by a sequence of operands. In computer algebra software, the expressions are usually represented in this way. This representation is very flexible, and many things, that seem not to be mathematical expressions at first glance, may be represented and manipulated as such. For example, an equation is an expression with "=" as an operator, a matrix may be represented as an expression with "matrix" as an operator and its rows as operands.

Even programs may be considered and represented as expressions with operator "procedure" and, at least, two operands, the list of parameters and the body, which is itself an expression with "body" as an operator and a sequence of instructions as operands. Conversely, any mathematical expression may be viewed as a program. For example, the expression $a + b$ may be viewed as a program for the addition, with a and b as parameters. Executing this program consists in *evaluating* the expression for given values of a and b; if they do not have any value—that is they are indeterminates—, the result of the evaluation is simply its input.

This process of delayed evaluation is fundamental in computer algebra. For example, the operator "=" of the equations is also, in most computer algebra systems, the name of the program of the equality test: normally, the evaluation of an equation results in an equation, but, when an equality test is needed,—either explicitly asked by the user through an "evaluation to a Boolean" command, or automatically started by the system in the case of a test inside a program—then the evaluation to a boolean 0 or 1 is executed.

As the size of the operands of an expression is unpredictable and may change during a working session, the sequence of the operands is usually represented as a sequence of either pointers (like in Macsyma) or entries in a hash table (like in Maple).

Simplification

The raw application of the basic rules of differentiation with respect to x on the expression a^x gives the result $x \cdot a^{x-1} \cdot 0 + a^x \cdot \left(1 \cdot \log a + x \cdot \dfrac{0}{a}\right)$. Such a complicated expression is clearly not acceptable, and a procedure of simplification is needed as soon as one works with general expressions.

This simplification is normally done through rewriting rules. There are several classes of rewriting rules that have to be considered. The simplest consists in the rewriting rules that always reduce the size of the expression, like $E - E \to 0$ or $\sin(0) \to 0$. They are systematically applied in the computer algebra systems.

The first difficulty occurs with associative operations like addition and multiplication. The standard way to deal with associativity is to consider that addition and multiplication have an arbitrary number of operands, that is that $a + b + c$ is represented as "+"(a, b, c). Thus $a + (b + c)$ and $(a + b) + c$ are both simplified to "+"(a, b, c), which is displayed $a + b + c$. What about $a - b + c$? To deal with this problem, the simplest way is to rewrite systematically $-E, E - F, E/F$ as, respectively, $(-1) \cdot E, E + (-1) \cdot F, E \cdot F^{-1}$. In other words, in the internal representation of the expressions, there is no subtraction nor division nor unary minus, outside the representation of the numbers.

A second difficulty occurs with the commutativity of addition and multiplication. The problem is to recognize quickly the like terms in order to combine or canceling them. In fact, the method for finding like terms, consisting of testing every pair of terms, is too costly for being practicable with very long sums and products. For solving this problem, Macsyma sorts the operands of sums and products with a function of comparison that is designed in order that like terms are in consecutive places, and thus easily detected. In Maple, the hash function is designed for generating collisions when like terms are entered, allowing to combine them as soon as they are introduced. This design of the hash function allows also to recognize immediately the expressions or subexpressions that appear several times in a computation and to store them only once. This allows not only to save some memory space, but also to speed up computation, by avoiding repetition of the same operations on several identical expressions.

Some rewriting rules sometimes increase and sometimes decrease the size of the expressions to which they are applied. This is the case of distributivity or trigonometric identities. For example, the distributivity law allows rewriting $(x+1)^4 \to x^4 + 4x^3 + 6x^2 + 4x + 1$ and $(x-1)(x^4 + x^3 + x^2 + x + 1) \to x^5 - 1$. As there is no way to make a good general choice of applying or not such a rewriting rule, such rewritings are done only when explicitly asked by the user. For the distributivity, the computer function that apply this rewriting rule is generally called "expand". The reverse rewriting rule, called "factor", requires a non-trivial algorithm, which is thus a key function in computer algebra systems.

Mathematical Aspects

In this section we consider some fundamental mathematical questions that arise as soon as one wants to manipulate mathematical expressions in a computer. We consider mainly the case of the multivariate rational fractions. This is not a real restriction, because, as soon as the irrational functions appearing in an expression are simplified, they are usually considered as new indeterminates. For example, $(\sin(x+y)^2 + \log(z^2 - 5))^3$ is viewed as a polynomial in $\sin(x+y)$ and $\log(z^2 - 5)$

Equality

There are two notions of equality for mathematical expressions. The *syntactic equality* is the equality of the expressions which means that they are written (or represented in a computer) in the same way. As trivial, it is rarely considered by mathematicians, but it is the only equality that is easy to test with a program. The *semantic equality* is when two expressions represent the same mathematical object, like in

It is known from Richardson's theorem that there may not exist an algorithm that decides if two expressions representing numbers are semantically equal, if exponentials and logarithms are allowed in the expressions. Therefore, (semantical) equality may be tested only on some classes of expressions such as the polynomials and the rational fractions.

To test the equality of two expressions, instead to design a specific algorithm, it is usual to put them in some *canonical form* or to put their difference in a *normal form* and to test the syntactic equality of the result.

Unlike in usual mathematics, "canonical form" and "normal form" are not synonymous in computer algebra. A *canonical form* is such that two expressions in canonical form are semantically equal if and only if they are syntactically equal, while a *normal form* is such that an expression in normal form is semantically zero only if it is syntactically zero. In other words, zero has a unique representation by expressions in normal form.

Normal forms are usually preferred in computer algebra for several reasons. Firstly, canonical forms may be more costly to compute than normal forms. For example, to put a polynomial in canonical form, one has to expand by distributivity every product, while it is not necessary with a normal form. Secondly, It may be the case, like for expressions involving radicals, that a canonical form, if it exists, depends on some arbitrary choices and that these choices may be different for two expressions that have been computed independently. This may make impracticable the use of a canonical form.

Permutation

In mathematics, the notion of permutation relates to the act of arranging all the members of a set into some sequence or order, or if the set is already ordered, rearranging (reordering) its elements, a process called permuting. These differ from combinations, which are selections of some members of a set where order is disregarded. For example, written as tuples, there are six permutations of the set {1,2,3}, namely: (1,2,3), (1,3,2), (2,1,3), (2,3,1), (3,1,2), and (3,2,1). These are all the possible orderings of this three element set. As another example, an anagram of a word, all of whose letters are different, is a permutation of its letters. In this example, the letters are already ordered in the original word and the anagram is a reordering of the letters. The study of permutations of finite sets is a topic in the field of combinatorics.

Each of the six rows is a different permutation of three distinct balls

Permutations occur, in more or less prominent ways, in almost every area of mathematics. They often arise when different orderings on certain finite sets are considered, possibly only because one wants to ignore such orderings and needs to know how many configurations are thus identified. For similar reasons permutations arise in the study of sorting algorithms in computer science.

The number of permutations of n distinct objects is n factorial, usually written as $n!$, which means the product of all positive integers less than or equal to n.

In algebra and particularly in group theory, a permutation of a set S is defined as a bijection from S to itself. That is, it is a function from S to S for which every element occurs exactly once as an image value. This is related to the rearrangement of the elements of S in which each element s is replaced by the corresponding $f(s)$. The collection of such permutations form a group called the symmetric group of S. The key to this group's structure is the fact that the composition of two permutations (performing two given rearrangements in succession) results in another rearrangement. Permutations may *act* on structured objects by rearranging their components, or by certain replacements (substitutions) of symbols.

In elementary combinatorics, the k-permutations, or partial permutations, are the ordered arrangements of k distinct elements selected from a set. When k is equal to the size of the set, these are the permutations of the set.

In the popular puzzle Rubik's cube invented in 1974 by Ernő Rubik, each turn of the puzzle faces creates a permutation of the surface colors.

History

The rule to determine the number of permutations of n objects was known in Indian culture at least as early as around 1150: the Lilavati by the Indian mathematician Bhaskara II contains a passage that translates to

The product of multiplication of the arithmetical series beginning and increasing by unity and continued to the number of places, will be the variations of number with specific figures.

Fabian Stedman in 1677 described factorials when explaining the number of permutations of bells in change ringing. Starting from two bells: "first, *two* must be admitted to be varied in two ways" which he illustrates by showing 1 2 and 2 1. He then explains that with three bells there are "three times two figures to be produced out of three" which again is illustrated. His explanation involves "cast away 3, and 1.2 will remain; cast away 2, and 1.3 will remain; cast away 1, and 2.3 will remain". He then moves on to four bells and repeats the casting away argument showing that there will be four different sets of three. Effectively this is an recursive process. He continues with five bells using the "casting away" method and tabulates the resulting 120 combinations. At this point he gives up and remarks:

Now the nature of these methods is such, that the changes on one number comprehends the changes on all lesser numbers, ... insomuch that a compleat Peal of changes on one number seemeth to be formed by uniting of the compleat Peals on all lesser numbers into one entire body;

Stedman widens the consideration of permutations; he goes on to consider the number of permutations of the letters of the alphabet and horses from a stable of 20.

A first case in which seemingly unrelated mathematical questions were studied with the help of permutations occurred around 1770, when Joseph Louis Lagrange, in the study of polynomial equations, observed that properties of the permutations of the roots of an equation are related to the possibilities to solve it. This line of work ultimately resulted, through the work of Évariste Galois, in Galois theory, which gives a complete description of what is possible and impossible with respect to solving polynomial equations (in one unknown) by radicals. In modern mathematics there are many similar situations in which understanding a problem requires studying certain permutations related to it.

Definition and Notations

There are two equivalent common ways of regarding permutations, sometimes called the "active" and "passive" forms, or in older terminology "substitutions" and "permutations". Which form is preferable depends on the type of questions being asked in a given discipline.

The "active" way to regard permutations of a set S (finite or infinite) is to define them as the bijections from S to itself. Thus, the permutations are thought of as functions which can be composed with each other, forming groups of permutations. From this viewpoint, the elements of S have no internal structure and are just labels for the objects being moved: one may refer to permutations of any set of n elements as "permutations on n letters".

In Cauchy's *two-line notation*, one lists the elements of S in the first row, and for each one its image below it in the second row. For instance, a particular permutation of the set $S = \{1,2,3,4,5\}$ can be written as:

$$\sigma = \begin{pmatrix} 1 & 2 & 3 & 4 & 5 \\ 2 & 5 & 4 & 3 & 1 \end{pmatrix};$$

this means that σ satisfies $\sigma(1)=2$, $\sigma(2)=5$, $\sigma(3)=4$, $\sigma(4)=3$, and $\sigma(5)=1$. The elements of S may appear in any order in the first row. This permutation could also be written as:

$$\sigma = \begin{pmatrix} 3 & 2 & 5 & 1 & 4 \\ 4 & 5 & 1 & 2 & 3 \end{pmatrix}.$$

The "passive" way to regard a permutation of the set S is an *ordered arrangement* (or listing, or linearly ordered arrangement, or sequence without repetition) of the elements of S. This is related to the active form as follows. If there is a "natural" order for the elements of S, say $x_1, x_2, \ldots, x_n$, then one uses this for the first row of the two-line notation:

$$\sigma = \begin{pmatrix} x_1 & x_2 & x_3 & \cdots & x_n \\ \sigma(x_1) & \sigma(x_2) & \sigma(x_3) & \cdots & \sigma(x_n) \end{pmatrix}.$$

Under this assumption, one may omit the first row and write the permutation in *one-line notation* as $\sigma(x_1)\,\sigma(x_2)\,\sigma(x_3)\quad\sigma(x\,)$,, that is, an ordered arrangement of S. Care must be taken to distinguish one-line notation from the cycle notation described later. In mathematics literature, a common usage is to omit parentheses for one-line notation, while using them for cycle notation. The one-line notation is also called the *word representation* of a permutation. The example above would then be 2 5 4 3 1 since the natural order 1 2 3 4 5 would be assumed for the first row. (It is typical to use commas to separate these entries only if some have two or more digits.) This form is more compact, and is common in elementary combinatorics and computer science. It is especially useful in applications where the elements of S or the permutations are to be compared as larger or smaller.

Other uses of the Term *Permutation*

The concept of a permutation as an ordered arrangement admits several generalizations that are not permutations but have been called permutations in the literature.

k-permutations of *n*

A weaker meaning of the term "permutation", sometimes used in elementary combinatorics texts, designates those ordered arrangements in which no element occurs more than once, but without the requirement of using all the elements from a given set. These are not permutations except in special cases, but are natural generalizations of the ordered arrangement concept. Indeed, this use often involves considering arrangements of a fixed length k of elements taken from a given set of size n, in other words, these **k-permutations of n** are the different ordered arrangements of a k-element subset of an n-set (sometimes called variations in the older literature.) These objects are also known as partial permutations or as sequences without repetition, terms that avoid confusion with the other, more common, meaning of "permutation". The number of such k-permutations of n is denoted variously by such symbols as P_k^n, $_nP_k$, ^nP_k, $P_{n,k}$, or $P(n,k)$, and its value is given by the product

$$P(n,k) = \underbrace{n \cdot (n-1) \cdot (n-2) \cdots (n-k+1)}_{k \text{ factors}},$$

which is 0 when $k > n$, and otherwise is equal to

$$\frac{n!}{(n-k)!}.$$

The product is well defined without the assumption that n is a non-negative integer and is of importance outside combinatorics as well; it is known as the Pochhammer symbol $(n)_k$ or as the $n^{\underline{k}}$ -th falling factorial power $n^{\underline{k}}$ of n.

This usage of the term "permutation" is closely related to the term "combination". A k-element combination of an n-set S is a k element subset of S, the elements of which are not ordered. By taking all the k element subsets of S and ordering each of them in all possible ways we obtain all the k-permutations of S. The number of k-combinations of an n-set, $C(n,k)$, is therefore related to the number of k-permutations of n by:

$$C(n,k) = \frac{P(n,k)}{P(k,k)} = \frac{n!}{(n-k)!k!}.$$

These numbers are also known as binomial coefficients and denoted $\binom{n}{k}$.

Permutations with Repetition

Ordered arrangements of the elements of a set S of length n where repetition is allowed are called n-tuples, but have sometimes been referred to as permutations with repetition although they are not permutations in general. They are also called words over the alphabet S in some contexts. If the set S has k elements, the number of n-tuples over S is:

$$k^n.$$

There is no restriction on how often an element can appear in an n-tuple, but if restrictions are placed on how often an element can appear, this formula is no longer valid.

Permutations of Multisets

If M is a finite multiset, then a multiset permutation is an ordered arrangement of elements of M in which each element appears exactly as often as is its multiplicity in M. An anagram of a word having some repeated letters is an example of a multiset permutation. If the multiplicities of the elements of M (taken in some order) are m_1, m_2, ..., m_l and their sum (i.e., the size of M) is n, then the number of multiset permutations of M is given by the multinomial coefficient,

$$\binom{n}{m_1, m_2, \ldots, m_l} = \frac{n!}{m_1! m_2! \cdots m_l!}.$$

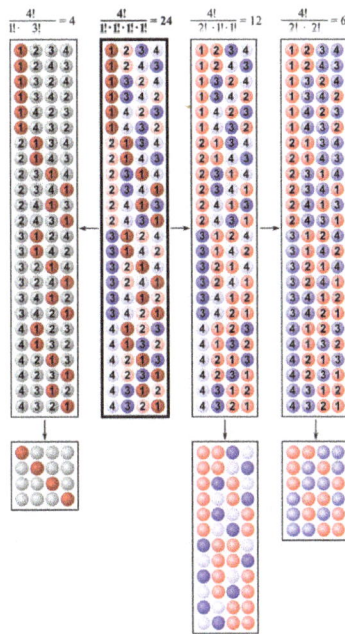

Permutations of multisets

For example, the number of distinct anagrams of the word MISSISSIPPI is:

$$\frac{11!}{1!4!4!2!}.$$

A k-permutation of a multiset M is a sequence of length k of elements of M in which each element appears *at most* its multiplicity in M times (an element's *repetition number*).

Circular Permutations

Permutations, when considered as arrangements, are sometimes referred to as *linearly ordered* arrangements. In these arrangements there is a first element, a second element, and so on. If, however, the objects are arranged in a circular manner this distinguished ordering no longer exists, that is, there is no "first element" in the arrangement, any element can be considered as the start of the arrangement. The arrangements of objects in a circular manner are called circular permutations. These can be formally defined as equivalence classes of ordinary permutations of the objects, for the equivalence relation generated by moving the final element of the linear arrangement to its front.

Two circular permutations are equivalent if one can be rotated into the other (that is, cycled without changing the relative positions of the elements). The following two circular permutations on four letters are considered to be the same.

The circular arrangements are to be read counterclockwise, so the following two are not equivalent since no rotation can bring one to the other.

$$
\begin{matrix}
 & 1 & & & & 1 & \\
4 & & 3 & & 3 & & 4 \\
 & 2 & & & & 2 &
\end{matrix}
$$

The number of circular permutations of a set S with n elements is $(n - 1)!$.

Permutations in Group Theory

The set of all permutations of any given set S forms a group, with the composition of maps as the product operation and the identity function as the neutral element of the group. This is the symmetric group of S, denoted by Sym(S). Up to isomorphism, this symmetric group only depends on the cardinality of the set (called the *degree* of the group), so the nature of elements of S is irrelevant for the structure of the group. Symmetric groups have been studied mostly in the case of finite sets, so, confined to this case, one can assume without loss of generality that $S = \{1,2,...,n\}$ for some natural number n. This is then the symmetric group of degree n, usually written as S_n.

Any subgroup of a symmetric group is called a permutation group. By Cayley's theorem any group is isomorphic to some permutation group, and every finite group to a subgroup of some finite symmetric group.

Cycle Notation

This alternative notation describes the effect of repeatedly applying the permutation, thought of as a function from a set onto itself. It expresses the permutation as a product of cycles corresponding to the orbits of the permutation; since distinct orbits are disjoint, this is referred to as "decomposition into disjoint cycles".

Starting from some element x of S, one writes the sequence $(x\ \sigma(x)\ \sigma(\sigma(x))\ ...)$ of successive images under σ, until the image returns to x, at which point one closes the parenthesis rather than repeat x. The set of values written down forms the orbit (under σ) of x, and the parenthesized expression gives the corresponding cycle of σ. One then continues by choosing a new element y of S outside the previous orbit and writing down the cycle starting at y; and so on until all elements of S are written in cycles. Since for every new cycle the starting point can be chosen in different ways, there are in general many different cycle notations for the same permutation; for the example above one has:

$$
\begin{pmatrix} 1 & 2 & 3 & 4 & 5 \\ 2 & 5 & 4 & 3 & 1 \end{pmatrix} = (1\ \ 2\ \ 5)(3\ \ 4) = (3\ \ 4)(1\ \ 2\ \ 5) = (3\ \ 4)(5\ \ 1\ \ 2).
$$

A cycle (x) of length 1 occurs when $\sigma(x) = x$, and is commonly omitted from the cycle notation, provided the set S is clear: for any element $x \in S$ not appearing in a cycle, one implicitly assumes $\sigma(x) = x$. The identity permutation, which consists only of 1-cycles, can be denoted by a single 1-cycle (x), by the number 1, or by *id*.

A cycle $(x_1\,x_2\,...\,x_k)$ of length k is called a k-cycle. Written by itself, it denotes a permutation in its own right, which maps x_i to x_{i+1} for $i < k$, and x_k to x_1, while implicitly mapping all other elements of S to themselves (omitted 1-cycles). Therefore, the individual cycles in the cycle notation can be interpreted as factors in a composition product. Since the orbits are disjoint, the corresponding cycles commute under composition, and so can be written in any order. The cycle decomposition is essentially unique: apart from the reordering the cycles in the product, there are no other ways to write σ as a product of cycles. Each individual cycle can be written in different ways, as in the example above where (5 1 2) and (1 2 5) and (2 5 1) all denote the same cycle, though note that (5 2 1) denotes a different cycle.

An element in a 1-cycle (x), corresponding to $\sigma(x) = x$, is called a fixed point of the permutation σ. A permutation with no fixed points is called a derangement. Cycles of length two are called transpositions; such permutations merely exchange the place of two elements, implicitly leaving the others fixed. Since the orbits of a permutation partition the set S, for a finite set of size n, the lengths of the cycles of a permutation σ form a partition of n called the cycle type of σ. There is a "1" in the cycle type for every fixed point of σ, a "2" for every transposition, and so on. The cycle type of $\beta = (1\,2\,5)(3\,4)(6\,8)(7)$, is $(3,2,2,1)$ which is sometimes written in a more compact form as $(1^1,2^2,3^1)$.

The number of n-permutations with k disjoint cycles is the signless Stirling number of the first kind, denoted by $c(n,k)$.

Abstract Groups vs. Permutations vs. Group Actions

Permutation groups have more structure than abstract groups, and different realizations of a group as a permutation group need not be equivalent as permutations. For instance S_3 is naturally a permutation group, in which any transposition has cycle type $(2,1)$; but the proof of Cayley's theorem realizes S_3 as a subgroup of S_6 (namely the permutations of the 6 elements of S_3 itself), in which permutation group transpositions have cycle type $(2,2,2)$. Finding the minimal-order symmetric group containing a subgroup isomorphic to a given group (sometimes called minimal faithful degree representation) is a rather difficult problem. So in spite of Cayley's theorem, the study of permutation groups differs from the study of abstract groups, being a branch of representation theory.

Much of the power of permutations can be regained in an abstract setting by considering group actions instead. A group action actually permutes the elements of a set according to the recipe provided by the abstract group. For example, S_3 acts faithfully and transitively on a set with exactly three elements (by permuting them).

Product and Inverse

The product of two permutations is defined as their composition as functions, in other words $\sigma \cdot \pi$ is the function that maps any element x of the set to $\sigma(\pi(x))$. Note that the rightmost permutation is applied to the argument first, because of the way function application is written. Some authors prefer the leftmost factor acting first, but to that end permutations must be written to the *right* of their argument, for instance as an exponent, where σ acting on x is written x^σ; then the product is defined by $x^{\sigma \cdot \pi} = (x^\sigma)^\pi$. However this gives a *different* rule for multiplying permutations; this article uses the definition where the rightmost permutation is applied first.

Since the composition of two bijections always gives another bijection, the product of two permutations is again a permutation. In two-line notation, the product of two permutations is obtained by rearranging the columns of the second (leftmost) permutation so that its first row is identical with the second row of the first (rightmost) permutation. The product can then be written as the first row of the first permutation over the second row of the modified second permutation. For example, given the permutations,

$$P = \begin{pmatrix} 1 & 2 & 3 & 4 & 5 \\ 2 & 4 & 1 & 3 & 5 \end{pmatrix} \quad \text{and} \quad Q = \begin{pmatrix} 1 & 2 & 3 & 4 & 5 \\ 5 & 4 & 3 & 2 & 1 \end{pmatrix},$$

the product QP is:

$$QP = \begin{pmatrix} 1 & 2 & 3 & 4 & 5 \\ 5 & 4 & 3 & 2 & 1 \end{pmatrix}\begin{pmatrix} 1 & 2 & 3 & 4 & 5 \\ 2 & 4 & 1 & 3 & 5 \end{pmatrix} = \begin{pmatrix} 2 & 4 & 1 & 3 & 5 \\ 4 & 2 & 5 & 3 & 1 \end{pmatrix}\begin{pmatrix} 1 & 2 & 3 & 4 & 5 \\ 2 & 4 & 1 & 3 & 5 \end{pmatrix} = \begin{pmatrix} 1 & 2 & 3 & 4 & 5 \\ 4 & 2 & 5 & 3 & 1 \end{pmatrix}.$$

In cycle notation this same product would be given by:

$$Q \cdot P = (15)(24) \cdot (1243) = (1435).$$

In general the composition of two permutations is not commutative, so that PQ can be different from QP as in the above example:

$$Q = \begin{pmatrix} 1 & 2 & 3 & 4 & 5 \\ 2 & 4 & 1 & 3 & 5 \end{pmatrix}\begin{pmatrix} 1 & 2 & 3 & 4 & 5 \\ 5 & 4 & 3 & 2 & 1 \end{pmatrix} = \begin{pmatrix} 1 & 2 & 3 & 4 & 5 \\ 5 & 3 & 1 & 4 & 2 \end{pmatrix} \neq QP.$$

Since function composition is associative, so is the product operation on permutations: $(\sigma \cdot \pi) \cdot \rho = \sigma \cdot (\pi \cdot \rho)$. Therefore, products of more than two permutations are usually written without adding parentheses to express grouping; they are also usually written without a dot or other sign to indicate multiplication.

The identity permutation, which maps every element of the set to itself, is the neutral element for this product. In two-line notation, the identity is

$$\begin{pmatrix} 1 & 2 & 3 & \cdots & n \\ 1 & 2 & 3 & \cdots & n \end{pmatrix}.$$

Since bijections have inverses, so do permutations, and the inverse σ^{-1} of σ is again a permutation. Explicitly, whenever $\sigma(x)=y$ one also has $\sigma^{-1}(y)=x$. In two-line notation the inverse can be obtained by interchanging the two lines (and sorting the columns if one wishes the first line to be in a given order). For instance

$$\begin{pmatrix} 1 & 2 & 3 & 4 & 5 \\ 2 & 5 & 4 & 3 & 1 \end{pmatrix}^{-1} = \begin{pmatrix} 2 & 5 & 4 & 3 & 1 \\ 1 & 2 & 3 & 4 & 5 \end{pmatrix} = \begin{pmatrix} 1 & 2 & 3 & 4 & 5 \\ 5 & 1 & 4 & 3 & 2 \end{pmatrix}.$$

In cycle notation one can reverse the order of the elements in each cycle to obtain a cycle notation for its inverse. Thus,

$$[(125)(34)]^{-1} = (521)(43) = (152)(34).$$

Having an associative product, a neutral element, and inverses for all its elements, makes the set of all permutations of S into a group, called the symmetric group of S.

Properties

Every permutation of a finite set can be expressed as the product of transpositions. Moreover, although many such expressions for a given permutation may exist, there can never be among them both expressions with an even number and expressions with an odd number of transpositions. All permutations are then classified as even or odd, according to the parity of the number of transpositions in any such expression.

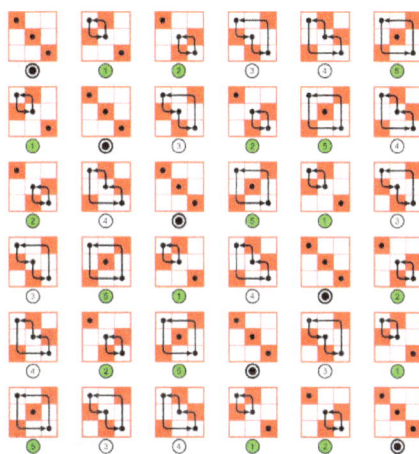

Composition of permutations corresponds to multiplication of permutation matrices.

Multiplying permutations written in cycle notation follows no easily described pattern, and the cycles of the product can be entirely different from those of the permutations being composed. However the cycle structure is preserved in the special case of conjugating a permutation σ by another permutation π, which means forming the product $\pi \cdot \sigma \cdot \pi^{-1}$. Here the cycle notation of the result can be obtained by taking the cycle notation for σ and applying π to all the entries in it.

Matrix Representation

One can represent a permutation of $\{1, 2, ..., n\}$ as an $n \times n$ matrix. There are two natural ways to do so, but only one for which multiplications of matrices corresponds to multiplication of permutations in the same order: this is the one that associates to σ the matrix M whose entry $M_{i,j}$ is 1 if $i = \sigma(j)$, and 0 otherwise. The resulting matrix has exactly one entry 1 in each column and in each row, and is called a *permutation matrix*. Here (file) is a list of these matrices for permutations of 4 elements. The Cayley table on the right shows these matrices for permutations of 3 elements.

Permutation of Components of a Sequence

As with any group, one can consider actions of a symmetric group on a set, and there are many ways in which such an action can be defined. For the symmetric group of $\{1, 2, ..., n\}$ there is one particularly natural action, namely the action by permutation on the set X^n of sequences of n symbols taken from some set X. As with the matrix representation, there are two natural ways in which the result of permuting a sequence $(x_1, x_2, ..., x_n)$ by σ can be defined, but only one is compatible with the multiplication of permutations (so as to give a left action of the symmetric group on X^n); with the multiplication rule used in this article this is the one given by

$$\sigma \cdot (x_1, ..., x_n) = (x_{\sigma^{-1}(1)}, ..., x_{\sigma^{-1}(n)}).$$

This means that each component x_i ends up at position $\sigma(i)$ in the sequence permuted by σ.

Permutations of Totally Ordered Sets

In some applications, the elements of the set being permuted will be compared with each other. This requires that the set S has a total order so that any two elements can be compared. The set $\{1, 2, ..., n\}$ is totally ordered by the usual "$\leq$" relation and so it is the most frequently used set in these applications, but in general, any totally ordered set will do. In these applications, the ordered arrangement view of a permutation is needed to talk about the *positions* in a permutation.

Here are a number of properties that are directly related to the total ordering of S.

Ascents, Descents, Runs and Excedances

An *ascent* of a permutation σ of n is any position $i < n$ where the following value is bigger than the current one. That is, if $\sigma = \sigma_1 \sigma_2 ... \sigma_n$, then i is an ascent if $\sigma_i < \sigma_{i+1}$.

For example, the permutation 3452167 has ascents (at positions) 1,2,5,6.

Similarly, a *descent* is a position $i < n$ with $\sigma_i > \sigma_{i+1}$, so every i with $1 \leq i < n$ either is an ascent or is a descent of σ.

An *ascending run* of a permutation is a nonempty increasing contiguous subsequence of the permutation that cannot be extended at either end; it corresponds to a maximal sequence of successive ascents (the latter may be empty: between two successive descents there is still an ascending run of length 1). By contrast an *increasing subsequence* of a permutation is not necessarily contiguous: it is an increasing sequence of elements obtained from the permutation by omitting the values at some positions. For example, the permutation 2453167 has the ascending runs 245, 3, and 167, while it has an increasing subsequence 2367.

If a permutation has $k - 1$ descents, then it must be the union of k ascending runs.

The number of permutations of n with k ascents is (by definition) the Eulerian number $\left\langle {n \atop k} \right\rangle$; this is also the number of permutations of n with k descents. Some authors however define the Eulerian number $\left\langle {n \atop k} \right\rangle$ as the number of permutations with k ascending runs, which corresponds to k-1 descents.

An excedance of a permutation $\sigma_1 \sigma_2 ... \sigma_n$ in an index j such that $\sigma_j > j$. If the inequality is not strict, i.e. $\sigma_j \geq j$, then j is called a *weak excedance*. The number of n-permutations with k excedances coincides with the number of n-permutations with k descents.

Canonical Cycle Notation (Aka Standard Form)

In some combinatorial contexts it useful to fix a certain order or the elements in the cycles and of the (disjoint) cycles themselves. Miklós Bóna calls the following ordering choices the *canonical cycle notation*:

- in each cycle the *largest* element is listed first
- the cycles are sorted in *increasing* order of their first element

For example, (312)(54)(8)(976) is a permutation in canonical cycle notation. Note that the canonical cycle notation does not omit one-cycles.

Richard P. Stanley calls the same choice of representation the "standard representation" of a permutation. and Martin Aigner uses the term "standard form" for the same notion. Sergey Kitaev also uses the "standard form" terminology, but reverses both choices, i.e. each cycle lists its least element first and the cycles are sorted in decreasing order of their least/first elements. In the remainder of this article, we use the first of these dual forms as the standard/canonical one.

Foata's Transition Lemma (or the Fundamental Bijection)

A natural question arises as to the relationship of the one-line and the canonical cycle notation. For example, considering the permutation (2)(31), which is in canonical cycle notation, if we erase its cycle parentheses, we obtain a different permutation in one-line notation, namely 231. (The permutation (2)(31) is actually 321 in one-line notation.) Foata's transition lemma establishes the nature of this correspondence as a bijection on the set of n-permutations (to itself). Richard P. Stanley calls this correspondence the *fundamental bijection*.

Let f be the parentheses-erasing transformation. The inverse of $q = f(p)$ is a bit less intuitive. Starting with the one-line notation $q = q_1 q_2 \cdots q_n$, the first cycle in canonical cycle notation must start with q_1. As long as the subsequent elements are smaller than q_1, we are in the same cycle. The second cycle starts at the smallest index j such that $q_j > q_1$. In other words, q_j is larger than everything else to its left, so it is called a *left-to-right maximum*. Every cycle in the canonical cycle notation starts with a left-to-right maximum.

For example, in the one-line notation 312548976, 5 is the first element larger than 3, so the first cycle must be (312). Then 8 is the next element larger than 5, so the second cycle is (54). Since 9 is larger than 8, (8) is a cycle by itself. Finally, 9 is larger than all the remaining elements to its right, so the last cycle is (976).

As a first corollary, the number of n-permutations with exactly k left-to-right maxima is also equal to the signless Stirling number of the first kind, $c(n,k)$.. Furthermore, Foata's mapping takes an n-permutation with k-weak excedances to an n-permutations with $k - 1$ ascents. For example, (2)(31) = 321 has two weak excedances (at index 1 and 2), whereas $f(321)=231$ has one ascent (at index 1, i.e. from 2 to 3).

Inversions

An *inversion* of a permutation σ is a pair (i,j) of positions where the entries of a permutation are in the opposite order: $i < j$ and $\sigma_i > \sigma_j$. So a descent is just an inversion at two adjacent positions. For example, the permutation $\sigma = 23154$ has three inversions: $(1,3), (2,3), (4,5)$, for the pairs of entries $(2,1), (3,1), (5,4)$.

In the 15 puzzle the goal is to get the squares in ascending order. Initial positions which have an odd number of inversions are impossible to solve.

Sometimes an inversion is defined as the pair of values (σ_i, σ_j) itself whose order is reversed; this makes no difference for the *number* of inversions, and this pair (reversed) is also an inversion in the above sense for the inverse permutation σ^{-1}. The number of inversions is an important measure for the degree to which the entries of a permutation are out of order; it is the same for σ and for σ^{-1}. To bring a permutation with k inversions into order (i.e., transform it into the identity permutation), by successively applying (right-multiplication by) adjacent transpositions, is always possible and requires a sequence of k such operations. Moreover, any reasonable choice for the adjacent transpositions will work: it suffices to choose at each step a transposition of i and $i + 1$ where i is a descent of the permutation as modified so far (so that the transposition will remove this particular descent, although it might create other descents). This is so because applying such a transposition reduces the number of inversions by 1; also note that as long as this number is not zero, the permutation is not the identity, so it has at least one descent. Bubble sort and insertion sort can be interpreted as particular instances of this procedure to put a sequence into order. Incidentally this procedure proves that any permutation σ can be written as a product of adjacent transpositions; for this one may simply reverse any sequence of such transpositions that transforms σ into the identity. In fact, by enumerating all sequences of adjacent transpositions that would transform σ into the identity, one obtains (after reversal) a *complete* list of all expressions of minimal length writing σ as a product of adjacent transpositions.

The number of permutations of n with k inversions is expressed by a Mahonian number, it is the coefficient of X^k in the expansion of the product

$$\prod_{m=1}^{n} \sum_{i=0}^{m-1} X^i = 1(1+X)(1+X+X^2)\cdots(1+X+X^2+\cdots+X^{n-1}),$$

which is also known (with q substituted for X) as the q-factorial $[n]_q!$. The expansion of the product appears in Necklace (combinatorics).

Permutations in Computing

Numbering Permutations

One way to represent permutations of n is by an integer N with $0 \leq N < n!$, provided convenient methods are given to convert between the number and the representation of a permutation as an ordered arrangement (sequence). This gives the most compact representation of arbitrary permutations, and in computing is particularly attractive when n is small enough that N can be held in a machine word; for 32-bit words this means $n \leq 12$, and for 64-bit words this means $n \leq 20$. The conversion can be done via the intermediate form of a sequence of numbers $d_n, d_{n-1}, \ldots, d_2, d_1$, where d_i is a non-negative integer less than i (one may omit d_1, as it is always 0, but its presence makes the subsequent conversion to a permutation easier to describe). The first step then is to simply express N in the factorial number system, which is just a particular mixed radix representation, where for numbers up to $n!$ the bases for successive digits are $n, n-1, \ldots, 2, 1$. The second step interprets this sequence as a Lehmer code or (almost equivalently) as an inversion table.

Rothe diagram for $\sigma = (6,3,8,1,4,9,7,2,5)$

$i \backslash \sigma_i$	1	2	3	4	5	6	7	8	9	Lehmer code
1	×	×	×	×	×	•				$d_9 = 5$
2	×	×	•							$d_8 = 2$
3	×	×		×	×		×	•		$d_7 = 5$
4	•									$d_6 = 0$
5		×		•						$d_5 = 1$
6		×			×		×		•	$d_4 = 3$
7		×			×		•			$d_3 = 2$
8		•								$d_2 = 0$
9					•					$d_1 = 0$
inversion table	3	6	1	2	4	0	2	0	0	

In the Lehmer code for a permutation σ, the number d_n represents the choice made for the first term σ_1, the number d_{n-1} represents the choice made for the second term σ_2 among the remaining $n-1$ elements of the set, and so forth. More precisely, each d_{n+1-i} gives the number of *remaining* elements strictly less than the term σ_i. Since those remaining elements are bound to turn up as some later term σ_j, the digit d_{n+1-i} counts the *inversions* (i,j) involving i as smaller index (the number of values j for which $i < j$ and $\sigma_i > \sigma_j$). The inversion table for σ is quite similar, but here d_{n+1-k} counts the number of inversions (i,j) where $k = \sigma_j$ occurs as the smaller of the two values appearing in inverted order. Both encodings can be visualized by an n by n Rothe diagram (named after Heinrich August Rothe) in which dots at (i,σ_i) mark the entries of the permutation, and a cross at (i,σ_j) marks the inversion (i,j); by the definition of inversions a cross appears in any square that comes both before the dot (j,σ_j) in its column, and before the dot (i,σ_i) in its row. The Lehmer code lists the numbers of crosses in successive rows, while the inversion table lists the numbers of crosses in successive columns; it is just the Lehmer code for the inverse permutation, and vice versa.

To effectively convert a Lehmer code $d_n, d_{n-1}, ..., d_2, d_1$ into a permutation of an ordered set S, one can start with a list of the elements of S in increasing order, and for i increasing from 1 to n set σ_i to the element in the list that is preceded by d_{n+1-i} other ones, and remove that element from the list. To convert an inversion table $d_n, d_{n-1}, ..., d_2, d_1$ into the corresponding permutation, one can traverse the numbers from d_1 to d_n while inserting the elements of S from largest to smallest into an initially empty sequence; at the step using the number d from the inversion table, the element from S inserted into the sequence at the point where it is preceded by d elements already present. Alternatively one could process the numbers from the inversion table and the elements of S both in the opposite order, starting with a row of n empty slots, and at each step place the element from S into the empty slot that is preceded by d other empty slots.

Converting successive natural numbers to the factorial number system produces those sequences in lexicographic order (as is the case with any mixed radix number system), and further converting them to permutations preserves the lexicographic ordering, provided the Lehmer code interpretation is used (using inversion tables, one gets a different ordering, where one starts by comparing permutations by the *place* of their entries 1 rather than by the value of their first entries). The sum of the numbers in the factorial number system representation gives the number of inversions of the permutation, and the parity of that sum gives the signature of the permutation. Moreover, the positions of the zeroes in the inversion table give the values of left-to-right maxima of the permutation (in the example 6, 8, 9) while the positions of the zeroes in the Lehmer code are the positions of the right-to-left minima (in the example positions the 4, 8, 9 of the values 1, 2, 5); this allows computing the distribution of such extrema among all permutations. A permutation with Lehmer code $d_n, d_{n-1}, ..., d_2, d_1$ has an ascent $n - i$ if and only if $d_i \geq d_{i+1}$.

Algorithms to Generate Permutations

In computing it may be required to generate permutations of a given sequence of values. The methods best adapted to do this depend on whether one wants some randomly chosen permutations, or all permutations, and in the latter case if a specific ordering is required. Another question is whether possible equality among entries in the given sequence is to be taken into account; if so, one should only generate distinct multiset permutations of the sequence.

An obvious way to generate permutations of n is to generate values for the Lehmer code (possibly using the factorial number system representation of integers up to $n!$), and convert those into the corresponding permutations. However, the latter step, while straightforward, is hard to implement efficiently, because it requires n operations each of selection from a sequence and deletion from it, at an arbitrary position; of the obvious representations of the sequence as an array or a linked list, both require (for different reasons) about $n^2/4$ operations to perform the conversion. With n likely to be rather small (especially if generation of all permutations is needed) that is not too much of a problem, but it turns out that both for random and for systematic generation there are simple alternatives that do considerably better. For this reason it does not seem useful, although certainly possible, to employ a special data structure that would allow performing the conversion from Lehmer code to permutation in $O(n \log n)$ time.

Random Generation of Permutations

For generating random permutations of a given sequence of n values, it makes no difference whether one applies a randomly selected permutation of n to the sequence, or chooses a random element from the set of distinct (multiset) permutations of the sequence. This is because, even though in case of repeated values there can be many distinct permutations of n that result in the same permuted sequence, the number of such permutations is the same for each possible result. Unlike for systematic generation, which becomes unfeasible for large n due to the growth of the number $n!$, there is no reason to assume that n will be small for random generation.

Biologist and statistician Ronald Fisher

The basic idea to generate a random permutation is to generate at random one of the $n!$ sequences of integers $d_1, d_2, ..., d_n$ satisfying $0 \leq d_i < i$ (since d_1 is always zero it may be omitted) and to convert it to a permutation through a bijective correspondence. For the latter correspondence one could interpret the (reverse) sequence as a Lehmer code, and this gives a generation method first published in 1938 by Ronald Fisher and Frank Yates. While at the time computer implementation was not an issue, this method suffers from the difficulty sketched above to convert from Lehmer code to permutation efficiently. This can be remedied by using a different bijective correspondence: after using d_i to select an element among i remaining elements of the sequence (for decreasing values of i), rather than removing the element and compacting the sequence by shifting down further elements one place, one swaps the element with the final remaining element. Thus the elements remaining for selection form a consecutive range at each point in time, even though they may not occur in the same order as they did in the original sequence. The mapping from sequence of integers to permutations is somewhat complicated, but it can be seen to produce each permutation in exactly one way, by an immediate induction. When the selected element happens to be the final remaining element, the swap operation can be omitted. This does not occur sufficiently often to warrant testing for the condition, but the final element must be included among the candidates of the selection, to guarantee that all permutations can be generated.

The resulting algorithm for generating a random permutation of $a, a, ..., a[n - 1]$ can be described as follows in pseudocode:

    for i from n downto 2

do $\quad d_i \leftarrow$ random element of $\{\, 0, ..., i-1 \,\}$

swap $a[d_i]$ and $a[i-1]$

This can be combined with the initialization of the array $a[i] = i$ as follows:

for i from 0 to $n-1$

do $\quad d_{i+1} \leftarrow$ random element of $\{\, 0, ..., i \,\}$

$a[i] \leftarrow a[d_{i+1}]$

$a[d_{i+1}] \leftarrow i$

If $d_{i+1} = i$, the first assignment will copy an uninitialized value, but the second will overwrite it with the correct value i.

Generation in Lexicographic Order

There are many ways to systematically generate all permutations of a given sequence. One classical algorithm, which is both simple and flexible, is based on finding the next permutation in lexicographic ordering, if it exists. It can handle repeated values, for which case it generates the distinct multiset permutations each once. Even for ordinary permutations it is significantly more efficient than generating values for the Lehmer code in lexicographic order (possibly using the factorial number system) and converting those to permutations. To use it, one starts by sorting the sequence in (weakly) increasing order (which gives its lexicographically minimal permutation), and then repeats advancing to the next permutation as long as one is found. The method goes back to Narayana Pandita in 14th century India, and has been frequently rediscovered ever since.

The following algorithm generates the next permutation lexicographically after a given permutation. It changes the given permutation in-place.

1. Find the largest index k such that $a[k] < a[k+1]$. If no such index exists, the permutation is the last permutation.

2. Find the largest index l greater than k such that $a[k] < a[l]$.

3. Swap the value of $a[k]$ with that of $a[l]$.

4. Reverse the sequence from $a[k+1]$ up to and including the final element $a[n]$.

For example, given the sequence [1, 2, 3, 4] (which is in increasing order), and given that the index is zero-based, the steps are as follows:

1. Index $k = 2$, because 3 is placed at an index that satisfies condition of being the largest index that is still less than $a[k+1]$ which is 4.

2. Index $l = 3$, because 4 is the only value in the sequence that is greater than 3 in order to satisfy the condition $a[k] < a[l]$.

3. The values of a and a are swapped to form the new sequence [1,2,4,3].

4. The sequence after k-index a to the final element is reversed. Because only one value lies after this index (the 3), the sequence remains unchanged in this instance. Thus the lexicographic successor of the initial state is permuted: [1,2,4,3].

Following this algorithm, the next lexicographic permutation will be [1,3,2,4], and the 24th permutation will be [4,3,2,1] at which point $a[k] < a[k + 1]$ does not exist, indicating that this is the last permutation.

Generation with Minimal Changes

An alternative to the above algorithm, the Steinhaus–Johnson–Trotter algorithm, generates an ordering on all the permutations of a given sequence with the property that any two consecutive permutations in its output differ by swapping two adjacent values. This ordering on the permutations was known to 17th-century English bell ringers, among whom it was known as "plain changes". One advantage of this method is that the small amount of change from one permutation to the next allows the method to be implemented in constant time per permutation. The same can also easily generate the subset of even permutations, again in constant time per permutation, by skipping every other output permutation.

An alternative to Steinhaus–Johnson–Trotter is Heap's algorithm, said by Robert Sedgewick in 1977 to be the fastest algorithm of generating permutations in applications.

Meandric Permutations

Meandric systems give rise to *meandric permutations*, a special subset of *alternate permutations*. An alternate permutation of the set {1,2,...,2n} is a cyclic permutation (with no fixed points) such that the digits in the cyclic notation form alternate between odd and even integers. Meandric permutations are useful in the analysis of RNA secondary structure. Not all alternate permutations are meandric. A modification of Heap's algorithm has been used to generate all alternate permutations of order n (that is, of length $2n$) without generating all $(2n)!$ permutations. Generation of these alternate permutations is needed before they are analyzed to determine if they are meandric or not.

The algorithm is recursive. The following table exhibits a step in the procedure. In the previous step, all alternate permutations of length 5 have been generated. Three copies of each of these have a "6" added to the right end, and then a different transposition involving this last entry and a previous entry in an even position is applied (including the identity, i.e., no transposition).

Previous sets	Transposition of digits	Alternate permutations
1-2-3-4-5-6		1-2-3-4-5-6
1-2-3-4-5-6	4,6	1-2-3-6-5-4
1-2-3-4-5-6	2,6	1-6-3-4-5-2
1-2-5-4-3-6		1-2-5-4-3-6
1-2-5-4-3-6	4,6	1-2-5-6-3-4
1-2-5-4-3-6	2,6	1-6-5-4-3-2

1-4-3-2-5-6			1-4-3-2-5-6
1-4-3-2-5-6	2,6		1-4-3-6-5-2
1-4-3-2-5-6	4,6		1-6-3-2-5-4
1-4-5-2-3-6			1-4-5-2-3-6
1-4-5-2-3-6	2,6		1-4-5-6-3-2
1-4-5-2-3-6	4,6		1-6-5-2-3-4

Software Implementations

```
// Insert into all positions solution
static List<string> Permute(string input)
{
  if (input.Length == 1)
  {
    return new List<string>() { input };
  }

  var permutations = new List<string>();

  var toInsert = input[0].ToString();
  foreach (var item in Permute(input.Substring(1)))
  {
    for (int i = 0; i <= item.Length; ++i)
    {
      string newPermutation = item.Insert(i, toInsert);
      permutations.Add(newPermutation);
    }
  }

  return permutations;
}

// Fixed-head solution
static void Permute(string prefix, string suffix, ref int count)
{
```

```
  if (suffix.Length == 0)
  {
    ++count;

    Console.WriteLine(prefix);
    return;
  }

  for (int i = 0; i < suffix.Length; ++i)
  {
    string newSuffix = suffix.Substring(0, i) + suffix.Substring(i + 1, suffix.Length - i - 1);

    Permute(prefix + suffix[i], newSuffix, ref count);
  }
}
```

Calculator Functions

- Casio and TI calculators: nPr

- HP calculators: PERM

- Mathematica: FactorialPower

Spreadsheet Functions

Most spreadsheet software also provides a built-in function for calculating the number of k-permutations of n, called PERMUT in many popular spreadsheets.

Applications

Permutations are used in the interleaver component of the error detection and correction algorithms, such as turbo codes, for example 3GPP Long Term Evolution mobile telecommunication standard uses these ideas. Such applications raise the question of fast generation of permutations satisfying certain desirable properties. One of the methods is based on the permutation polynomials. Also as a base for optimal hashing in Unique Permutation Hashing.

Algebraic Number Theory

Algebraic number theory is a major branch of number theory that studies algebraic structures related to algebraic integers. This is generally accomplished by considering a ring of algebraic integers O in an algebraic number field $K/\mathbb{Q}$, and studying their algebraic properties such as factorization, the behaviour of ideals, and field extensions. In this setting, the familiar features of the integers—such as unique factorization—need not hold. The virtue of the primary machinery employed—Galois theory, group cohomology, group representations, and L-functions—is that it allows one to deal with new phenomena and yet partially recover the behaviour of the usual integers.

DISQVISITIONES

ARITHMETICAE

AVCTORE

D. CAROLO FRIDERICO GAVSS

LIPSIAE

IN COMMISSIS APVD GERH. FLEISCHER, JVR.

1801.

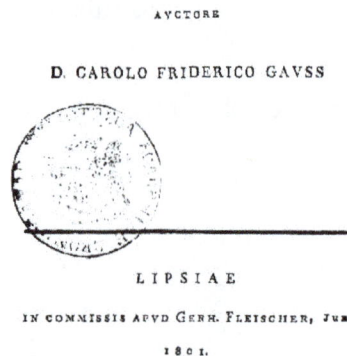

Title page of the first edition of Disquisitiones Arithmeticae, one of the founding works of modern algebraic number theory.

History of Algebraic Number Theory

Diophantus

The beginnings of algebraic number theory can be traced to Diophantine equations, named after the 3rd-century Alexandrian mathematician, Diophantus, who studied them and developed methods for the solution of some kinds of Diophantine equations. A typical Diophantine problem is to find two integers x and y such that their sum, and the sum of their squares, equal two given numbers A and B, respectively:

$$A = x + y$$

$$B = x^2 + y^2.$$

Diophantine equations have been studied for thousands of years. For example, the solutions to the quadratic Diophantine equation $x^2 + y^2 = z^2$ are given by the Pythagorean triples, originally solved by the Babylonians (c. 1800 BC). Solutions to linear Diophantine equations, such as $26x + 65y = 13$, may be found using the Euclidean algorithm (c. 5th century BC).

Diophantus's major work was the *Arithmetica*, of which only a portion has survived.

Fermat

Fermat's last theorem was first conjectured by Pierre de Fermat in 1637, famously in the margin of a copy of *Arithmetica* where he claimed he had a proof that was too large to fit in the margin. No successful proof was published until 1995 despite the efforts of countless mathematicians during the 358 intervening years. The unsolved problem stimulated the development of algebraic number theory in the 19th century and the proof of the modularity theorem in the 20th century.

Gauss

One of the founding works of algebraic number theory, the *Disquisitiones Arithmeticae* (Latin: *Arithmetical Investigations*) is a textbook of number theory written in Latin by Carl Friedrich Gauss in 1798 when Gauss was 21 and first published in 1801 when he was 24. In this book Gauss brings together results in number theory obtained by mathematicians such as Fermat, Euler, Lagrange and Legendre and adds important new results of his own. Before the *Disquisitiones* was published, number theory consisted of a collection of isolated theorems and conjectures. Gauss brought the work of his predecessors together with his own original work into a systematic framework, filled in gaps, corrected unsound proofs, and extended the subject in numerous ways.

The *Disquisitiones* was the starting point for the work of other nineteenth century European mathematicians including Ernst Kummer, Peter Gustav Lejeune Dirichlet and Richard Dedekind. Many of the annotations given by Gauss are in effect announcements of further research of his own, some of which remained unpublished. They must have appeared particularly cryptic to his contemporaries; we can now read them as containing the germs of the theories of L-functions and complex multiplication, in particular.

Dirichlet

In a couple of papers in 1838 and 1839 Peter Gustav Lejeune Dirichlet proved the first class number formula, for quadratic forms (later refined by his student Kronecker). The formula, which Jacobi called a result "touching the utmost of human acumen", opened the way for similar results regarding more general number fields. Based on his research of the structure of the unit group of quadratic fields, he proved the Dirichlet unit theorem, a fundamental result in algebraic number theory.

He first used the pigeonhole principle, a basic counting argument, in the proof of a theorem in diophantine approximation, later named after him Dirichlet's approximation theorem. He published important contributions to Fermat's last theorem, for which he proved the cases $n = 5$ and $n = 14$, and to the biquadratic reciprocity law. The Dirichlet divisor problem, for which he found the first results, is still an unsolved problem in number theory despite later contributions by other researchers.

Dedekind

Richard Dedekind's study of Lejeune Dirichlet's work was what led him to his later study of algebraic number fields and ideals. In 1863, he published Lejeune Dirichlet's lectures on number theory as *Vorlesungen über Zahlentheorie* ("Lectures on Number Theory") about which it has been written that:

"Although the book is assuredly based on Dirichlet's lectures, and although Dedekind himself referred to the book throughout his life as Dirichlet's, the book itself was entirely written by Dedekind, for the most part after Dirichlet's death." (Edwards 1983)

1879 and 1894 editions of the *Vorlesungen* included supplements introducing the notion of an ideal, fundamental to ring theory. (The word "Ring", introduced later by Hilbert, does not appear in Dedekind's work.) Dedekind defined an ideal as a subset of a set of numbers, composed of algebraic integers that satisfy polynomial equations with integer coefficients. The concept underwent further development in the hands of Hilbert and, especially, of Emmy Noether. Ideals generalize Ernst Eduard Kummer's ideal numbers, devised as part of Kummer's 1843 attempt to prove Fermat's Last Theorem.

Hilbert

David Hilbert unified the field of algebraic number theory with his 1897 treatise *Zahlbericht* (literally "report on numbers"). He also resolved a significant number-theory problem formulated by Waring in 1770. As with the finiteness theorem, he used an existence proof that shows there must be solutions for the problem rather than providing a mechanism to produce the answers. He then had little more to publish on the subject; but the emergence of Hilbert modular forms in the dissertation of a student means his name is further attached to a major area.

He made a series of conjectures on class field theory. The concepts were highly influential, and his own contribution lives on in the names of the Hilbert class field and of the Hilbert symbol of local class field theory. Results were mostly proved by 1930, after work by Teiji Takagi.

Artin

Emil Artin established the Artin reciprocity law in a series of papers (1924; 1927; 1930). This law is a general theorem in number theory that forms a central part of global class field theory. The term "reciprocity law" refers to a long line of more concrete number theoretic statements which it generalized, from the quadratic reciprocity law and the reciprocity laws of Eisenstein and Kummer to Hilbert's product formula for the norm symbol. Artin's result provided a partial solution to Hilbert's ninth problem.

Modern Theory

Around 1955, Japanese mathematicians Goro Shimura and Yutaka Taniyama observed a possible link between two apparently completely distinct, branches of mathematics, elliptic curves and modular forms. The resulting modularity theorem (at the time known as the Taniyama–Shimura conjecture) states that every elliptic curve is modular, meaning that it can be associated with a unique modular form.

It was initially dismissed as unlikely or highly speculative, and was taken more seriously when number theorist André Weil found evidence supporting it, but no proof; as a result the "astounding" conjecture was often known as the Taniyama–Shimura-Weil conjecture. It became a part of the Langlands programme, a list of important conjectures needing proof or disproof.

From 1993 to 1994, Andrew Wiles provided a proof of the modularity theorem for semistable elliptic curves, which, together with Ribet's theorem, provides a proof for Fermat's Last Theorem. Both Fermat's Last Theorem and the Modularity Theorem were almost universally considered inaccessible to proof by contemporaneous mathematicians (meaning, impossible or virtually impossible to prove using current knowledge). Wiles first announced his proof in June 1993 in a version that was soon recognized as having a serious gap in a key point. The proof was corrected by Wiles, in part via collaboration with Richard Taylor, and the final, widely accepted, version was released in September 1994, and formally published in 1995. The proof uses many techniques from algebraic geometry and number theory, and has many ramifications in these branches of mathematics. It also uses standard constructions of modern algebraic geometry, such as the category of schemes and Iwasawa theory, and other 20th-century techniques not available to Fermat.

Basic Notions

Failure of unique factorization

An important property of the ring of integers is that it satisfies the fundamental theorem of arithmetic, that every (positive) integer has a factorization into a product of prime numbers, and this factorization is unique up to the ordering of the factors. This may no longer be true in the ring of integers O of an algebraic number field K.

A prime element is an element p of O such that if p divides a product ab, then it divides one of the factors a or b. This property is closely related to primality in the integers, because any positive integer satisfying this property is either 1 or a prime number. However, it is strictly weaker. For example, −2 is not a prime number because it is negative, but it is a prime element. If factorizations into prime elements are permitted, then, even in the integers, there are alternative factorizations such as

$$6 = 2 \cdot 3 = (-2) \cdot (-3).$$

In general, if u is a unit, meaning a number with a multiplicative inverse in O, and if p is a prime element, then up is also a prime element. Numbers such as p and up are said to be associate. In the integers, the primes p and −p are associate, but only one of these is positive. Requiring that prime numbers be positive selects a unique element from among a set of associated prime elements. When K is not the rational numbers, however, there is no analog of positivity. For example, in the Gaussian integers Z[i], the numbers 1 + 2i and −2 + i are associate because the latter is the product of the former by i, but there is no way to single out one as being more canonical than the other. This leads to equations such as

$$5 = (1+2i)(1-2i) = (2+i)(2-i),$$

which prove that in Z[i], it is not true that factorizations are unique up to the order of the factors. For this reason, one adopts the definition of unique factorization used in unique factorization domains (UFDs). In a UFD, the prime elements occurring in a factorization are only expected to be unique up to units and their ordering.

However, even with this weaker definition, many rings of integers in algebraic number fields do not admit unique factorization. There is an algebraic obstruction called the ideal class group. When the ideal class group is trivial, the ring is a UFD. When it is not, there is a distinction between a prime element and an irreducible element. An irreducible element x is an element such that if x = yz, then either y or z is a unit. These are the elements that cannot be factored any further. Every element in O admits a factorization into irreducible elements, but it may admit more than one. This is because, while all prime elements are irreducible, some irreducible elements may not be prime. For example, consider the ring $Z[\sqrt{-5}]$. In this ring, the numbers 3, $2 + \sqrt{-5}$ and $2 - \sqrt{-5}$ are irreducible. This means that the number 9 has two factorizations into irreducible elements,

$$9 = 3^2 = (2 + \sqrt{-5})(2 - \sqrt{-5}).$$

This equation shows that 3 divides the product $(2 + \sqrt{-5})(2 - \sqrt{-5}) = 9$. If 3 were a prime element, then it would divide $2 + \sqrt{-5}$ or $2 - \sqrt{-5}$, but it does not, because all elements divisible by 3 are of the form $3a + 3b\sqrt{-5}$. Similarly, $2 + \sqrt{-5}$ and $2 - \sqrt{-5}$ divide the product 32, but neither of these elements divides 3 itself, so neither of them are prime. As there is no sense in which the elements 3, $2 + \sqrt{-5}$ and $2 - \sqrt{-5}$ can be made equivalent, unique factorization fails in $Z[\sqrt{-5}]$. Unlike the situation with units, where uniqueness could be repaired by weakening the definition, overcoming this failure requires a new perspective.

Factorization into Prime Ideals

If I is an ideal in O, then there is always a factorization

$$I = \mathfrak{p}_1^{e_1} \cdots \mathfrak{p}_t^{e_t},$$

where each $\mathfrak{p}_i$ is a prime ideal, and where this expression is unique up to the order of the factors. In particular, this is true if I is the principal ideal generated by a single element. This is the strongest sense in which the ring of integers of a general number field admits unique factorization. In the language of ring theory, it says that rings of integers are Dedekind domains.

When O is a UFD, every prime ideal is generated by a prime element. Otherwise, there are prime ideals which are not generated by prime elements. In $Z[\sqrt{-5}]$, for instance, the ideal $(2, 1 + \sqrt{-5})$ is a prime ideal which cannot be generated by a single element.

Historically, the idea of factoring ideals into prime ideals was preceded by Ernst Kummer's introduction of ideal numbers. These are numbers lying in an extension field E of K. This extension field is now known as the Hilbert class field. By the principal ideal theorem, every prime ideal of O generates a principal ideal of the ring of integers of E. A generator of this principal ideal is called an ideal number. Kummer used these as a substitute for the failure of unique factorization in cyclotomic fields. These eventually led Richard Dedekind to introduce a forerunner of ideals and to prove unique factorization of ideals.

An ideal which is prime in the ring of integers in one number field may fail to be prime when extended to a larger number field. Consider, for example, the prime numbers. The corresponding ideals $p\mathbf{Z}$ are prime ideals of the ring $\mathbf{Z}$. However, when this ideal is extended to the Gaussian integers to get $p\mathbf{Z}[i]$, it may or may not be prime. For example, the factorization $2 = (1 + i)(1 - i)$ implies that

$$2\mathbf{Z}[i] = (1+i)\mathbf{Z}[i] \cdot (1-i)\mathbf{Z}[i] = ((1+i)\mathbf{Z}[i])^2;$$

note that because $1 + i = (1 - i) \cdot i$, the ideals generated by $1 + i$ and $1 - i$ are the same. A complete answer to the question of which ideals remain prime in the Gaussian integers is provided by Fermat's theorem on sums of two squares. It implies that for an odd prime number p, $p\mathbf{Z}[i]$ is a prime ideal if $p \equiv 3 \pmod 4$ and is not a prime ideal if $p \equiv 1 \pmod 4$. This, together with the observation that the ideal $(1 + i)\mathbf{Z}[i]$ is prime, provides a complete description of the prime ideals in the Gaussian integers. Generalizing this simple result to more general rings of integers is a basic problem in algebraic number theory. Class field theory accomplishes this goal when K is an abelian extension of $\mathbf{Q}$ (i.e. a Galois extension with abelian Galois group).

Ideal Class Group

Unique factorization fails if and only if there are prime ideals that fail to be principal. The object which measures the failure of prime ideals to be principal is called the ideal class group. Defining the ideal class group requires enlarging the set of ideals in a ring of algebraic integers so that they admit a group structure. This is done by generalizing ideals to fractional ideals. A fractional ideal is an additive subgroup J of K which is closed under multiplication by elements of O, meaning that $xJ \subseteq J$ if $x \in O$. All ideals of O are also fractional ideals. If I and J are fractional ideals, then the set IJ of all products of an element in I and an element in J is also a fractional ideal. This operation makes the set of non-zero fractional ideals into a group. The group identity is the ideal $(1) = O$, and the inverse of J is a (generalized) ideal quotient, $J^{-1} = (O: J) = \{\, x \in K : xJ \subseteq O \,\}$.

The principal fractional ideals, meaning the ones of the form Ox where $x \in K^\times$, form a subgroup of the group of all non-zero fractional ideals. The quotient of the group of non-zero fractional ideals by this subgroup is the ideal class group. Two fractional ideals I and J represent the same element of the ideal class group if and only if there exists an element $x \in K$ such that $xI = J$. Therefore the ideal class group makes two fractional ideals equivalent if one is as close to being principal as the other is. The ideal class group is generally denoted $\mathrm{Cl}\,K$, $\mathrm{Cl}\,O$, or $\mathrm{Pic}\,O$ (with the last notation identifying it with the Picard group in algebraic geometry).

The number of elements in the class group is called the class number of K. The class number of $\mathbf{Q}(\sqrt{-5})$ is 2. This means that there are only two ideal classes, the class of principal fractional ideals, and the class of a non-principal fractional ideal such as $(2, 1 + \sqrt{-5})$.

The ideal class group has another description in terms of divisors. These are formal objects which represent possible factorizations of numbers. The divisor group $\mathrm{Div}\,K$ is defined to be the free abelian group generated by the prime ideals of O. There is a group homomorphism from $K^\times$, the non-zero elements of K up to multiplication, to $\mathrm{Div}\,K$. Suppose that $x \in K$ satisfies

$$(x) = \mathfrak{p}_1^{e_1} \cdots \mathfrak{p}_t^{e_t}.$$

Then $\mathrm{div}\,x$ is defined to be the divisor

$$\mathrm{div}\,x = \sum_{i=1}^{t} e_i [\mathfrak{p}_i].$$

The kernel of div is the group of units in O, while the cokernel is the ideal class group. In the language of homological algebra, this says that there is an exact sequence of abelian groups (written multiplicatively),

$$1 \to O^\times \to K^\times \xrightarrow{\ \mathrm{div}\ } \mathrm{Div}\, K \to \mathrm{Cl}\, K \to 1.$$

Real and Complex Embeddings

Some number fields, such as $Q(\sqrt{2})$, can be specified as subfields of the real numbers. Others, such as $Q(\sqrt{-1})$, cannot. Abstractly, such a specification corresponds to a field homomorphism $K \to R$ or $K \to C$. These are called real embeddings and complex embeddings, respectively.

A real quadratic field $Q(\sqrt{d})$ is so-called because it admits two real embeddings and no complex embeddings. These are the field homomorphisms which send $\sqrt{d}$ to $\sqrt{d}$ and to $-\sqrt{d}$, respectively. Dually, an imaginary quadratic field $Q(\sqrt{-d})$ admits no real embeddings and a conjugate pair of complex embeddings. One of these embeddings sends $\sqrt{-d}$ to $\sqrt{-d}$, while the other sends it to its complex conjugate.

Conventionally, the number of real embeddings of K is denoted r_1, while the number of conjugate pairs of complex embeddings is denoted r_2. The signature of K is the pair (r_1, r_2). It is a theorem that $r_1 + 2r_2 = d$, where d is the degree of K.

Considering all embeddings at once determines a function

$$M : K \to \mathbf{R}^{r_1} \oplus \mathbf{C}^{2r_2}.$$

This is called the Minkowski embedding. The subspace of the codomain fixed by complex conjugation is a real vector space of dimension d called Minkowski space. Because the Minkowski embedding is defined by field homomorphisms, multiplication of elements of K by an element $x \in K$ corresponds to multiplication by a diagonal matrix in the Minkowski embedding. The dot product on Minkowski space corresponds to the trace form $\langle x, y \rangle = \mathrm{Tr}(xy)$.

The image of O in Minkowski space is a d-dimensional lattice. If B is a basis for this lattice, then $\det B^{\mathrm{T}}B$ is the discriminant of O. The discriminant is denoted Δ or D. The covolume of the image of O is $\sqrt{|\Delta|}$..

Places

Real and complex embeddings can be put on the same footing as prime ideals by adopting a perspective based on valuations. Consider, for example, the integers. In addition to the usual absolute value function $|\cdot| : Q \to R$, there are p-adic absolute value functions $|\cdot|_p : Q \to R$, defined for each prime number p, which measure divisibility by p. Ostrowski's theorem states that these are all possible absolute value functions on Q (up to equivalence). Therefore absolute values are a common language to describe both the real embedding of Q and the prime numbers.

A place of an algebraic number field is an equivalence class of absolute value functions on K. There are two types of places. There is a $\mathfrak{p}$-adic absolute value for each prime ideal $\mathfrak{p}$ of O, and, like the p-adic absolute values, it measures divisibility. These are called finite places. The other type of place is specified using a real or complex embedding of K and the standard absolute value function on R or C. These are infinite places. Because absolute values are unable to distinguish between a complex embedding and its conjugate, a complex embedding and its conjugate determine the same place. Therefore there are r_1 real places and r_2 complex places. Because places encompass the primes, places are sometimes referred to as primes. When this is done, finite places are called finite primes and infinite places are called infinite primes. If v is a valuation corresponding to an absolute value, then one frequently writes $v \mid \infty$ to mean that v is an infinite place and $v \nmid \infty$ to mean that it is a finite place.

Considering all the places of the field together produces the adele ring of the number field. The adele ring allows one to simultaneously track all the data available using absolute values. This produces significant advantages in situations where the behavior at one place can affect the behavior at other places, as in the Artin reciprocity law.

Units

The integers have only two units, 1 and −1. Other rings of integers may admit more units. The Gaussian integers have four units, the previous two as well as $\pm i$. The Eisenstein integers $\mathbf{Z}[\exp(2\pi i / 3)]$ have six units. The integers in real quadratic number fields have infinitely many units. For example, in $\mathbf{Z}[\sqrt{3}]$, every power of $2 + \sqrt{3}$ is a unit, and all these powers are distinct.

In general, the group of units of O, denoted $O^{\times}$, is a finitely generated abelian group. The fundamental theorem of finitely generated abelian groups therefore implies that it is a direct sum of a torsion part and a free part. Reinterpreting this in the context of a number field, the torsion part consists of the roots of unity that lie in O. This group is cyclic. The free part is described by Dirichlet's unit theorem. This theorem says that rank of the free part is $r_1 + r_2 - 1$. Thus, for example, the only fields for which the rank of the free part is zero are Q and the imaginary quadratic fields. A more precise statement giving the structure of $O^{\times} \otimes_{\mathbf{Z}} \mathrm{Q}$ as a Galois module for the Galois group of $K/\mathbf{Q}$ is also possible.

The free part of the unit group can be studied using the infinite places of K. Consider the function

$$L : K^{\times} \to \mathbf{R}^{r_1 + r_2}$$

defined by

$$L(x) = (\log |x|_v)_v,$$

where v varies over the infinite places of K and $|\cdot|_v$ is the absolute value associated with v. The function L is a homomorphism from $K^{\times}$ to a real vector space. It can be shown that the image of $O^{\times}$ is a lattice that spans the hyperplane defined by $x_1 + \cdots + x_{r_1 + r_2} = 0$. The covolume of this lattice is the **regulator** of the number field. One of the simplifications made possible by working with the adele ring is that there is a single object, the idele class group, that describes both the quotient by this lattice and the ideal class group.

Zeta Function

The Dedekind zeta function of a number field, analogous to the Riemann zeta function is an analytic object which describes the behavior of prime ideals in K. When K is an abelian extension of $\mathbf{Q}$, Dedekind zeta functions are products of Dirichlet L-functions, with there being one factor for each Dirichlet character. The trivial character corresponds to the Riemann zeta function. When K is a Galois extension, the Dedekind zeta function is the Artin L-function of the regular representation of the Galois group of K, and it has a factorization in terms of irreducible Artin representations of the Galois group.

The zeta function is related to the other invariants described above by the class number formula.

Local Fields

Completing a number field K at a place w gives a complete field. If the valuation is archimedean, one gets R or C, if it is non-archimedean and lies over a prime p of the rationals, one gets a finite extension K_w / Q_p: a complete, discrete valued field with finite residue field. This process simplifies the arithmetic of the field and allows the local study of problems. For example, the Kronecker–Weber theorem can be deduced easily from the analogous local statement. The philosophy behind the study of local fields is largely motivated by geometric methods. In algebraic geometry, it is common to study varieties locally at a point by localizing to a maximal ideal. Global information can then be recovered by gluing together local data. This spirit is adopted in algebraic number theory. Given a prime in the ring of algebraic integers in a number field, it is desirable to study the field locally at that prime. Therefore, one localizes the ring of algebraic integers to that prime and then completes the fraction field much in the spirit of geometry.

Major Results

Finiteness of the Class Group

One of the classical results in algebraic number theory is that the ideal class group of an algebraic number field K is finite. The order of the class group is called the class number, and is often denoted by the letter h.

Dirichlet's Unit Theorem

Dirichlet's unit theorem provides a description of the structure of the multiplicative group of units $O^\times$ of the ring of integers O. Specifically, it states that $O^\times$ is isomorphic to $G \times Z^r$, where G is the finite cyclic group consisting of all the roots of unity in O, and $r = r_1 + r_2 - 1$ (where r_1 (respectively, r_2) denotes the number of real embeddings (respectively, pairs of conjugate non-real embeddings) of K). In other words, $O^\times$ is a finitely generated abelian group of rank $r_1 + r_2 - 1$ whose torsion consists of the roots of unity in O.

Reciprocity Laws

In terms of the Legendre symbol, the law of quadratic reciprocity for positive odd primes states

$$\left(\frac{p}{q}\right)\left(\frac{q}{p}\right) = (-1)^{\frac{p-1}{2}\frac{q-1}{2}}.$$

A reciprocity law is a generalization of the law of quadratic reciprocity.

There are several different ways to express reciprocity laws. The early reciprocity laws found in the 19th century were usually expressed in terms of a power residue symbol (p/q) generalizing the quadratic reciprocity symbol, that describes when a prime number is an nth power residue modulo another prime, and gave a relation between (p/q) and (q/p). Hilbert reformulated the reciprocity laws as saying that a product over p of Hilbert symbols $(a,b/p)$, taking values in roots of unity, is equal to 1. Artin's reformulated reciprocity law states that the Artin symbol from ideals (or ideles) to elements of a Galois group is trivial on a certain subgroup. Several more recent generalizations express reciprocity laws using cohomology of groups or representations of adelic groups or algebraic K-groups, and their relationship with the original quadratic reciprocity law can be hard to see.

Class Number Formula

The class number formula relates many important invariants of a number field to a special value of its Dedekind zeta function.

Related Areas

Algebraic number theory interacts with many other mathematical disciplines. It uses tools from homological algebra. Via the analogy of function fields vs. number fields, it relies on techniques and ideas from algebraic geometry. Moreover, the study of higher-dimensional schemes over Z instead of number rings is referred to as arithmetic geometry. Algebraic number theory is also used in the study of arithmetic hyperbolic 3-manifolds.

Algebraic K-theory

Algebraic K-theory is a subject area in mathematics with connections to geometry, topology, ring theory, and number theory. Geometric, algebraic, and arithmetic objects are assigned objects called K-groups. These are groups in the sense of abstract algebra. They contain detailed information about the original object but are notoriously difficult to compute; for example, an important outstanding problem is to compute the K-groups of the integers.

K-theory was invented in the late 1950s by Alexander Grothendieck in his study of intersection theory on algebraic varieties. Intersection theory is still a motivating force in the development of algebraic K-theory through its links with motivic cohomology and specifically Chow groups. The subject also includes classical number-theoretic topics like quadratic reciprocity and embeddings of number fields into the real numbers and complex numbers, as well as more modern concerns like the construction of higher regulators and special values of L-functions.

The lower K-groups were discovered first, in the sense that adequate descriptions of these groups in terms of other algebraic structures were found. For example, if F is a field, then $K_0(F)$ is isomorphic to the integers Z and is closely related to the notion of vector space dimension. For a commutative ring R, $K_0(R)$ is the Picard group of R, and when R is the ring of integers in a number field, this generalizes the classical construction of the class group. The group $K_1(R)$ is closely related to the group of units $R^\times$, and if R is a field, it is exactly the group of units. For a number field F, $K_2(F)$ is related to class field theory, the Hilbert symbol, and the solvability of quadratic equations over completions. In contrast, finding the correct definition of the higher K-groups of rings was a difficult achievement of Daniel Quillen, and many of the basic facts about the higher K-groups of algebraic varieties were not known until the work of Robert Thomason.

History

The history of K-theory was detailed by Weibel.

K_0, K_1, and K_2

In the 19th century, Bernhard Riemann and his student Gustav Roch proved what is now known as the Riemann–Roch theorem. If X is a Riemann surface, then the sets of meromorphic functions and meromorphic differential forms on X form vector spaces. A line bundle on X determines subspaces of these vector spaces, and if X is projective, then these subspaces are finite dimensional. The Riemann–Roch theorem states that the difference in dimensions between these subspaces is equal to the degree of the line bundle (a measure of twistedness) plus one minus the genus of X. In the mid-20th century, the Riemann–Roch theorem was generalized by Friedrich Hirzebruch to all algebraic varieties. In Hirzebruch's formulation, the Hirzebruch–Riemann–Roch theorem, the theorem became a statement about Euler characteristics: The Euler characteristic of a vector bundle on an algebraic variety (which is the alternating sum of the dimensions of its cohomology groups) equals the Euler characteristic of the trivial bundle plus a correction factor coming from characteristic classes of the vector bundle. This is a generalization because on a projective Riemann surface, the Euler characteristic of a line bundle equals the difference in dimensions mentioned previously, the Euler characteristic of the trivial bundle is one minus the genus, and the only non-trivial characteristic class is the degree.

The subject of K-theory takes its name from a 1957 construction of Alexander Grothendieck which appeared in the Grothendieck–Riemann–Roch theorem, his generalization of Hirzebruch's theorem. Let X be a smooth algebraic variety. To each vector bundle on X, Grothendieck associates an invariant, its *class*. The set of all classes on X was called $K(X)$ from the German *Klasse*. By definition, $K(X)$ is a quotient of the free abelian group on isomorphism classes of vector bundles on X, and so it is an abelian group. If the basis element corresponding to a vector bundle V is denoted $[V]$, then for each short exact sequence of vector bundles:

$$0 \to V' \to V \to V'' \to 0,$$

Grothendieck imposed the relation $[V] = [V'] + [V'']$. These generators and relations define $K(X)$, and they imply that it is the universal way to assign invariants to vector bundles in a way compatible with exact sequences.

Grothendieck took the perspective that the Riemann–Roch theorem is a statement about morphisms of varieties, not the varieties themselves. He proved that there is a homomorphism from $K(X)$ to the Chow groups of X coming from the Chern character and Todd class of X. Additionally, he proved that a proper morphism $f: X \to Y$ to a smooth variety Y determines a homomorphism $f_*: K(X) \to K(Y)$ called the *pushforward*. This gives two ways of determining an element in the Chow group of Y from a vector bundle on X: Starting from X, one can first compute the pushforward in K-theory and then apply the Chern character and Todd class of Y, or one can first apply the Chern character and Todd class of X and then compute the pushforward for Chow groups. The Grothendieck–Riemann–Roch theorem says that these are equal. When Y is a point, a vector bundle is a vector space, the class of a vector space is its dimension, and the Grothendieck–Riemann–Roch theorem specializes to Hirzebruch's theorem.

The group $K(X)$ is now known as $K_0(X)$. Upon replacing vector bundles by projective modules, K_0 also became defined for non-commutative rings, where it had applications to group representations. Atiyah and Hirzebruch quickly transported Grothendieck's construction to topology and used it to define topological K-theory. Topological K-theory was one of the first examples of an extraordinary cohomology theory: It associates to each topological space X (satisfying some mild technical constraints) a sequence of groups $K_n(X)$ which satisfy all the Eilenberg–Steenrod axioms except the normalization axiom. The setting of algebraic varieties, however, is much more rigid, and the flexible constructions used in topology were not available. While the group K_0 seemed to satisfy the necessary properties to be the beginning of a cohomology theory of algebraic varieties and of non-commutative rings, there was no clear definition of the higher $K_n(X)$. Even as such definitions were developed, technical issues surrounding restriction and gluing usually forced K_n to be defined only for rings, not for varieties.

While it was not initially known, a group related to K_1 had already been introduced in another context. Henri Poincaré had attempted to define the Betti numbers of a manifold in terms of a triangulation. His methods, however, had a serious gap: Poincaré could not prove that two triangulations of a manifold always yielded the same Betti numbers. It was clearly true that Betti numbers were unchanged by subdividing the triangulation, and therefore it was clear that any two triangulations that shared a common subdivision had the same Betti numbers. What was not known was that any two triangulations admitted a common subdivision. This hypothesis became a conjecture known as the *Hauptvermutung* (roughly "main conjecture"). The fact that triangulations were stable under subdivision led J.H.C. Whitehead to introduce the notion of simple homotopy type. A simple homotopy equivalence is defined in terms of adding simplices or cells to a simplicial complex or cell complex in such a way that each additional simplex or cell deformation retracts into a subdivision of the old space. Part of the motivation for this definition is that a subdivision of a triangulation is simple homotopy equivalent to the original triangulation, and therefore two triangulations that share a common subdivision must be simple homotopy equivalent. Whitehead proved that simple homotopy equivalence is a finer invariant than homotopy equivalence by introducing an invariant called the *torsion*. The torsion of a homotopy equivalence takes values in a group now called the *Whitehead group* and denoted $Wh(\pi)$, where π is the fundamental group of the two complexes. Whitehead found examples of non-trivial torsion and thereby proved that some homotopy equivalences were not simple. The Whitehead group was later discovered to be a quotient of $K_1(Z\pi)$, where $Z\pi$ is the integral group ring of π. Later John Milnor used Reidemeister torsion, an invariant related to Whitehead torsion, to disprove the Hauptvermutung.

The first adequate definition of K_1 of a ring was made by Hyman Bass and Stephen Schanuel. In topological K-theory, K_1 is defined using vector bundles on a suspension of the space. All such vector bundles come from the clutching construction, where two trivial vector bundles on two halves of a space are glued along a common strip of the space. This gluing data is expressed using the general linear group, but elements of that group coming from elementary matrices (matrices corresponding to elementary row or column operations) define equivalent gluings. Motivated by this, the Bass–Schanuel definition of K_1 of a ring R is $GL(R) / E(R)$, where $GL(R)$ is the infinite general linear group (the union of all $GL_n(R)$) and $E(R)$ is the subgroup of elementary matrices. They also provided a definition of K_0 of a homomorphism of rings and proved that K_0 and K_1 could be fit together into an exact sequence similar to the relative homology exact sequence.

Work in K-theory from this period culminated in Bass' book *Algebraic K-theory*. In addition to providing a coherent exposition of the results then known, Bass improved many of the statements of the theorems. Of particular note is that Bass, building on his earlier work with Murthy, provided the first proof of what is now known as the fundamental theorem of algebraic K-theory. This is a four-term exact sequence relating K_0 of a ring R to K_1 of R, the polynomial ring $R[t]$, and the localization $R[t, t^{-1}]$. Bass recognized that this theorem provided a description of K_0 entirely in terms of K_1. By applying this description recursively, he produced negative K-groups $K_{-n}(R)$. In independent work, Max Karoubi gave another definition of negative K-groups for certain categories and proved that his definitions yielded that same groups as those of Bass.

The next major development in the subject came with the definition of K_2. Steinberg studied the universal central extensions of a Chevalley group over a field and gave an explicit presentation of this group in terms of generators and relations. In the case of the group $E_n(k)$ of elementary matrices, the universal central extension is now written $St_n(k)$ and called the *Steinberg group*. In the spring of 1967, John Milnor defined $K_2(R)$ to be the kernel of the homomorphism $St(R) \to E(R)$. The group K_2 further extended some of the exact sequences known for K_1 and K_0, and it had striking applications to number theory. Matsumoto's 1968 thesis showed that for a field F, $K_2(F)$ was isomorphic to:

$$F^\times \otimes_{\mathbf{Z}} F^\times / \langle x \otimes (1-x) : x \in F \setminus \{0,1\}\rangle.$$

This relation is also satisfied by the Hilbert symbol, which expresses the solvability of quadratic equations over local fields. In particular, John Tate was able to prove that $K_2(\mathbf{Q})$ is essentially structured around the law of quadratic reciprocity.

Higher *K*-groups

In the late 1960s and early 1970s, several definitions of higher K-theory were proposed. Swan and Gersten both produced definitions of K_n for all n, and Gersten proved that his and Swan's theories were equivalent, but the two theories were not known to satisfy all the expected properties. Nobile and Villamayor also proposed a definition of higher K-groups. Karoubi and Villamayor defined well-behaved K-groups for all n, but their equivalent of K_1 was sometimes a proper quotient of the Bass–Schanuel K_1. Their K-groups are now called KV_n and are related to homotopy-invariant modifications of K-theory.

Inspired in part by Matsumoto's theorem, Milnor made a definition of the higher K-groups of a field. He referred to his definition as "purely *ad hoc*", and it neither appeared to generalize to all rings nor did it appear to be the correct definition of the higher K-theory of fields. Much later, it was discovered by Nesterenko and Suslin and by Totaro that Milnor K-theory is actually a direct summand of the true K-theory of the field. Specifically, K-groups have a filtration called the *weight filtration*, and the Milnor K-theory of a field is the highest weight-graded piece of the K-theory. Additionally, Thomason discovered that there is no analog of Milnor K-theory for a general variety.

The first definition of higher K-theory to be widely accepted was Daniel Quillen's. As part of Quillen's work on the Adams conjecture in topology, he had constructed maps from the classifying spaces $BGL(\mathbf{F}_q)$ to the homotopy fiber of $\psi^q - 1$, where ψ^q is the qth Adams operation acting on the classifying space BU. This map is acyclic, and after modifying $BGL(\mathbf{F}_q)$ slightly to produce a new space $BGL(\mathbf{F}_q)^+$, the map became a homotopy equivalence. This modification was called the plus construction. The Adams operations had been known to be related to Chern classes and to K-theory since the work of Grothendieck, and so Quillen was led to define the K-theory of R as the homotopy groups of $BGL(R)^+$. Not only did this recover K_1 and K_2, the relation of K-theory to the Adams operations allowed Quillen to compute the K-groups of finite fields.

The classifying space BGL is connected, so Quillen's definition failed to give the correct value for

K_0. Additionally, it did not give any negative K-groups. Since K_0 had a known and accepted definition it was possible to sidestep this difficulty, but it remained technically awkward. Conceptually, the problem was that the definition sprung from GL, which was classically the source of K_1. Because GL knows only about gluing vector bundles, not about the vector bundles themselves, it was impossible for it to describe K_0.

Inspired by conversations with Quillen, Segal soon introduced another approach to constructing algebraic K-theory under the name of Γ-objects. Segal's approach is a homotopy analog of Grothendieck's construction of K_0. Where Grothendieck worked with isomorphism classes of bundles, Segal worked with the bundles themselves and used isomorphisms of the bundles as part of his data. This results in a spectrum whose homotopy groups are the higher K-groups (including K_0). However, Segal's approach was only able to impose relations for split exact sequences, not general exact sequences. In the category of projective modules over a ring, every short exact sequence splits, and so Γ-objects could be used to define the K-theory of a ring. However, there are non-split short exact sequences in the category of vector bundles on a variety and in the category of all modules over a ring, so Segal's approach did not apply to all cases of interest.

In the spring of 1972, Quillen found another approach to the construction of higher K-theory which was to prove enormously successful. This new definition began with an exact category, a category satisfying certain formal properties similar to, but slightly weaker than, the properties satisfied by a category of modules or vector bundles. From this he constructed an auxiliary category using a new device called his "Q-construction." Like Segal's Γ-objects, the Q-construction has its roots in Grothendieck's definition of K_0. Unlike Grothendieck's definition, however, the Q-construction builds a category, not an abelian group, and unlike Segal's Γ-objects, the Q-construction works directly with short exact sequences. If C is an abelian category, then QC is a category with the same objects as C but whose morphisms are defined in terms of short exact sequences in C. The K-groups of the exact category are the homotopy groups of ΩBQC, the loop space of the geometric realization (taking the loop space corrects the indexing). Quillen additionally proved his "+ = Q theorem" that his two definitions of K-theory agreed with each other. This yielded the correct K_0 and led to simpler proofs, but still did not yield any negative K-groups.

All abelian categories are exact categories, but not all exact categories are abelian. Because Quillen was able to work in this more general situation, he was able to use exact categories as tools in his proofs. This technique allowed him to prove many of the basic theorems of algebraic K-theory. Additionally, it was possible to prove that the earlier definitions of Swan and Gersten were equivalent to Quillen's under certain conditions.

K-theory now appeared to be a homology theory for rings and a cohomology theory for varieties. However, many of its basic theorems carried the hypothesis that the ring or variety in question was regular. One of the basic expected relations was a long exact sequence (called the "localization sequence") relating the K-theory of a variety X and an open subset U. Quillen was unable to prove the existence of the localization sequence in full generality. He was, however, able to prove its existence for a related theory called G-theory (or sometimes K'-theory). G-theory had been defined early in the development of the subject by Grothendieck. Grothendieck defined $G_0(X)$ for a variety X to be the free abelian group on isomorphism classes of coherent sheaves on X, modulo relations coming from exact sequences of coherent sheaves. In the categorical framework adopted by later authors, the K-theory of a variety is the K-theory of its category of vector bundles, while its G-theory is the K-theory of its category of coherent sheaves. Not only could Quillen prove the existence of a localization exact sequence for G-theory, he could prove that for a regular ring or variety, K-theory equaled G-theory, and therefore K-theory of regular varieties had a localization exact sequence. Since this sequence was fundamental to many of the facts in the subject, regularity hypotheses pervaded early work on higher K-theory.

Applications of Algebraic K-theory in Topology

The earliest application of algebraic K-theory to topology was Whitehead's construction of Whitehead torsion. A closely related construction was found by C. T. C. Wall in 1963. Wall found that a space π dominated by a finite complex has a generalized Euler characteristic taking values in a quotient of $K_0(\mathbb{Z}\pi)$, where π is the fundamental group of the space. This invariant is called *Wall's finiteness obstruction* because X is homotopy equivalent to a finite complex if and only if the invariant vanishes. Laurent Siebenmann in his thesis found an invariant similar to Wall's that gives an obstruction to an open manifold being the interior of a compact manifold with boundary. If two manifolds with boundary M and N have isomorphic interiors (in TOP, PL, or DIFF as appropriate), then the isomorphism between them defines an h-cobordism between M and N.

Whitehead torsion was eventually reinterpreted in a more directly K-theoretic way. This reinterpretation happened through the study of h-cobordisms. Two n-dimensional manifolds M and N are h-cobordant if there exists an $(n+1)$-dimensional manifold with boundary W whose boundary is the disjoint union of M and N and for which the inclusions of M and N into W are homotopy equivalences (in the categories TOP, PL, or DIFF). Stephen Smale's h-cobordism theorem asserted that if $n \geq 5$, W is compact, and M, N, and W are simply connected, then W is isomorphic to the cylinder $M \times [0, 1]$ (in TOP, PL, or DIFF as appropriate). This theorem proved the Poincaré conjecture for $n \geq 5$.

If M and N are not assumed to be simply connected, then an h-cobordism need not be a cylinder. The s-cobordism theorem, due independently to Mazur, Stallings, and Barden, explains the general situation: An h-cobordism is a cylinder if and only if the Whitehead torsion of the inclusion $M \subset W$ vanishes. This generalizes the h-cobordism theorem because the simple connectedness hypotheses imply that the relevant Whitehead group is trivial. In fact the s-cobordism theorem implies that there is a bijective correspondence between isomorphism classes of h-cobordisms and elements of the Whitehead group.

An obvious question associated with the existence of h-cobordisms is their uniqueness. The natural notion of equivalence is isotopy. Jean Cerf proved that for simply connected smooth manifolds M of dimension at least 5, isotopy of h-cobordisms is the same as a weaker notion called pseudo-isotopy. Hatcher and Wagoner studied the components of the space of pseudo-isotopies and related it to a quotient of $K_2(\mathbb{Z}\pi)$.

The proper context for the s-cobordism theorem is the classifying space of h-cobordisms. If M is a CAT manifold, then $H^{\text{CAT}}(M)$ is a space that classifies bundles of h-cobordisms on M. The s-cobordism theorem can be reinterpreted as the statement that the set of connected components of this space is the Whitehead group of $\pi_1(M)$. This space contains strictly more information than the Whitehead group; for example, the connected component of the trivial cobordism describes the possible cylinders on M and in particular is the obstruction to the uniqueness of a homotopy between a manifold and $M \times [0, 1]$. Consideration of these questions led Waldhausen to introduced his *algebraic K-theory of spaces*. The algebraic K-theory of M is a space $A(M)$ which is defined so that it plays essentially the same role for higher K-groups as $K_1(Z\pi_1(M))$ does for M. In particular, Waldhausen showed that there is a map from $A(M)$ to a space $\text{Wh}(M)$ which generalizes the map $K_1(Z\pi_1(M)) \to \text{Wh}(\pi_1(M))$ and whose homotopy fiber is a homology theory.

In order to fully develop A-theory, Waldhausen made significant technical advances in the foundations of K-theory. Waldhausen introduced Waldhausen categories, and for a Waldhausen category C he introduced a simplicial category $S_.C$ (the S is for Segal) defined in terms of chains of cofibrations in C. This freed the foundations of K-theory from the need to invoke analogs of exact sequences.

Algebraic Topology and Algebraic Geometry in Algebraic K-theory

Quillen suggested to his student Kenneth Brown that it might be possible to create a theory of sheaves of spectra of which K-theory would provide an example. The sheaf of K-theory spectra would, to each open subset of a variety, associate the K-theory of that open subset. Brown developed such a theory for his thesis. Simultaneously, Gersten had the same idea. At a Seattle conference in autumn of 1972, they together discovered a spectral sequence converging from the sheaf cohomology of $\mathcal{K}_n$, the sheaf of K_n-groups on X, to the K-group of the total space. This is now called the Brown–Gersten spectral sequence.

Spencer Bloch, influenced by Gersten's work on sheaves of K-groups, proved that on a regular surface, the cohomology group $H^2(X, \mathcal{K}_2)$ is isomorphic to the Chow group $CH^2(X)$ of codimension 2 cycles on X. Inspired by this, Gersten conjectured that for a regular local ring R with fraction field F, $K_n(R)$ injects into $K_n(F)$ for all n. Soon Quillen proved that this is true when R contains a field, and using this he proved that $H^p(X, \mathcal{K}_p) \cong CH^p(X)$ for all p. This is known as *Bloch's formula*. While progress has been made on Gersten's conjecture since then, the general case remains open.

Lichtenbaum conjectured that special values of the zeta function of a number field could be expressed in terms of the K-groups of the ring of integers of the field. These special values were known to be related to the etale cohomology of the ring of integers. Quillen therefore generalized Lichtenbaum's conjecture, predicting the existence of a spectral sequence like the Atiyah–Hirzebruch spectral sequence in topological K-theory. Quillen's proposed spectral sequence would start from the etale cohomology of a ring R and, in high enough degrees and after completing at a prime ℓ invertible in R, abut to the ℓ-adic completion of the K-theory of R. In the case studied by Lichtenbaum, the spectral sequence would degenerate, yielding Lichtenbaum's conjecture.

The necessity of localizing at a prime ℓ suggested to Browder that there should be a variant of K-theory with finite coefficients. He introduced K-theory groups $K_n(R; Z/\ell Z)$ which were $Z/\ell Z$-vector spaces, and he found an analog of the Bott element in topological K-theory. Soule used this theory to construct "etale Chern classes", an analog of topological Chern classes which took elements of algebraic K-theory to classes in etale cohomology. Unlike algebraic K-theory, etale cohomology is highly computable, so etale Chern classes provided an effective tool for detecting the existence of elements in K-theory. Dwyer and Friedlander then invented an analog of K-theory for the etale topology called etale K-theory. For varieties defined over the complex numbers, etale K-theory is isomorphic to topological K-theory. Moreover, etale K-theory admitted a spectral sequence similar to the one conjectured by Quillen. Thomason proved around 1980 that after inverting the Bott element, algebraic K-theory with finite coefficients became isomorphic to etale K-theory.

Throughout the 1970s and early 1980s, K-theory on singular varieties still lacked adequate foundations. While it was believed that Quillen's K-theory gave the correct groups, it was not known that these groups had all of the envisaged properties. For this, algebraic K-theory had to be reformulated. This was done by Thomason in a lengthy monograph which he co-credited to his dead friend Thomas Trobaugh, whom he said gave him a key idea in a dream. Thomason combined Waldhausen's construction of K-theory with the foundations of intersection theory described in volume six of Grothendieck's Séminaire de Géométrie Algébrique du Bois Marie. There, K_0 was described in terms of complexes of sheaves on algebraic varieties. Thomason discovered that if one worked with in derived category of sheaves, there was a simple description of when a complex of sheaves could be extended from an open subset of a variety to the whole variety. By applying Waldhausen's construction of K-theory to derived categories, Thomason was able to prove that algebraic K-theory had all the expected properties of a cohomology theory.

In 1976, Keith Dennis discovered an entirely novel technique for computing K-theory based on Hochschild homology. This was based around the existence of the Dennis trace map, a homomorphism from K-theory to Hochschild homology. While the Dennis trace map seemed to be successful for calculations of K-theory with finite coefficients, it was less successful for rational calculations. Goodwillie, motivated by his "calculus of functors", conjectured the existence of a theory intermediate to K-theory and Hochschild homology. He called this theory topological Hochschild homology because its ground ring should be the sphere spectrum (considered as a ring whose operations are defined only up to homotopy). In the mid-80s, Bokstedt gave a definition of topological Hochschild homology that satisfied nearly all of Goodwillie's conjectural properties, and this made possible further computations of K-groups. Bokstedt's version of the Dennis trace map was a transformation of spectra $K \to THH$. This transformation factored through the fixed points of a circle action on THH, which suggested a relationship with cyclic homology. In the course of proving an algebraic K-theory analog of the Novikov conjecture, Bokstedt, Hsiang, and Madsen introduced topological cyclic homology, which bore the same relationship to topological Hochschild homology as cyclic homology did to Hochschild homology. The Dennis trace map to topological Hochschild homology facts through topological cyclic homology, providing an even more detailed tool for calculations. In 1996, Dundas, Goodwillie, and McCarthy proved that topological cyclic homology has in a precise sense the same local structure as algebraic K-theory, so that if a calculation in K-theory or topological cyclic homology is possible, then many other "nearby" calculations follow.

Lower K-groups

The lower K-groups were discovered first, and given various ad hoc descriptions, which remain useful. Throughout, let A be a ring.

K_0

The functor K_0 takes a ring A to the Grothendieck group of the set of isomorphism classes of its finitely generated projective modules, regarded as a monoid under direct sum. Any ring homomorphism $A \to B$ gives a map $K_0(A) \to K_0(B)$ by mapping (the class of) a projective A-module M to $M \otimes_A B$, making K_0 a covariant functor.

If the ring A is commutative, we can define a subgroup of $K_0(A)$ as the set

$$\tilde{K}_0(A) = \bigcap_{\mathfrak{p} \text{ prime ideal of } A} \operatorname{Ker} \dim_{\mathfrak{p}},$$

where :

$$\dim_{\mathfrak{p}} : K_0(A) \to \mathbf{Z}$$

is the map sending every (class of a) finitely generated projective A-module M to the rank of the free $A_{\mathfrak{p}}$-module $M_{\mathfrak{p}}$ (this module is indeed free, as any finitely generated projective module over a local ring is free). This subgroup $\tilde{K}_0(A)$ is known as the *reduced zeroth K-theory* of A.

If B is a ring without an identity element, we can extend the definition of K_0 as follows. Let $A = B \oplus Z$ be the extension of B to a ring with unity obtaining by adjoining an identity element (0,1). There is a short exact sequence $B \to A \to Z$ and we define $K_0(B)$ to be the kernel of the corresponding map $K_0(A) \to K_0(Z) = Z$.

Examples

- (Projective) modules over a field k are vector spaces and $K_0(k)$ is isomorphic to Z, by dimension.

- Finitely generated projective modules over a local ring A are free and so in this case again $K_0(A)$ is isomorphic to Z, by rank.

- For A a Dedekind domain,

$$K_0(A) = \operatorname{Pic}(A) \oplus Z,$$

where $\operatorname{Pic}(A)$ is the Picard group of A, and similarly the reduced K-theory is given by

$$\tilde{K}_0(A) = \operatorname{Pic} A.$$

An algebro-geometric variant of this construction is applied to the category of algebraic varieties; it associates with a given algebraic variety X the Grothendieck's K-group of the category of locally free sheaves (or coherent sheaves) on X. Given a compact topological space X, the topological K-theory $K^{top}(X)$ of (real) vector bundles over X coincides with K_0 of the ring of continuous real-valued functions on X.

Relative K_0

Let I be an ideal of A and define the "double" to be a subring of the Cartesian product $A \times A$:

$$D(A,I) = \{(x,y) \in A \times A : x - y \in I\}.$$

The *relative K-group* is defined in terms of the "double"

$$K_0(A,I) = \ker\left(K_0(D(A,I)) \to K_0(A)\right).$$

where the map is induced by projection along the first factor.

The relative $K_0(A,I)$ is isomorphic to $K_0(I)$, regarding I as a ring without identity. The independence from A is an analogue of the Excision theorem in homology.

K_0 as a Ring

If A is a commutative ring, then the tensor product of projective modules is again projective, and so tensor product induces a multiplication turning K_0 into a commutative ring with the class $[A]$ as identity. The exterior product similarly induces a λ-ring structure. The Picard group embeds as a subgroup of the group of units $K_0(A)^*$.

K_1

Hyman Bass provided this definition, which generalizes the group of units of a ring: $K_1(A)$ is the abelianization of the infinite general linear group:

$$K_1(A) = \mathrm{GL}(A)^{\mathrm{ab}} = \mathrm{GL}(A)\,/\,[\mathrm{GL}(A), \mathrm{GL}(A)]$$

Here

$$\mathrm{GL}(A) = \operatorname{colim} \mathrm{GL}(n, A)$$

is the direct limit of the GL(n), which embeds in GL(n + 1) as the upper left block matrix, and [GL (A) , GL (A)] is its commutator subgroup. Define an elementary matrix to be one which is the sum of an identity matrix and a single off-diagonal element (this is different from the definition used in linear algebra). Then Whitehead's lemma states that the group E(A) generated by elementary matrices equals the commutator subgroup [GL(A), GL(A)]. Indeed, the group GL(A)/E(A) was first defined and studied by Whitehead,[52] and is called the Whitehead group of the ring A.

Relative K_1

The *relative K-group* is defined in terms of the "double"
$$K_1(A, I) = \ker\left(K_1(D(A, I)) \to K_1(A)\right).$$

There is a natural exact sequence

$$K_1(A, I) \to K_1(A) \to K_1(A\,/\,I) \to K_0(A, I) \to K_0(A) \to K_0(A\,/\,I).$$

Commutative Rings and Fields

For A a commutative ring, one can define a determinant det: $\mathrm{GL}(A) \to A^*$ to the group of units of A, which vanishes on E(A) and thus descends to a map det: $K_1(A) \to A^*$. As E(A) ◁ SL(A), one can also define the special Whitehead group $SK_1(A) := SL(A)/E(A)$. This map splits via the map $A^* \to \mathrm{GL}(1, A) \to K_1(A)$ (unit in the upper left corner), and hence is onto, and has the special Whitehead group as kernel, yielding the split short exact sequence:

$$1 \to SK_1(A) \to K_1(A) \to A^* \to 1,$$

which is a quotient of the usual split short exact sequence defining the special linear group, namely

$$1 \to SL(A) \to GL(A) \to A^* \to 1.$$

The determinant is split by including the group of units $A^* = GL_1(A)$ into the general linear group $GL(A)$, so $K_1(A)$ splits as the direct sum of the group of units and the special Whitehead group: $K_1(A) \cong A^* \oplus SK_1(A)$.

When A is a Euclidean domain (e.g. a field, or the integers) $SK_1(A)$ vanishes, and the determinant map is an isomorphism from $K_1(A)$ to A^*. This is *false* in general for PIDs, thus providing one of the rare mathematical features of Euclidean domains that do not generalize to all PIDs. An explicit PID such that SK_1 is nonzero was given by Ischebeck in 1980 and by Grayson in 1981. If A is a Dedekind domain whose quotient field is an algebraic number field (a finite extension of the rationals) then Milnor (1971, corollary 16.3) shows that $SK_1(A)$ vanishes.

The vanishing of SK_1 can be interpreted as saying that K_1 is generated by the image of GL_1 in GL. When this fails, one can ask whether K_1 is generated by the image of GL_2. For a Dedekind domain, this is the case: indeed, K_1 is generated by the images of GL_1 and SL_2 in GL. The subgroup of SK_1 generated by SL_2 may be studied by Mennicke symbols. For Dedekind domains with all quotients by maximal ideals finite, SK_1 is a torsion group.

For a non-commutative ring, the determinant cannot in general be defined, but the map $GL(A) \to K_1(A)$ is a generalisation of the determinant.

Central Simple Algebras

In the case of a central simple algebra A over a field F, the reduced norm provides a generalisation of the determinant giving a map $K_1(A) \to F^*$ and $SK_1(A)$ may be defined as the kernel. Wang's theorem states that if A has prime degree then $SK_1(A)$ is trivial, and this may be extended to square-free degree. Wang also showed that $SK_1(A)$ is trivial for any central simple algebra over a number field, but Platonov has given examples of algebras of degree prime squared for which $SK_1(A)$ is non-trivial.

K_2

John Milnor found the right definition of K_2: it is the center of the Steinberg group St(A) of A.

It can also be defined as the kernel of the map

$$\varphi : \mathrm{St}(A) \to \mathrm{GL}(A),$$

or as the Schur multiplier of the group of elementary matrices.

For a field, K_2 is determined by Steinberg symbols: this leads to Matsumoto's theorem.

One can compute that K_2 is zero for any finite field. The computation of $K_2(\mathbf{Q})$ is complicated: Tate proved

$$K_2(\mathbf{Q}) = (\mathbf{Z}/4)^* \times \prod_{p \text{ odd prime}} (\mathbf{Z}/p)^*$$

and remarked that the proof followed Gauss's first proof of the Law of Quadratic Reciprocity.

For non-Archimedean local fields, the group $K_2(F)$ is the direct sum of a finite cyclic group of order m, say, and a divisible group $K_2(F)^m$.

We have $K_2(Z) = Z/2$, and in general K_2 is finite for the ring of integers of a number field.

We further have $K_2(Z/n) = Z/2$ if n is divisible by 4, and otherwise zero.

Matsumoto's Theorem

Matsumoto's theorem states that for a field k, the second K-group is given by

$$K_2(k) = k^\times \otimes_Z k^\times / \langle a \otimes (1-a) \mid a \neq 0, 1 \rangle.$$

Matsumoto's original theorem is even more general: For any root system, it gives a presentation for the unstable K-theory. This presentation is different from the one given here only for symplectic root systems. For non-symplectic root systems, the unstable second K-group with respect to the root system is exactly the stable K-group for GL(A). Unstable second K-groups (in this context) are defined by taking the kernel of the universal central extension of the Chevalley group of universal type for a given root system. This construction yields the kernel of the Steinberg extension for the root systems A_n ($n > 1$) and, in the limit, stable second K-groups.

Long Exact Sequences

If A is a Dedekind domain with field of fractions F then there is a long exact sequence

$$K_2 F \to \oplus_p K_1 A / \mathbf{p} \to K_1 A \to K_1 F \to \oplus_p K_0 A / \mathbf{p} \to K_0 A \to K_0 F \to 0$$

where $\mathbf{p}$ runs over all prime ideals of A.

There is also an extension of the exact sequence for relative K_1 and K_0:

$$K_2(A) \to K_2(A/I) \to K_1(A, I) \to K_1(A) \cdots.$$

Pairing

There is a pairing on K_1 with values in K_2. Given commuting matrices X and Y over A, take elements x and y in the Steinberg group with X, Y as images. The commutator $xyx^{-1}y^{-1}$ is an element of K_2. The map is not always surjective.

Milnor K-theory

The above expression for K_2 of a field k led Milnor to the following definition of "higher" K-groups by

$$K_*^M(k) := T^*(k^\times) / (a \otimes (1-a)),$$

thus as graded parts of a quotient of the tensor algebra of the multiplicative group $k^\times$ by the two-sided ideal, generated by the

$$\{a \otimes (1-a) : a \neq 0,1\}.$$

For $n=0,1,2$ these coincide with those below, but for $n \geq 3$ they differ in general. For example, we have KM $n(F_q) = o$ for $n \geq 2$ but $K_n F_q$ is nonzero for odd n.

The tensor product on the tensor algebra induces a product $K_m \times K_n \to K_{m+n}$ making $K_*^M(F)$ a graded ring which is graded-commutative.

The images of elements $a_1 \otimes \cdots \otimes a_n$ in $K_n^M(k)$ are termed *symbols*, denoted $\{a_1, \ldots, a_n\}$. For integer m invertible in k there is a map

$$\partial : k^* \to H^1(k, \mu_m)$$

where μ_m denotes the group of m-th roots of unity in some separable extension of k. This extends to

$$\partial^n : k^* \times \cdots \times k^* \to H^n\left(k, \mu_m^{\otimes n}\right)$$

satisfying the defining relations of the Milnor K-group. Hence ∂^n may be regarded as a map on $K_n^M(k)$, called the *Galois symbol* map.

The relation between étale (or Galois) cohomology of the field and Milnor K-theory modulo 2 is the Milnor conjecture, proven by Vladimir Voevodsky. The analogous statement for odd primes is the Bloch-Kato conjecture, proved by Voevodsky, Rost, and others.

Higher K-theory

The accepted definitions of higher K-groups were given by Quillen (1973), after a few years during which several incompatible definitions were suggested. The object of the program was to find definitions of K(R) and K(R,I) in terms of classifying spaces so that $R \Rightarrow K(R)$ and $(R,I) \Rightarrow K(R,I)$ are functors into a homotopy category of spaces and the long exact sequence for relative K-groups arises as the long exact homotopy sequence of a fibration K(R,I) $\to$ K(R) $\to$ K(R/I).

Quillen gave two constructions, the "plus-construction" and the "Q-construction", the latter subsequently modified in different ways. The two constructions yield the same K-groups.

The +-construction

One possible definition of higher algebraic K-theory of rings was given by Quillen
$$K_n(R) = \pi_n(BGL(R)^+),$$

Here π_n is a homotopy group, GL(R) is the direct limit of the general linear groups over R for the size of the matrix tending to infinity, B is the classifying space construction of homotopy theory, and the $^+$ is Quillen's plus construction.

This definition only holds for $n > 0$ so one often defines the higher algebraic K-theory via

$$K_n(R) = \pi_n(BGL(R)^+ \times K_0(R))$$

Since $BGL(R)^+$ is path connected and $K_0(R)$ discrete, this definition doesn't differ in higher degrees and also holds for $n = 0$.

The Q-construction

The Q-construction gives the same results as the +-construction, but it applies in more general situations. Moreover, the definition is more direct in the sense that the K-groups, defined via the Q-construction are functorial by definition. This fact is not automatic in the plus-construction.

Suppose P is an exact category; associated to P a new category QP is defined, objects of which are those of P and morphisms from M' to M'' are isomorphism classes of diagrams

$$M' \leftarrow N \rightarrow M'',$$

where the first arrow is an admissible epimorphism and the second arrow is an admissible monomorphism.

The i-th **K**-group of the exact category P is then defined as

$$K_i(P) = \pi_{i+1}(\mathrm{B}QP, 0)$$

with a fixed zero-object 0, where $\mathrm{B}QP$ is the *classifying space* of QP, which is defined to be the geometric realisation of the *nerve* of QP.

This definition coincides with the above definition of $K_0(P)$. If P is the category of finitely generated projective R-modules, this definition agrees with the above BGL^+ definition of $K_n(R)$ for all n. More generally, for a scheme X, the higher K-groups of X are defined to be the K-groups of (the exact category of) locally free coherent sheaves on X.

The following variant of this is also used: instead of finitely generated projective (= locally free) modules, take finitely generated modules. The resulting K-groups are usually written $G_n(R)$. When R is a noetherian regular ring, then G- and K-theory coincide. Indeed, the global dimension of regular rings is finite, i.e. any finitely generated module has a finite projective resolution $P_* \rightarrow M$, and a simple argument shows that the canonical map $K_0(\mathrm{R}) \rightarrow G_0(\mathrm{R})$ is an isomorphism, with $[M] = \Sigma \pm [P_n]$. This isomorphism extends to the higher K-groups, too.

The S-construction

A third construction of K-theory groups is the S-construction, due to Waldhausen. It applies to categories with cofibrations (also called Waldhausen categories). This is a more general concept than exact categories.

Examples

While the Quillen algebraic K-theory has provided deep insight into various aspects of algebraic geometry and topology, the K-groups have proved particularly difficult to compute except in a few isolated but interesting cases.

Algebraic K-groups of Finite Fields

The first and one of the most important calculations of the higher algebraic K-groups of a ring were made by Quillen himself for the case of finite fields:

If F_q is the finite field with q elements, then:

- $K_0(F_q) = Z$,
- $K_{2i}(F_q) = 0$ for $i \geq 1$,
- $K_{2i-1}(F_q) = Z/(q^i - 1)Z$ for $i \geq 1$.

Algebraic K-groups of Rings of Integers

Quillen proved that if A is the ring of algebraic integers in an algebraic number field F (a finite extension of the rationals), then the algebraic K-groups of A are finitely generated. Borel used this to calculate $K_i(A)$ and $K_i(F)$ modulo torsion. For example, for the integers Z, Borel proved that (modulo torsion)

- $K_i(Z)/\text{tors.} = 0$ for positive i unless $i = 4k+1$ with k positive
- $K_{4k+1}(Z)/\text{tors.} = Z$ for positive k.

The torsion subgroups of $K_{2i+1}(Z)$, and the orders of the finite groups $K_{4k+2}(Z)$ have recently been determined, but whether the latter groups are cyclic, and whether the groups $K_{4k}(Z)$ vanish depends upon Vandiver's conjecture about the class groups of cyclotomic integers.

Applications and Open Questions

Algebraic K-groups are used in conjectures on special values of L-functions and the formulation of an non-commutative main conjecture of Iwasawa theory and in construction of higher regulators.

Parshin's conjecture concerns the higher algebraic K-groups for smooth varieties over finite fields, and states that in this case the groups vanish up to torsion.

Another fundamental conjecture due to Hyman Bass (Bass' conjecture) says that all of the groups $G_n(A)$ are finitely generated when A is a finitely generated Z-algebra. (The groups $G_n(A)$ are the K-groups of the category of finitely generated A-modules)

References

- Falb, Peter (1990). Methods of Algebraic Geometry in Control Theory Part II Multivariable Linear Systems and Projective Algebraic Geometry. Springer. ISBN 978-0-8176-4113-9.

- Allen Tannenbaum (1982), Invariance and Systems Theory: Algebraic and Geometric Aspects, Lecture Notes in Mathematics, volume 845, Springer-Verlag, ISBN 9783540105657

- Tsfasman, Michael A.; Vlăduţ, Serge G.; Nogin, Dmitry (1990). Algebraic Geometric Codes Basic Notions. American Mathematical Soc. ISBN 978-0-8218-7520-9.

- Cox, David A.; Katz, Sheldon (1999). Mirror Symmetry and Algebraic Geometry. American Mathematical Soc. ISBN 978-0-8218-2127-5.

- Cox, David A.; Sturmfels, Bernd. Manocha, Dinesh N., ed. Applications of Computational Algebraic Geometry. American Mathematical Soc. ISBN 978-0-8218-6758-7.

- Cohen, Joel S. (2003). Computer Algebra and Symbolic Computation: Mathematical Methods. AK Peters, Ltd. p. 14. ISBN 978-1-56881-159-8.

- Donald Knuth. The Art of Computer Programming, Volume 4: Generating All Tuples and Permutations, Fascicle 2, first printing. Addison–Wesley, 2005. ISBN 0-201-85393-0.

- Donald Knuth. The Art of Computer Programming, Volume 3: Sorting and Searching, Second Edition. Addison–Wesley, 1998. ISBN 0-201-89685-0.

- Kanemitsu, Shigeru; Chaohua Jia (2002). Number theoretic methods: future trends. Springer. pp. 271–274. ISBN 978-1-4020-1080-4.

- Neukirch, Jürgen; Schmidt, Alexander; Wingberg, Kay (2000), Cohomology of Number Fields, Grundlehren der Mathematischen Wissenschaften, 323, Berlin: Springer-Verlag, ISBN 978-3-540-66671-4.

- Friedlander, Eric; Grayson, Daniel, eds. (2005), Handbook of K-Theory, Berlin, New York: Springer-Verlag, ISBN 978-3-540-30436-4, MR 2182598

- Gille, Philippe; Szamuely, Tamás (2006), Central simple algebras and Galois cohomology, Cambridge Studies in Advanced Mathematics, 101, Cambridge: Cambridge University Press, ISBN 0-521-86103-9, Zbl 1137.12001

- Gras, Georges (2003), Class field theory. From theory to practice, Springer Monographs in Mathematics, Berlin: Springer-Verlag, ISBN 3-540-44133-6.

- Lam, Tsit-Yuen (2005), Introduction to Quadratic Forms over Fields, Graduate Studies in Mathematics, 67, American Mathematical Society, ISBN 0-8218-1095-2, MR 2104929, Zbl 1068.11023

- Lemmermeyer, Franz (2000), Reciprocity laws. From Euler to Eisenstein, Springer Monographs in Mathematics, Berlin: Springer-Verlag, doi:10.1007/978-3-662-12893-0, ISBN 3-540-66957-4.

- Rosenberg, Jonathan (1994), Algebraic K-theory and its applications, Graduate Texts in Mathematics, 147, Berlin, New York: Springer-Verlag, ISBN 978-0-387-94248-3.

- Silvester, John R. (1981), Introduction to algebraic K-theory, Chapman and Hall Mathematics Series, London, New York: Chapman and Hall, ISBN 0-412-22700-2.

Evolution of Algebra

Algebra is one of the basic branches of mathematics. It emerged in the 16th century and since then has continuously been advanced as a branch of mathematics. In order to completely understand algebra, it is vital to understand the evolution of algebra. The aspects elucidated in this chapter are of vital importance, and provide a better understanding of algebra.

History of Algebra

As a branch of mathematics, algebra emerged at the end of 16th century in Europe, with the work of François Viète. Algebra can essentially be considered as doing computations similar to those of arithmetic but with non-numerical mathematical objects. However, until the 19th century, algebra consisted essentially of the theory of equations. For example, the fundamental theorem of algebra belongs to the theory of equations and is not, nowadays, considered as belonging to algebra.

This chapter describes the history of the theory of equations, called here "algebra", from the origins to the emergence of algebra as a separate area of mathematics.

Etymology

The word "algebra" is derived from the Arabic word الجبر *al-jabr*, and this comes from the treatise written in the year 830 by the medieval Persian mathematician, Muhammad ibn Mūsā al-Khwārizmī, whose Arabic title. *The Compendious Book on Calculation by Completion and Balancing*. The treatise provided for the systematic solution of linear and quadratic equations. According to one history, "[i]t is not certain just what the terms *al-jabr* and *muqabalah* mean, but the usual interpretation is similar to that implied in the [above] translation. The word 'al-jabr' presumably meant something like 'restoration' or 'completion' and seems to refer to the transposition of subtracted terms to the other side of an equation; the word 'muqabalah' is said to refer to 'reduction' or 'balancing'—that is, the cancellation of like terms on opposite sides of the equation. Arabic influence in Spain long after the time of al-Khwarizmi is found in *Don Quixote*, where the word 'algebrista' is used for a bone-setter, that is, a 'restorer'." The term is used by al-Khwarizmi to describe the operations that he introduced, "reduction" and "balancing", referring to the transposition of subtracted terms to the other side of an equation, that is, the cancellation of like terms on opposite sides of the equation.

Stages of Algebra

Algebraic Expression

Algebra did not always make use of the symbolism that is now ubiquitous in mathematics; instead, it went through three distinct stages. The stages in the development of symbolic algebra are approximately as follows:

- Rhetorical algebra, in which equations are written in full sentences. For example, the rhetorical form of $x + 1 = 2$ is "The thing plus one equals two" or possibly "The thing plus 1 equals 2". Rhetorical algebra was first developed by the ancient Babylonians and remained dominant up to the 16th century.

- Syncopated algebra, in which some symbolism is used, but which does not contain all of the characteristics of symbolic algebra. For instance, there may be a restriction that subtraction may be used only once within one side of an equation, which is not the case with symbolic algebra. Syncopated algebraic expression first appeared in Diophantus' *Arithmetica* (3rd century AD), followed by Brahmagupta's *Brahma Sphuta Siddhanta* (7th century).

- Symbolic algebra, in which full symbolism is used. Early steps toward this can be seen in the work of several Islamic mathematicians such as Ibn al-Banna (13th-14th centuries) and al-Qalasadi (15th century), although fully symbolic algebra was developed by François Viète (16th century). Later, René Descartes (17th century) introduced the modern notation and showed that the problems occurring in geometry can be expressed and solved in terms of algebra (Cartesian geometry).

Equally important as the use or lack of symbolism in algebra was the degree of the equations that were addressed. Quadratic equations played an important role in early algebra; and throughout most of history, until the early modern period, all quadratic equations were classified as belonging to one of three categories.

$$x^2 + px = q$$

$$x^2 = px + q$$

$$x^2 + q = px$$

where p and q are positive. This trichotomy comes about because quadratic equations of the form $x^2 + px + q = 0$, with p and q positive, have no positive roots.

In between the rhetorical and syncopated stages of symbolic algebra, a geometric constructive algebra was developed by classical Greek and Vedic Indian mathematicians in which algebraic equations were solved through geometry. For instance, an equation of the form $x^2 = A$ was solved by finding the side of a square of area A.

Conceptual Stages

In addition to the three stages of expressing algebraic ideas, there were four conceptual stages in the development of algebra that occurred alongside the changes in expression. These four stages were as follows:

- Geometric stage, where the concepts of algebra are largely geometric. This dates back to the Babylonians and continued with the Greeks, and was later revived by Omar Khayyám.

- Static equation-solving stage, where the objective is to find numbers satisfying certain relationships. The move away from geometric algebra dates back to Diophantus and Brahmagupta, but algebra didn't decisively move to the static equation-solving stage until Al-Khwarizmi's *Al-Jabr*.

- Dynamic function stage, where motion is an underlying idea. The idea of a function began emerging with Sharaf al-Dīn al-Tūsī, but algebra did not decisively move to the dynamic function stage until Gottfried Leibniz.

- Abstract stage, where mathematical structure plays a central role. Abstract algebra is largely a product of the 19th and 20th centuries.

Babylonian Algebra

The Plimpton 322 tablet.

The origins of algebra can be traced to the ancient Babylonians, who developed a positional number system that greatly aided them in solving their rhetorical algebraic equations. The Babylonians were not interested in exact solutions but approximations, and so they would commonly use linear interpolation to approximate intermediate values. One of the most famous tablets is the Plimpton 322 tablet, created around 1900–1600 BCE, which gives a table of Pythagorean triples and represents some of the most advanced mathematics prior to Greek mathematics.

Babylonian algebra was much more advanced than the Egyptian algebra of the time; whereas the Egyptians were mainly concerned with linear equations the Babylonians were more concerned with quadratic and cubic equations. The Babylonians had developed flexible algebraic operations with which they were able to add equals to equals and multiply both sides of an equation by like quantities so as to eliminate fractions and factors. They were familiar with many simple forms of factoring, three-term quadratic equations with positive roots, and many cubic equations although it is not known if they were able to reduce the general cubic equation.

Egyptian Algebra

Ancient Egyptian algebra dealt mainly with linear equations while the Babylonians found these equations too elementary and developed mathematics to a higher level than the Egyptians.

The Rhind Papyrus, also known as the Ahmes Papyrus, is an ancient Egyptian papyrus written c. 1650 BCE by Ahmes, who transcribed it from an earlier work that he dated to between 2000 and 1800 BCE. It is the most extensive ancient Egyptian mathematical document known to historians. The Rhind Papyrus contains problems where linear equations of the form $cx + ax = b$ and $x + ax + bx = c$ are solved, where a, b, and c are known and x, which is referred to as "aha" or heap, is the unknown. The solutions were possibly, but not likely, arrived at by using the "method of false position", or *regula falsi*, where first a specific value is substituted into the left hand side of the equation, then the required arithmetic calculations are done, thirdly the result is compared to the right hand side of the equation, and finally the correct answer is found through the use of proportions. In some of the problems the author "checks" his solution, thereby writing one of the earliest known simple proofs.

A portion of the Rhind Papyrus.

Greek Geometric Algebra

It is sometimes alleged that the Greeks had no algebra, but this is inaccurate. By the time of Plato, Greek mathematics had undergone a drastic change. The Greeks created a geometric algebra where terms were represented by sides of geometric objects, usually lines, that had letters associated with them, and with this new form of algebra they were able to find solutions to equations by using a process that they invented, known as "the application of areas". "The application of areas" is only a part of geometric algebra and it is thoroughly covered in Euclid's *Elements*.

An example of geometric algebra would be solving the linear equation ax = bc. The ancient Greeks would solve this equation by looking at it as an equality of areas rather than as an equality between the ratios a:b and c:x. The Greeks would construct a rectangle with sides of length b and c, then extend a side of the rectangle to length a, and finally they would complete the extended rectangle so as to find the side of the rectangle that is the solution.

Bloom of Thymaridas

Iamblichus in *Introductio arithmatica* tells us that Thymaridas (c. 400 BCE – c. 350 BCE) worked with simultaneous linear equations. In particular, he created the then famous rule that was known as the "bloom of Thymaridas" or as the "flower of Thymaridas", which states that:

If the sum of *n* quantities be given, and also the sum of every pair containing a particular quantity, then this particular quantity is equal to 1/ (n - 2) of the difference between the sums of these pairs and the first given sum.

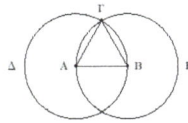

A proof from Euclid's *Elements* that, given a line segment, an equilateral triangle exists that includes the segment as one of its sides.

or using modern notion, the solution of the following system of n linear equations in n unknowns,

$$x + x_1 + x_2 + \ldots + x_{n-1} = s$$
$$x + x_1 = m_1$$
$$x + x_2 = m_2$$

.

.

.

$$x + x_{n-1} = m_{n-1}$$

is,

$$x = \frac{(m_1 + m_2 + \ldots + m_{n-1}) - s}{n - 2} = \frac{(\sum_{i=1}^{n-1} m_i) - s}{n - 2}$$

Iamblichus goes on to describe how some systems of linear equations that are not in this form can be placed into this form.

Euclid of Alexandria

Hellenistic mathematician Euclid details geometrical algebra.

Euclid was a Greek mathematician who flourished in Alexandria, Egypt, almost certainly during the reign of Ptolemy I (323–283 BCE). Neither the year nor place of his birth have been established, nor the circumstances of his death.

Euclid is regarded as the "father of geometry". His *Elements* is the most successful textbook in the history of mathematics. Although he is one of the most famous mathematicians in history there are no new discoveries attributed to him, rather he is remembered for his great explanatory skills. The *Elements* is not, as is sometimes thought, a collection of all Greek mathematical knowledge to its date, rather, it is an elementary introduction to it.

Elements

The geometric work of the Greeks, typified in Euclid's *Elements*, provided the framework for generalizing formulae beyond the solution of particular problems into more general systems of stating and solving equations.

Book II of the *Elements* contains fourteen propositions, which in Euclid's time were extremely significant for doing geometric algebra. These propositions and their results are the geometric equivalents of our modern symbolic algebra and trigonometry. Today, using modern symbolic algebra, we let symbols represent known and unknown magnitudes (i.e. numbers) and then apply algebraic operations on them. While in Euclid's time magnitudes were viewed as line segments and then results were deduced using the axioms or theorems of geometry.

Many basic laws of addition and multiplication are included or proved geometrically in the *Elements*. For instance, proposition 1 of Book II states:

> If there be two straight lines, and one of them be cut into any number of segments whatever, the rectangle contained by the two straight lines is equal to the rectangles contained by the uncut straight line and each of the segments.

But this is nothing more than the geometric version of the (left) distributive law, $a(b + c + d) = ab + ac + ad$; and in Books V and VII of the *Elements* the commutative and associative laws for multiplication are demonstrated.

Many basic equations were also proved geometrically. For instance, proposition 5 in Book II proves that $a^2 - b^2 = (a+b)(a-b)$, and proposition 4 in Book II proves that $(a+b)^2 = a^2 + 2ab + b^2$.

Furthermore, there are also geometric solutions given to many equations. For instance, proposition 6 of Book II gives the solution to the quadratic equation $ax + x^2 = b^2$, and proposition 11 of Book II gives a solution to $ax + x^2 = a^2$.

Data

Data is a work written by Euclid for use at the schools of Alexandria and it was meant to be used as a companion volume to the first six books of the *Elements*. The book contains some fifteen definitions and ninety-five statements, of which there are about two dozen statements that serve as algebraic rules or formulas. Some of these statements are geometric equivalents to solutions of quadratic equations. For instance, *Data* contains the solutions to the equations $dx^2 - adx + b^2c = 0$ and the familiar Babylonian equation $xy = a^2, x \pm y = b$.

Conic Sections

A conic section is a curve that results from the intersection of a cone with a plane. There are three primary types of conic sections: ellipses (including circles), parabolas, and hyperbolas. The conic sections are reputed to have been discovered by Menaechmus (c. 380 BCE – c. 320 BCE) and since dealing with conic sections is equivalent to dealing with their respective equations, they played geometric roles equivalent to cubic equations and other higher order equations.

Menaechmus knew that in a parabola, the equation $y^2 = lx$ holds, where l is a constant called the latus rectum, although he was not aware of the fact that any equation in two unknowns determines a curve. He apparently derived these properties of conic sections and others as well. Using this information it was now possible to find a solution to the problem of the duplication of the cube by solving for the points at which two parabolas intersect, a solution equivalent to solving a cubic equation.

We are informed by Eutocius that the method he used to solve the cubic equation was due to Dionysodorus (250 BCE – 190 BCE). Dionysodorus solved the cubic by means of the intersection of a rectangular hyperbola and a parabola. This was related to a problem in Archimedes' *On the Sphere and Cylinder*. Conic sections would be studied and used for thousands of years by Greek, and later Islamic and European, mathematicians. In particular Apollonius of Perga's famous *Conics* deals with conic sections, among other topics.

Chinese Algebra

Chinese Mathematics dates to at least 300 BCE with the *Chou Pei Suan Ching*, generally considered to be one of the oldest Chinese mathematical documents.

Nine Chapters on the Mathematical Art

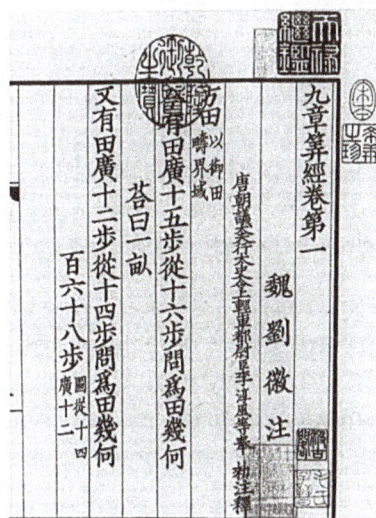

Nine Chapters on the Mathematical Art

Chiu-chang suan-shu or *The Nine Chapters on the Mathematical Art*, written around 250 BCE, is one of the most influential of all Chinese math books and it is composed of some 246 problems.

Chapter eight deals with solving determinate and indeterminate simultaneous linear equations using positive and negative numbers, with one problem dealing with solving four equations in five unknowns.

Sea-Mirror of the Circle Measurements

Ts'e-yuan hai-ching, or *Sea-Mirror of the Circle Measurements*, is a collection of some 170 problems written by Li Zhi (or Li Ye) (1192 – 1279 CE). He used *fan fa*, or Horner's method, to solve equations of degree as high as six, although he did not describe his method of solving equations.

Mathematical Treatise in Nine Sections

Shu-shu chiu-chang, or *Mathematical Treatise in Nine Sections*, was written by the wealthy governor and minister Ch'in Chiu-shao (c. 1202 – c. 1261 CE) and with the invention of a method of solving simultaneous congruences, now called Chinese remainder theorem, it marks the high point in Chinese indeterminate analysis.

Magic Squares

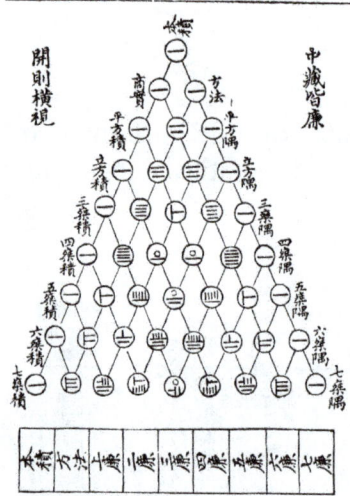

Yang Hui (Pascal's) triangle, as depicted by the ancient Chinese using rod numerals.

The earliest known magic squares appeared in China. In *Nine Chapters* the author solves a system of simultaneous linear equations by placing the coefficients and constant terms of the linear equations into a magic square (i.e. a matrix) and performing column reducing operations on the magic square. The earliest known magic squares of order greater than three are attributed to Yang Hui (fl. c. 1261 – 1275), who worked with magic squares of order as high as ten.

Precious Mirror of the Four Elements

Ssy-yüan yü-chien 《四元玉鑒》, or *Precious Mirror of the Four Elements*, was written by Chu Shih-chieh in 1303 and it marks the peak in the development of Chinese algebra. The four elements, called heaven, earth, man and matter, represented the four unknown quantities in his algebraic equations. The *Ssy-yüan yü-chien* deals with simultaneous equations and with equations of degrees as high as fourteen. The author uses the method of *fan fa*, today called Horner's method, to solve these equations.

The *Precious Mirror* opens with a diagram of the arithmetic triangle (Pascal's triangle) using a round zero symbol, but Chu Shih-chieh denies credit for it. A similar triangle appears in Yang Hui's work, but without the zero symbol.

There are many summation series equations given without proof in the *Precious mirror*. A few of the summation series are:

$$1^2 + 2^2 + 3^2 + \cdots + n^2 = \frac{n(n+1)(2n+1)}{3!}$$

$$1 + 8 + 30 + 80 + \cdots + \frac{n^2(n+1)(n+2)}{3!} = \frac{n(n+1)(n+2)(n+3)(4n+1)}{5!}$$

Diophantine Algebra

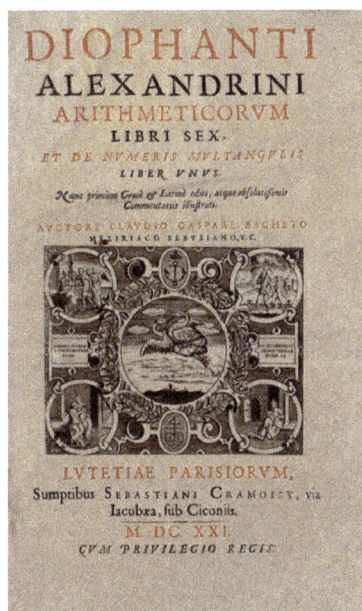

Cover of the 1621 edition of Diophantus' *Arithmetica*, translated into Latin by Claude Gaspard Bachet de Méziriac.

Diophantus was a Hellenistic mathematician who lived c. 250 CE, but the uncertainty of this date is so great that it may be off by more than a century. He is known for having written *Arithmetica*, a treatise that was originally thirteen books but of which only the first six have survived. *Arithmetica* has very little in common with traditional Greek mathematics since it is divorced from geometric methods, and it is different from Babylonian mathematics in that Diophantus is concerned primarily with exact solutions, both determinate and indeterminate, instead of simple approximations.

In *Arithmetica*, Diophantus is the first to use symbols for unknown numbers as well as abbreviations for powers of numbers, relationships, and operations; thus he used what is now known as *syncopated* algebra. The main difference between Diophantine syncopated algebra and modern algebraic notation is that the former lacked special symbols for operations, relations, and exponentials. So, for example, what we would write as

$$x^3 - 2x^2 + 10x - 1 = 5$$

Diophantus would have written this as

K^Y ᾱς ῑ ⌑ Δ^Y β M ᾱ ῐσ M ε̄

where the symbols represent the following:

Symbol	Representation
ᾱ	represents 1
β	represents 2
ε̄	represents 5
ῑ	represents 10
ς	represents the unknown quantity (i.e. the variable)
ῐσ	(short for ῐσος) represents "equals"
⌑	represents the subtraction of everything that follows it up to ῐσ
M	represents the zeroth power of the variable (i.e. a constant term)
Δ^Y	represents the second power of the variable
K^Y	represents the third power of the variable
Δ^YΔ	represents the fourth power of the variable
ΔK^Y	represents the fifth power of the variable
K^YK	represents the sixth power of the variable

Note that the coefficients come after the variables and that addition is represented by the juxtaposition of terms. A literal symbol-for-symbol translation of Diophantus's syncopated equation into a modern symbolic equation would be the following:

$$x^3 1 x 10 - x^2 2 x^0 1 = x^0 5$$

and, to clarify, if the modern parentheses and plus are used then the above equation can be rewritten as:

$$(x^3 1 + x 10) - (x^2 2 + x^0 1) = x^0 5$$

Arithmetica is a collection of some 150 solved problems with specific numbers and there is no postulational development nor is a general method explicitly explained, although generality of method may have been intended and there is no attempt to find all of the solutions to the equations. *Arithmetica* does contain solved problems involving several unknown quantities, which are solved, if possible, by expressing the unknown quantities in terms of only one of them. *Arithmetica* also makes use of the identities:

$$(a^2 + b^2)(c^2 + d^2) = (ac + db)^2 + (bc - ad)^2$$
$$= (ad + bc)^2 + (ac - bd)^2$$

Indian Algebra

The Indian mathematicians were active in studying about number systems. The earliest known Indian mathematical documents are dated to around the middle of the first millennium BCE (around the 6th century BCE).

The recurring themes in Indian mathematics are, among others, determinate and indeterminate linear and quadratic equations, simple mensuration, and Pythagorean triples.

Aryabhata

Aryabhata (476–550 CE) was an Indian mathematician who authored *Aryabhatiya*. In it he gave the rules,

$$1^2 + 2^2 + \cdots + n^2 = \frac{n(n+1)(2n+1)}{6}$$

and

$$1^3 + 2^3 + \cdots + n^3 = (1 + 2 + \cdots + n)^2$$

Brahma Sphuta Siddhanta

Brahmagupta (fl. 628) was an Indian mathematician who authored *Brahma Sphuta Siddhanta*. In his work Brahmagupta solves the general quadratic equation for both positive and negative roots.

In indeterminate analysis Brahmagupta gives the Pythagorean triads m, $\frac{1}{2}(\frac{m^2}{n} - n)$, $\frac{1}{2}(\frac{m^2}{n} + n)$, but this is a modified form of an old Babylonian rule that Brahmagupta may have been familiar with. He was the first to give a general solution to the linear Diophantine equation ax + by = c, where a, b, and c are integers. Unlike Diophantus who only gave one solution to an indeterminate equation, Brahmagupta gave *all* integer solutions; but that Brahmagupta used some of the same examples as Diophantus has led some historians to consider the possibility of a Greek influence on Brahmagupta's work, or at least a common Babylonian source.

Like the algebra of Diophantus, the algebra of Brahmagupta was syncopated. Addition was indicated by placing the numbers side by side, subtraction by placing a dot over the subtrahend, and division by placing the divisor below the dividend, similar to our notation but without the bar. Multiplication, evolution, and unknown quantities were represented by abbreviations of appropriate terms. The extent of Greek influence on this syncopation, if any, is not known and it is possible that both Greek and Indian syncopation may be derived from a common Babylonian source.

Bhāskara II

Bhāskara II (1114 – c. 1185) was the leading mathematician of the 12th century. In Algebra, he gave the general solution of Pell's equation. He is the author of *Lilavati* and *Vija-Ganita*, which contain problems dealing with determinate and indeterminate linear and quadratic equations, and Pythagorean triples and he fails to distinguish between exact and approximate statements. Many of the problems in *Lilavati* and *Vija-Ganita* are derived from other Hindu sources, and so Bhaskara is at his best in dealing with indeterminate analysis.

Bhaskara uses the initial symbols of the names for colors as the symbols of unknown variables. So, for example, what we would write today as

$$(-x-1)+(2x-8)=x-9$$

Bhaskara would have written as

> *ya* 1 *ru* 1
>
> *ya* 2 *ru* 8
>
> Sum *ya* 1 ru 9

where *ya* indicates the first syllable of the word for *black*, and *ru* is taken from the word *species*. The dots over the numbers indicate subtraction.

Islamic Algebra

The first century of the Islamic Arab Empire saw almost no scientific or mathematical achievements since the Arabs, with their newly conquered empire, had not yet gained any intellectual drive and research in other parts of the world had faded. In the second half of the 8th century, Islam had a cultural awakening, and research in mathematics and the sciences increased. The Muslim Abbasid caliph al-Mamun (809–833) is said to have had a dream where Aristotle appeared to him, and as a consequence al-Mamun ordered that Arabic translation be made of as many Greek works as possible, including Ptolemy's *Almagest* and Euclid's *Elements*. Greek works would be given to the Muslims by the Byzantine Empire in exchange for treaties, as the two empires held an uneasy peace. Many of these Greek works were translated by Thabit ibn Qurra (826–901), who translated books written by Euclid, Archimedes, Apollonius, Ptolemy, and Eutocius.

There are three theories about the origins of Arabic Algebra. The first emphasizes Hindu influence, the second emphasizes Mesopotamian or Persian-Syriac influence and the third emphasizes Greek influence. Many scholars believe that it is the result of a combination of all three sources.

Throughout their time in power, before the fall of Islamic civilization, the Arabs used a fully rhetorical algebra, where often even the numbers were spelled out in words. The Arabs would eventually replace spelled out numbers (e.g. twenty-two) with Arabic numerals (e.g. 22), but the Arabs did not adopt or develop a syncopated or symbolic algebra until the work of Ibn al-Banna in the 13th century and Abū al-Hasan ibn Alī al-Qalasādī in the 15th century.

Al-jabr wa'l muqabalah

Left: The original Arabic print manuscript of the Book of Algebra by Al-Khwarizmi. Right: A page from The Algebra of Al-Khwarizmi by Fredrick Rosen, in English.

The Muslim Persian mathematician Muhammad ibn Mūsā al-Khwārizmī was a faculty member of the "House of Wisdom" (*Bait al-Hikma*) in Baghdad, which was established by Al-Mamun. Al-Khwarizmi, who died around 850 CE, wrote more than half a dozen mathematical and astronomical works, some of which were based on the Indian *Sindhind*. One of al-Khwarizmi's most famous books is entitled *Al-jabr wa'l muqabalah* or *The Compendious Book on Calculation by Completion and Balancing*, and it gives an exhaustive account of solving polynomials up to the second degree. The book also introduced the fundamental concept of "reduction" and "balancing", referring to the transposition of subtracted terms to the other side of an equation, that is, the cancellation of like terms on opposite sides of the equation. This is the operation which Al-Khwarizmi originally described as *al-jabr*.

R. Rashed and Angela Armstrong write:

"Al-Khwarizmi's text can be seen to be distinct not only from the Babylonian tablets, but also from Diophantus' *Arithmetica*. It no longer concerns a series of problems to be resolved, but an exposition which starts with primitive terms in which the combinations must give all possible prototypes for equations, which henceforward explicitly constitute the true object of study. On the other hand, the idea of an equation for its own sake appears from the beginning and, one could say, in a generic manner, insofar as it does not simply emerge in the course of solving a problem, but is specifically called on to define an infinite class of problems."

Al-Jabr is divided into six chapters, each of which deals with a different type of formula. The first chapter of *Al-Jabr* deals with equations whose squares equal its roots ($ax^2 = bx$), the second chapter deals with squares equal to number ($ax^2 = c$), the third chapter deals with roots equal to a number ($bx = c$), the fourth chapter deals with squares and roots equal a number ($ax^2 + bx = c$), the fifth chapter deals with squares and number equal roots ($ax^2 + c = bx$), and the sixth and final chapter deals with roots and number equal to squares ($bx + c = ax^2$).

Pages from a 14th-century Arabic copy of the book, showing geometric solutions to two quadratic equations

In *Al-Jabr*, al-Khwarizmi uses geometric proofs, he does not recognize the root x = 0, and he only deals with positive roots. He also recognizes that the discriminant must be positive and described the method of completing the square, though he does not justify the procedure. The Greek influence is shown by *Al-Jabr's* geometric foundations and by one problem taken from Heron. He makes use of lettered diagrams but all of the coefficients in all of his equations are specific numbers since he had no way of expressing with parameters what he could express geometrically; although generality of method is intended.

Al-Khwarizmi most likely did not know of Diophantus's *Arithmetica*, which became known to the Arabs sometime before the 10th century. And even though al-Khwarizmi most likely knew of Brahmagupta's work, *Al-Jabr* is fully rhetorical with the numbers even being spelled out in words. So, for example, what we would write as

$$x^2 + 10x = 39$$

Diophantus would have written as

$$\Delta^Y \bar{\alpha} \, \varsigma \bar{\iota} \, \text{'}\iota\sigma \, M \, \overline{\lambda\theta}$$

And al-Khwarizmi would have written as

> One square and ten roots of the same amount to thirty-nine *dirhems*; that is to say, what must be the square which, when increased by ten of its own roots, amounts to thirty-nine?

Logical Necessities in Mixed Equations

'Abd al-Hamīd ibn Turk authored a manuscript entitled *Logical Necessities in Mixed Equations*, which is very similar to al-Khwarizimi's *Al-Jabr* and was published at around the same time as, or even possibly earlier than, *Al-Jabr*. The manuscript gives exactly the same geometric demonstration as is found in *Al-Jabr*, and in one case the same example as found in *Al-Jabr*, and even goes beyond *Al-Jabr* by giving a geometric proof that if the discriminant is negative then the quadratic equation has no solution. The similarity between these two works has led some historians to conclude that Arabic algebra may have been well developed by the time of al-Khwarizmi and 'Abd al-Hamid.

Abu Kamil and al-Karkhi

Arabic mathematicians treated irrational numbers as algebraic objects. The Egyptian mathematician Abū Kāmil Shujā ibn Aslam (c. 850–930) was the first to accept irrational numbers (often in the form of a square root, cube root or fourth root) as solutions to quadratic equations or as coefficients in an equation. He was also the first to solve three non-linear simultaneous equations with three unknown variables.

Al-Karkhi (953–1029), also known as Al-Karaji, was the successor of Abū al-Wafā' al-Būzjānī (940–998) and he discovered the first numerical solution to equations of the form $ax^{2n} + bx^n = c$. Al-Karkhi only considered positive roots. Al-Karkhi is also regarded as the first person to free algebra from geometrical operations and replace them with the type of arithmetic operations which are at the core of algebra today. His work on algebra and polynomials, gave the rules for arithmetic operations to manipulate polynomials. The historian of mathematics F. Woepcke, in *Extrait du Fakhri, traité d'Algèbre par Abou Bekr Mohammed Ben Alhacan Alkarkhi* (Paris, 1853), praised Al-Karaji for being "the first who introduced the theory of algebraic calculus". Stemming from this, Al-Karaji investigated binomial coefficients and Pascal's triangle.

Omar Khayyám, Sharaf al-Dīn, and al-Kashi

Omar Khayyam is credited with identifying the foundations of algebraic geometry and found the general geometric solution of the cubic equation.

Omar Khayyám (c. 1050 – 1123) wrote a book on Algebra that went beyond *Al-Jabr* to include equations of the third degree. Omar Khayyám provided both arithmetic and geometric solutions for quadratic equations, but he only gave geometric solutions for general cubic equations since he mistakenly believed that arithmetic solutions were impossible. His method of solving cubic equations by using intersecting conics had been used by Menaechmus, Archimedes, and Ibn al-Haytham (Alhazen), but Omar Khayyám generalized the method to cover all cubic equations with positive roots. He only considered positive roots and he did not go past the third degree. He also saw a strong relationship between Geometry and Algebra.

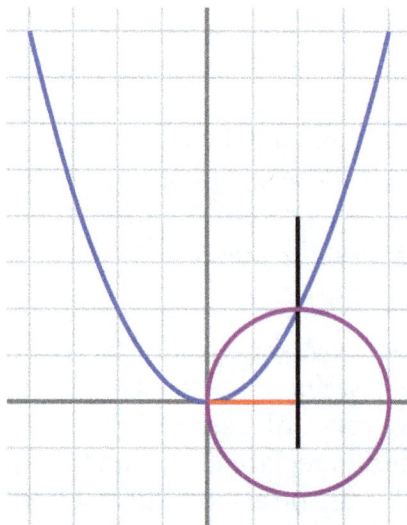

To solve the third-degree equation $x^3 + a^2x = b$ Khayyám constructed the parabola $x^2 = ay$, a circle with diameter b/a^2, and a vertical line through the intersection point. The solution is given by the length of the horizontal line segment from the origin to the intersection of the vertical line and the x-axis.

In the 12th century, Sharaf al-Dīn al-Tūsī (1135–1213) wrote the *Al-Mu'adalat* (*Treatise on Equations*), which dealt with eight types of cubic equations with positive solutions and five types of cubic equations which may not have positive solutions. He used what would later be known as the "Ruffini-Horner method" to numerically approximate the root of a cubic equation. He also developed the concepts of the maxima and minima of curves in order to solve cubic equations which may not have positive solutions. He understood the importance of the discriminant of the cubic equation and used an early version of Cardano's formula to find algebraic solutions to certain types of cubic equations. Some scholars, such as Roshdi Rashed, argue that Sharaf al-Din discovered the derivative of cubic polynomials and realized its significance, while other scholars connect his solution to the ideas of Euclid and Archimedes.

Sharaf al-Din also developed the concept of a function. In his analysis of the equation $x^3 + d = bx^2$ for example, he begins by changing the equation's form to $x^2(b - x) = d$. He then states that the question of whether the equation has a solution depends on whether or not the "function" on the left side reaches the value d, To determine this, he finds a maximum value for the function. He proves that the maximum value occurs when $x = \dfrac{2b}{3}$;, which gives the functional value $\dfrac{4b^3}{27}$. Sharaf al-Din then states that if this value is less than d, there are no positive solutions; if it is equal to d, then there is one solution at $x = \dfrac{2b}{3}$; and if it is greater than d, then there are two solutions, one between 0 and $\dfrac{2b}{3}$ and one between $\dfrac{2b}{3}$ and b.

In the early 15th century, Jamshīd al-Kāshī developed an early form of Newton's method to numerically solve the equation to find roots of . Al-Kāshī also developed decimal fractions and claimed to have discovered it himself. However, J. Lennart Berggrenn notes that he was mistaken, as decimal fractions were first used five centuries before him by the Baghdadi mathematician Abu'l-Hasan al-Uqlidisi as early as the 10th century.

Al-Hassār, Ibn al-Banna, and al-Qalasadi

Al-Hassār, a mathematician from Morocco specializing in Islamic inheritance jurisprudence during the 12th century, developed the modern symbolic mathematical notation for fractions, where the numerator and denominator are separated by a horizontal bar. This same fractional notation appeared soon after in the work of Fibonacci in the 13th century.

Abū al-Hasan ibn Alī al-Qalasādī (1412–1486) was the last major medieval Arab algebraist, who made the first attempt at creating an algebraic notation since Ibn al-Banna two centuries earlier, who was himself the first to make such an attempt since Diophantus and Brahmagupta in ancient times. The syncopated notations of his predecessors, however, lacked symbols for mathematical operations. Al-Qalasadi "took the first steps toward the introduction of algebraic symbolism by using letters in place of numbers" and by "using short Arabic words, or just their initial letters, as mathematical symbols."

European Algebra

Dark Ages

Just as the death of Hypatia signals the close of the Library of Alexandria as a mathematical center, so does the death of Boethius signal the end of mathematics in the Western Roman Empire. Although there was some work being done at Athens, it came to a close when in 529 the Byzantine emperor Justinian closed the pagan philosophical schools. The year 529 is now taken to be the beginning of the medieval period. Scholars fled the West towards the more hospitable East, particularly towards Persia, where they found haven under King Chosroes and established what might be termed an "Athenian Academy in Exile". Under a treaty with Justinian, Chosroes would eventually return the scholars to the Eastern Empire. During the Dark Ages, European mathematics was at its nadir with mathematical research consisting mainly of commentaries on ancient treatises; and most of this research was centered in the Byzantine Empire. The end of the medieval period is set as the fall of Constantinople to the Turks in 1453.

Late Middle Ages

The 12th century saw a flood of translations from Arabic into Latin and by the 13th century, European mathematics was beginning to rival the mathematics of other lands. In the 13th century, the solution of a cubic equation by Fibonacci is representative of the beginning of a revival in European algebra.

As the Islamic world was declining after the 15th century, the European world was ascending. And it is here that Algebra was further developed.

Modern Algebra

Modern notation for arithmetic operations was introduced between the end of the 15th century and the beginning of the 16th century by Johannes Widmann and Michael Stifel.

Another key event in the further development of algebra was the general algebraic solution of the cubic and quartic equations, developed in the mid-16th century. The idea of a determinant was developed by Japanese mathematician Kowa Seki in the 17th century, followed by Gottfried Leibniz ten years later, for the purpose of solving systems of simultaneous linear equations using matrices. Gabriel Cramer also did some work on matrices and determinants in the 18th century.

The symbol x

By tradition, the first unknown variable in an algebraic problem is nowadays represented by the symbol x; if there is a second or a third unknown, these are labeled y and z respectively. Algebraic x is conventionally printed in italic type to distinguish it from the sign of multiplication.

Mathematical historians generally agree that the use of x in algebra was introduced by René Descartes and was first published in his treatise *La Géométrie* (1637). In that work, he used letters from the beginning of the alphabet (a, b, c,...) for known quantities, and letters from the end of the alphabet (z, y, x,...) for unknowns. It has been suggested that he later settled on x (in place of z) for the first unknown because of its relatively greater abundance in the French and Latin typographical fonts of the time.

Three alternative theories of the origin of algebraic x were suggested in the 19th century: (1) a symbol used by German algebraists and thought to be derived from a cursive letter r, mistaken for x; (2) the numeral 1 with oblique strikethrough; and (3) an Arabic/Spanish source. But the Swiss-American historian of mathematics Florian Cajori examined these and found all three lacking in concrete evidence; Cajori credited Descartes as the originator, and described his x, y, and z as "free from tradition[,] and their choice purely arbitrary."

Nevertheless, the Hispano-Arabic hypothesis continues to have a presence in popular culture today. It is the claim that algebraic x is the abbreviation of a supposed loanword from Arabic in Old Spanish. The theory originated in 1884 with the German "orientalist" Paul de Lagarde, shortly after he published his edition of a 1505 Spanish/Arabic bilingual glossary in which Spanish *cosa* ("thing") was paired with its Arabic equivalent, شيء (*shayʔ*), transcribed as *xei*. (The "sh" sound in Old Spanish was routinely spelled x.) Evidently Lagarde was aware that Arab mathematicians, in the "rhetorical" stage of algebra's development, often used that word to represent the unknown quantity. He surmised that "nothing could be more natural" (Nichts war also natürlicher...) than for the initial of the Arabic word—romanized as the Old Spanish x—to be adopted for use in algebra. A later reader reinterpreted Lagarde's conjecture as having "proven" the point. Lagarde was unaware that early Spanish mathematicians used, not a *transcription* of the Arabic word, but rather its *translation* in their own language, "cosa". There is no instance of *xei* or similar forms in several compiled historical vocabularies of Spanish.

Gottfried Leibniz

Although the mathematical notion of function was implicit in trigonometric and logarithmic tables, which existed in his day, Gottfried Leibniz was the first, in 1692 and 1694, to employ it explicitly, to denote any of several geometric concepts derived from a curve, such as abscissa, ordinate, tangent, chord, and the perpendicular. In the 18th century, "function" lost these geometrical associations.

Leibniz realized that the coefficients of a system of linear equations could be arranged into an array, now called a matrix, which can be manipulated to find the solution of the system, if any. This method was later called Gaussian elimination. Leibniz also discovered Boolean algebra and symbolic logic, also relevant to algebra.

Abstract Algebra

The ability to do algebra is a skill cultivated in mathematics education. As explained by Andrew Warwick, Cambridge University students in the early 19th century practiced "mixed mathematics", doing exercises based on physical variables such as space, time, and weight. Over time the association of variables with physical quantities faded away as mathematical technique grew. Eventually mathematics was concerned completely with abstract polynomials, complex numbers, hypercomplex numbers and other concepts. Application to physical situations was then called applied mathematics or mathematical physics, and the field of mathematics expanded to include abstract algebra. For instance, the issue of constructible numbers showed some mathematical limitations, and the field of Galois theory was developed.

The Father of Algebra

The Hellenistic mathematician Diophantus has traditionally been known as "the father of algebra" but debate now exists as to whether or not Al-Khwarizmi deserves this title instead. Those who support Diophantus point to the fact that the algebra found in *Al-Jabr* is more elementary than the algebra found in *Arithmetica* and that *Arithmetica* is syncopated while *Al-Jabr* is fully rhetorical.

Those who support Al-Khwarizmi point to the fact that he gave an exhaustive explanation for the algebraic solution of quadratic equations with positive roots, and was the first to teach algebra in an elementary form and for its own sake, whereas Diophantus was primarily concerned with the theory of numbers. Al-Khwarizmi also introduced the fundamental concept of "reduction" and "balancing" (which he originally used the term *al-jabr* to refer to), referring to the transposition of subtracted terms to the other side of an equation, that is, the cancellation of like terms on opposite sides of the equation. Other supporters of Al-Khwarizmi point to his algebra no longer being concerned "with a series of problems to be resolved, but an exposition which starts with primitive terms in which the combinations must give all possible prototypes for equations, which henceforward explicitly constitute the true object of study." They also point to his treatment of an equation for its own sake and "in a generic manner, insofar as it does not simply emerge in the course of solving a problem, but is specifically called on to define an infinite class of problems."

References

- Burton, David M. (1997), The History of Mathematics: An Introduction (Third ed.), The McGraw-Hill Companies, Inc., ISBN 0-07-009465-9

- Derbyshire, John (2006), Unknown Quantity: A Real And Imaginary History of Algebra, Washington, DC: Joseph Henry Press, ISBN 0-309-09657-X

- Katz, Victor J.; Parshall, Karen Hunger (2014), Taming the Unknown: A History of Algebra from Antiquity to the Early Twentieth Century, Princeton, NJ: Princeton University Press, ISBN 978-1-400-85052-5

- Stillwell, John (2004), Mathematics and its History (Second ed.), Springer Science + Business Media Inc., ISBN 0-387-95336-1

Permissions

Index